Theory of Integro-Differential Equations

Stability and Control: Theory, Methods and Applications
A series of books and monographs on the theory of stability and control
Edited by A. Martynyuk, Institute of Mechanics, Kiev, Ukraine and V. Lakshmikantham, Florida Institute of Technology, USA

Volume 1

Theory of Integro-Differential Equations
V. Lakshmikantham and M. Rama Mohana Rao

Additional Volumes in Preparation

Stability Analysis: Nonlinear Mechanics Equations
A.A. Martynyuk

Stability and Motion of Nonautonomous Systems (Method of Limiting Equations)
J. Kato, A.A. Martynyuk and A.A. Shestakov

This book is part of a series. The publisher will accept continuation orders which may be cancelled at any time and which provide for automatic billing and shipping of each title in the series upon publication. Please write for details.

Theory of Integro-Differential Equations

V. Lakshmikantham
Florida Institute of Technology, USA

and

M. Rama Mohana Rao
*Indian Institute of Technology,
Kanpur, India*

GORDON AND BREACH SCIENCE PUBLISHERS
Switzerland•Australia•Belgium•France•Germany•Great Britain
India•Japan•Malaysia•Netherlands•Russia•Singapore•USA

Copyright ©1995 by OPA (Amsterdam) B.V. All rights reserved. Published under license by Gordon and Breach Science Publishers S.A.

Gordon and Breach Science Publishers

World Trade Centre
Case Portale 531
1000 Lausanne 30 Grey
Switzerland

British Library Cataloguing in Publication Data

Lakshmikantham, V.
 Theory of Integro-differential
 Equations. - (Stability & Control:
 Theory, Methods & Applications Series;
 Vol. 1)
 I. Title II. Rao, M. Rama Mohana
 III. Series
 515.353

ISBN 2-88449-000-0

No part of this book may be reproduced or utilized in any form or by any means, electronic or mechanical, including photocopying and recording, or by any information storage or retrieval system, without permission in writing from the publisher. Printed in Singapore.

CONTENTS

Preface ix

Chapter 1. Basic Theory
       1.0   Introduction 1
       1.1   Local and Global Existence 2
       1.2   Integro-differential Inequalities 7
       1.3   Existence of Extremal Solutions 9
       1.4   Comparison Results 13
       1.5   Convergence of Successive Approximations 19
       1.6   Continuous Dependence 22
       1.7   Linear Variation of Parameters 24
       1.8   Nonlinear Variation of Parameters 31
       1.9   Monotone Iterative Technique 36
       1.10 Interval Analytic Method 43
       1.11 Notes and Comments 48

Chapter 2. Linear Analysis
       2.0   Introduction 49
       2.1   Basic Properties of Linear Systems 50
       2.2   Stability of Linear Convolution Systems 51
       2.3   Stability Criteria for General Linear Systems 58
       2.4   Stability by Method of Reduction 64
       2.5   Stability in Variation 68

	2.6	Lipschitz Stability	75
	2.7	Asymptotic Equivalence	80
	2.8	Ultimate Behavior of Solutions	93
	2.9	Difference Equations	99
	2.10	Impulsive Integro-differential Systems	106
	2.11	Periodic Solutions	115
	2.12	Notes and Comments	122

Chapter 3. Lyapunov Stability

	3.0	Introduction	125
	3.1	Method of Lyapunov Functionals	126
	3.2	Equations with Unbounded Delay	137
	3.3	Perturbed systems	145
	3.4	Method of Lyapunov Functions	158
	3.5	Lyapunov Functions on Product Spaces	171
	3.6	Impulsive Integro-differential Equations	181
	3.7	Impulsive Integro-differential Equations (continued)	191
	3.8	Notes and Comments	199

Chapter 4. Equations in Abstract Spaces

	4.0	Introduction	201
	4.1	Existence and Uniqueness	202
	4.2	Existence of Maximal and Minimal Solutions	210
	4.3	Well-Posedness of Linear Equations	218
	4.4	Semigroups and Resolvent Operators	229
	4.5	Evolution Operators and Resolvents	238
	4.6	Asymptotic Behavior and Perturbations	250
	4.7	Stability of Solutions	254
	4.8	Notes and Comments	268

Chapter 5. Applications

	5.0	Introduction	271
	5.1	Biological Population	272
	5.2	Grazing Systems	281

5.3	Wave Propagation	293
5.4	Nuclear Reactors	301
5.5	Viscoelasticity	310
5.6	Large-Scale Systems	324
5.7	Notes and Comments	337

References 339

Index 357

PREFACE

Integro-differential equations arise quite frequently as mathematical models in diverse disciplines. The origins of the study of integral and integro-differential equations may be traced to the work of Abel, Lotka, Fredholm, Malthus, Verhulst and Volterra on problems in mechanics, mathematical biology and economics. The work of Volterra on the problem of competing species is of fundamental importance for the development of mathematical modeling of real world problems. From those beginnings, the theory and applications of Volterra integro-differential equations with bounded and unbounded delays have emerged as new areas of investigation. This continuous process of development during the past few decades is reflected in the large number of research papers and books covering some of these areas. None of the available books, however, is exclusively devoted to the theory and applications of integro-differential equations as a subject in itself.

The present monograph is intended to fulfill this existing need. It provides the basic theory and qualitative properties of solutions of Volterra integro-differential equations, together with a large number of applications, and consists of five chapters. Chapter 1 is concerned with the basic theory, namely existence, uniqueness, theory of inequalities, comparison results, continuous dependence and differentiability of solutions with respect to the initial data, linear and nonlinear variation of parameters, and monotone iterative techniques.

Chapter 2 deals with linear and weakly nonlinear systems, for which fundamental properties such as stability, boundedness and periodicity of solutions are discussed. This chapter also covers equations with impulse effects, and difference equations resulting from Volterra integro-differential equations and of interest to applied scientists and engineers.

Lyapunov stability theory for Volterra integro-differential equations is discussed extensively in Chapter 3. In addition, this chapter deals with stability analysis of nonlinear equations with impulse effects, construction of Lyapunov functions and functionals, the Lyapunov–Razumukhin technique, equations with unbounded delay, and Lyapunov functions on product spaces.

Chapter 4 is concerned with equations in abstract spaces. The main ingredients of this chapter are the basic existence theory, well-posedness of linear equations, semigroups, resolvents, linear evolution operators, asymptotic behavior of solutions, and stability analysis.

Finally, in Chapter 5, we investigate various qualitative properties of solutions of integro-differential equations that arise in problems of biological population, grazing systems, wave propagation, nuclear reactors and viscoelasticity. Stability analysis of engineering systems such as input–output systems, multiloop systems and large scale systems are also covered in this chapter.

We are immensely grateful to Professors Leela and Sivasundaram for their valuable comments and suggestions during the preparation of the manuscript and to Ms Donn Harnish for the excellent typing of the monograph.

V. Lakshmikantham
M. Rama Mohana Rao

1 BASIC THEORY

1.0 Introduction

This chapter provides the basic theory of integro-differential equations of the form

$$x'(t) = f(t, x(t)) + \int_{t_0}^{t} K(t, s, x(s)) \, ds, \quad x(t_0) = x_0,$$

where $t_0 \geq 0$ and $dx(t)/dt = x'(t)$. The theory developed is close in spirit to that of classical ordinary differential equations. It exhibits the common features of existence theory and also provides insight into certain problems where there are differences.

We begin Section 1.1 by proving local and global existence results using Schauder and Tychonoff fixed point theorems. In Section 1.2, we consider the theory of integro-differential inequalities, which is essential for later purposes. Section 1.3 investigates the existence of maximal and minimal solutions and proves some convergence results that are useful for later developments. Necessary comparison results are studied in Section 1.4. In Section 1.5, we discuss the convergence of successive approximations and uniqueness in a general setup, while in Section 1.6, continuous dependence with respect to initial values is investigated. In Section 1.7, we first obtain a linear variation of parameters formula, and then develop a method of finding an equivalent linear differential system corresponding to the given linear integro-differential system. Section 1.8 considers first the differentiability of solutions with respect to initial values and then, utilizing these results, proves the nonlinear variation of parameters formula. In Section 1.9, we discuss the monotone iterative technique and the method of upper and lower solutions to study the periodic boundary value problem. Finally, in Section 1.10, employing the interval analytic method, we obtain simultaneous interval bounds for solutions.

1.1 Local and Global Existence

This section is devoted to the study of the initial value problem (IVP) for integro-differential systems of the type

$$(1.1.1) \qquad x'(t) = f(t, x(t)) + \int_{t_0}^{t} K(t, s, x(s)) \, ds, \quad x(t_0) = x_0, \quad x' = \frac{dx}{dt},$$

where $f \in C[J \times R^n, R^n]$, $K \in C[J \times J \times R^n, R^n]$ and $J = [t_0, t_0 + a]$. It is easy to show that the IVP (1.1.1) is equivalent to the integral equation

$$(1.1.2) \qquad x(t) = x_0 + \int_{t_0}^{t} \left[f(s, x(s)) + \int_{s}^{t} K(\sigma, s, x(s)) \, d\sigma \right] ds,$$

which can be seen by integrating (1.1.1) from t_0 to t and changing the order of integration. Since f and K are continuous, on differentiating (1.1.2), we obtain (1.1.1). Let us begin by proving the following local existence result by applying Schauder's fixed point theorem, which we state here in a suitable form.

Theorem (Schauder) *If E is a closed, bounded, convex subset of a Banach space B and $T: E \to E$ is completely continuous then T has a fixed point.*

Theorem 1.1.1 *Assume that $f \in C[J \times R^n, R^n]$, $K \in C[J \times J \times R^n, R^n]$ and $\int_{s}^{t} |K(\sigma, s, x(s))| \, d\sigma \leq N$, for $t_0 \leq s \leq t \leq t_0 + a$, $x \in \Omega = \{\phi \in C[J, R^n] : \phi(t_0) = x_0$ and $|\phi(t) - x_0| \leq b\}$. Then the IVP (1.1.1) possesses at least one solution $x(t)$ on $t_0 \leq t \leq t_0 + \alpha$, for some $0 < \alpha \leq a$.*

Proof Consider the set $D = \{[(t,x) : t \in J$ and $|x - x_0| \leq b\}$ and let $|f(t,x)| \leq M$ on D. Choose $\alpha = \min[a, b/M+N]$ and let $\Omega_0 = \{\phi \in C[J_0, R^n] : \phi(t_0) = x_0$ and $|\phi - x_0|_0 \leq b\}$, where $|\phi|_0 = \max_{t_0 \leq t \leq t_0 + \alpha} |\phi(t)|$ and $J_0 = [t_0, t_0 + \alpha]$. Clearly the set Ω_0 is closed, convex and bounded. For any $\phi \in \Omega_0$, define the function $T\phi$ by

$$T\phi(t) = x_0 + \int_{t_0}^{t}\left[f(s, \phi(s)) + \int_{s}^{t} K(\sigma, s, \phi(s))\, d\sigma\right] ds, \quad t \in [t_0, t_0 + \alpha].$$

We may apply Schauder's fixed point theorem to prove the existence of a fixed point of T in Ω_0, which is equivalent to solving the IVP (1.1.1). Clearly $T\phi(t_0) = x_0$, and for $t \in J_0$

$$|T\phi(t) - x_0| \leq \int_{t_0}^{t}\left[|f(s, \phi(s))| + \int_{s}^{t}|K(\sigma, s, \phi(s)|\, d\sigma\right] ds$$

$$\leq (M+N)\alpha \leq b,$$

which implies that $T\Omega_0 \subset \Omega_0$. Furthermore, for any $t_1, t_2 \in J_0$, such that $t_2 > t_1$, by changing the order of integration, we obtain

$$|T\phi(t_2) - T\phi(t_1)| \leq \int_{t_1}^{t_2}\left[|f(s, \phi(s))| + \int_{t_0}^{s}|K(s, \sigma, x(\sigma))|\, d\sigma\right] ds$$

$$\leq (M+N)|t_2 - t_1|.$$

This shows that the set $T(\Omega_0)$ is an equicontinuous family, and consequently the closure of $T(\Omega_0)$ is compact. For any ϕ, ψ in Ω_0, it follows, using uniform continuity of f and K, that for any $\epsilon > 0$, there exist $\delta > 0$ such that

$$|T\phi(t) - T\psi(t)|$$

$$\leq \int_{t_0}^{t}\left[|f(s, \phi(s)) - f(s, \psi(s))| + \int_{s}^{t}|K(\sigma, s, \phi(s)) - K(\sigma, s, \psi(s))|\, d\sigma\right] ds$$

$$\leq \epsilon(\alpha + \frac{\alpha^2}{2}) \quad \text{for all } t \in J_0,$$

provided that $|\phi(s) - \psi(s)| < \delta$ for all $s \in J_0$. This implies that T is a continuous mapping. By Schauder's fixed point theorem, there is a fixed point of T in Ω_0, which completes the proof. □

Sometime, one needs to consider $IVPs$ of integro-differential equations of the form

(1.1.3) $$x' = f(t, x, Kx), \quad x(t_0) = x_0,$$

where $(Kx)(t) = \int_{t_0}^{t} K(t,s)x(s)\,ds$, $K(t,s)$ is an $n \times n$ continuous matrix on $J \times J$, and $f \in C[J \times R^n \times R^n, R^n]$. A local existence result for the IVP (1.1.3) can be proved using arguments similar to Theorem 1.1.1. We merely state such a result.

Theorem 1.1.2 *Assume that*

$$|K(t,s)| \leq K, \quad (t,s) \in J \times J \text{ and } |f(t,x,y)| \leq M \text{ for } t \in J,$$

$x, y \in \Omega = \{x \in R^n : |x - x_0| \leq b\}$. *Then there exists a solution $x(t)$ of (1.1.3) on $[t_0, t_0 + \alpha]$ for some $\alpha > 0$.*

We shall next discuss a global existence result for IVP (1.1.1) using Tychonoff's fixed point theorem, which we state in the following form.

Theorem (Tychonoff) *Let B be a complete, locally convex, linear space and B_0 a closed convex subset of B. Let the mapping $T: B \to B$ be continuous and $T(B_0) \subset B_0$. If the closure of $T(B_0)$ is compact then T has a fixed point in B_0.*

Theorem 1.1.3 *Assume that*

(i) $f \in C[R_+ \times R^n, R^n]$, $g \in C[R_+^2, R_+]$, $g(t,u)$ is monotone nondecreasing in u for each $T \in J$ and

$$|f(t,x)| \leq g(t, |x|), \quad (t,x) \in R_+ \times R^n;$$

(ii) $K \in C[R_+^2 \times R^n, R^n]$, $G \in C[R_+^3, R_+]$, $G(t,s,u)$ is monotone nondecreasing in u for each $(t,s) \in R_+^2$ and

$$|K(t,s,x)| \leq G(t,s,|x|), \quad (t,s,x) \in R_+^2 \times R^n;$$

(iii) for every $u_0 > 0$, the scalar integro-differential equation

(1.1.4) $$u'(t) = g(t, u(t)) + \int_{t_0}^{t} G(t, s, u(s)) \, ds, \qquad u(t_0) = u_0$$

has a solution $u(t)$ existing for $t \geq t_0$,

(iv) $\int_{s}^{t} |K(\sigma, s, x(s))| \, d\sigma \leq N$ for $t, s \in R_+$, $x \in C[R_+, R^n]$.

Then for every $x_0 \in R^n$ such that $|x_0| \leq u_0$, there exists a solution $x(t)$ of (1.1.1) for $t \geq t_0$ satisfying $|x(t)| \leq u(t)$, $t \geq t_0$.

Proof Let us consider the real vector space B of all continuous functions from $[t_0, \infty)$ into R^n, the topology on B being that induced by the family of pseudo-norms $\{V_n(x)\}_{n=1}^{\infty}$ where for $x \in B$, $V_n(x) = \sup_{t_0 \leq t \leq n} |x(t)|$.

A fundamental system of neighborhoods is then given by $\{S_n\}_{n=1}^{\infty}$, where $S_n = \{x \in B : V_n(x) \leq 1\}$. Under this topology, B is a complete, locally convex linear space.

Now define a subset B_0 of B as follows:

$$B_0 = \{x \in B : |x(t)| \leq u(t), t \geq t_0\},$$

where $u(t)$ is a solution of (1.1.4) existing for $t \geq t_0$. It is clear that in the topology of B, B_0 is closed, convex and bounded. Consider the integral operator defined by

$$Tx(t) = x_0 + \int_{t_0}^{t} \left[f(s, x(s)) + \int_{s}^{t} K(\sigma, s, x(s)) \, d\sigma \right] ds$$

whose fixed point corresponds to a solution of (1.1.1). Evidently, the operator T is compact in the topology of B, and hence the closure of $T(B_0)$ is compact in view of the boundedness of B_0. Now to prove $T(B_0) \subset B_0$, observe that for any $x \in B_0$,

$$|Tx(t)| \leq |x_0| + \int_{t_0}^{t} |f(s, x(s))| \, ds + \int_{t_0}^{t} \int_{s}^{t} |K(\sigma, s, x(s))| \, d\sigma \, ds$$

$$\leq |x_0| + \int_{t_0}^{t} g(s, |x(s)|) \, ds + \int_{t_0}^{t} \int_{s}^{t} G(\sigma, s, |x(s)|) \, d\sigma \, ds$$

$$\leq u_0 + \int_{t_0}^{t} g(s, u(s)) \, ds + \int_{t_0}^{t} \int_{s}^{t} G(\sigma, s, u(s)) \, d\sigma \, ds = u(t),$$

using the monotonicity of g and G, the definition of B_0, and the fact $u(t)$ is a solution of (1.1.4). Therefore $|Tx(t)| \leq u(t)$, which implies $T(B_0) \subset B_0$. Hence by Tychonoff's fixed point theorem, T has a fixed point in B_0, which completes the proof of the theorem. □

If, instead of the form (1.1.1), an integro-differential equation is of the type

(1.1.5) $$x'(t) = f(t, x(t)) + \int_0^t K(t, s, x(s)) \, ds, \quad t \in R_+,$$

then it is sometimes convenient to specify an initial function $\phi_0(t)$ on the interval $0 \leq t \leq t_0$, $t_0 \geq 0$, namely

$$x(t) = \phi_0(t), \quad 0 \leq t \leq t_0,$$

and look for solutions of

(1.1.6) $$x'(t) = f(t, x(t)) + \int_0^{t_0} K(t, s, \phi_0(s)) \, ds + \int_{t_0}^t K(t, s, x(s)) \, ds,$$

for $t \geq t_0$, that depend on (t_0, ϕ_0). Naturally, (1.1.6) can also be expressed in the form (1.1.5). To see this, set $y(t) = x(t + t_0)$, so that (1.1.6) is transformed into

(1.1.7) $$y'(t) = F(t, y(t)) + \int_0^t G(t, s, y(s)) \, ds,$$

where $F(t, y(t)) = f(t + t_0, y(t)) + \int_0^{t_0} K(t + t_0, s, \phi_0(s)) \, ds$ and $G(t, s, y) = K(t + t_0, s + t_0, y)$. Here the initial function $\phi_0(t)$ is absorbed into the source function F.

We shall also sometimes consider integro-differential equations of the form

$$x'(t) = f(t, x(t)) + \int_{-\infty}^t K(t, s, x(s)) \, ds,$$

and

$$x'(t) = f(t, x(t)) + \int_{t-\tau}^t K(t, s, x(s)) \, ds, \quad \tau > 0,$$

BASIC THEORY

to which the methods developed here can also be applied.

1.2 Integro-differential Inequalities

In this section, we shall consider basic integro-differential inequalities that are needed for later use.

Theorem 1.2.1 *Assume that*

(A1) $g \in C[R_+ \times R, R]$, $H \in C[R_+^2 \times R, R]$ *and* $H(t, s, u)$ *is monotone nondecreasing in* u *for each fixed* $(t, s) \in R_+^2$;

(A2) $v' \leq g(t, v) + \int_{t_0}^{t} H(t, s, v(s))\,ds$, *and* $w' \geq g(t, w) + \int_{t_0}^{t} H(t, s, w(s))\,ds$;

(A3) *for* $(t, s) \in R_+^2$, $x \geq y$ *and* $L \geq 0$,

$$g(t, x) - g(t, y) \leq L(x - y), \quad H(t, s, x) - H(t, s, y) \leq L^2(x - y).$$

Then we have

(1.2.1) $\qquad v(t) \leq w(t) \quad \text{for } t \geq t_0, \quad \text{provided } v(t_0) \leq w(t_0).$

Proof We shall first consider the result for strict inequalities. Assume that the conclusion of the theorem is false; then there exists a t_1 such that

(1.2.2) $\qquad v(t_1) = w(t_1), \quad v(t) < w(t), \quad t_0 \leq t < t_1.$

Clearly $t_1 > t_0$, because in this case $v(t_0) < w(t_0)$. Since H is monotone nondecreasing in u, it follows from (1.2.2) that $H(t_1, s, v(s)) \leq H(t_1, s, w(s))$, $t_0 \leq s \leq t_1$. Also from (1.2.2), we get $v'(t_1) \geq w'(t_1)$. Hence, if one of the inequalities in (A2) is strict, we have

$$w'(t_1) \leq v'(t_1) = g(t_1, v(t_1)) + \int_{t_0}^{t_1} H(t_1, s, v(s))\,ds$$

$$\leq g(t_1, w(t_1)) + \int_{t_0}^{t_1} H(t_1, s, w(s))\,ds < w'(t_1),$$

which is a contradiction. Hence it follows that $v(t) < w(t)$, $t \geq t_0$ is valid. To prove the claim of the theorem, we set $w_\epsilon(t) = w(t) + \epsilon e^{2Lt}$, where $\epsilon > 0$. Then $v(t_0) \leq w(t_0) < w_\epsilon(t_0)$. Also

$$w'_\epsilon(t) = w'(t) + 2L\epsilon e^{2Lt} \geq g(t, w(t)) + \int_{t_0}^{t} H(t, s, w(s)) \, ds + 2L\epsilon e^{2Lt}$$

$$\geq g(t, w_\epsilon(t)) + \int_{t_0}^{t} H(t, s, w_\epsilon(s)) \, ds - \left(L\epsilon e^{2Lt} + \int_{t_0}^{t} L^2 \epsilon e^{2Ls} \, ds \right) + 2L\epsilon e^{2Lt}$$

$$> g(t, w_\epsilon(t)) + \int_{t_0}^{t} H(t, s, w_\epsilon(s)) \, ds, \quad t \geq t_0.$$

Now by the first part of the proof, we get $v(t) < w_\epsilon(t)$ for $t \geq t_0$, and letting $\epsilon \to 0$, we obtain the state results. The proof is therefore complete. □

The next result is concerned with implicit integro-differential inequalities which are of interest in some cases.

Theorem 1.2.2 Assume that
(i) $F \in C[R_+ \times R^3, R]$ and $F(t, x, y, z)$ is nondecreasing in x for each (t, y, z) and nonincreasing in z for each (t, x, y);
(ii) T maps $C[R_+, R]$ into $C[R_+, R]$ and for $u_1, u_2 \in C[R_+, R]$, the inequality $u_1(t) \leq u_2(t)$, $t_0 \leq t \leq t_1$, $t_0 \geq 0$, implies $Tu_1 \leq Tu_2$ for $t = t_1$;
(iii) $v, w \in C^1[R_+, R]$ and

$$F(t, v', v, Tv) \leq 0,$$
$$F(t, w', w, Tw) \geq 0, \quad t \geq t_0,$$

one of the inequalities being strict.

Then $v(t_0) < w(t_0)$ implies

(1.2.3) $$v(t) < w(t), \quad t \geq t_0.$$

Proof If the claim (1.2.3) is false, there exists a $t_1 > t_0$ such that $v(t_1) = w(t_1)$ and $v(t) < w(t)$, $t_0 \leq t < t_1$. This gives $v'(t_1) \geq w'(t_1)$.

It then follows from (ii) that $Tv \leq Tw$ at $t = t_1$. Using the monotone character of F and (iii), we then get $0 \geq F(t_1, v'(t_1), v(t_1), Tv) \geq F(t_1, w'(t_1), w(t_1), Tw) \geq 0$, which is a contradiction. Hence (1.2.3) is true and the proof is complete. □

BASIC THEORY

As a simple application of Theorem 1.2.2, we can obtain Gronwall's inequality.

Corollary 1.2.1 *Let $v, \lambda \in C[R_+, R_+]$ and suppose that*

$$v(t) \leq v_0 + \int_{t_0}^{t} \lambda(s)v(s)\,ds, \quad v_0 \geq 0,$$

then

$$v(t) \leq v_0 \exp\left[\int_{t_0}^{t} \lambda(s)\,ds\right].$$

Proof Set $F(t,x,y,z) = y - z - v_0$ and $Tu = \int_{t_0}^{t} \lambda(s)u(s)\,ds$. Consider the function $w(t) = (v_0 + \epsilon)\exp[\int_{t_0}^{t} \lambda(s)\,ds]$ for arbitrarily small $\epsilon > 0$. Then it is easy to check that

$$F(t, v, Tv) \leq 0, \quad F(t, w, Tw) > 0 \quad \text{and} \quad v_0 < w(0).$$

Then we get, by Theorem 1.2.2, $v(t) < w(t)$, $t \geq t_0$, and letting $\epsilon \to 0$, it follows that $v(t) \leq v_0 \exp[\int_{t_0}^{t} \lambda(s)\,ds]$, $t \geq t_0$, and we are done. □

1.3 Existence of Extremal Solutions

Having the existence and the inequality theory at our disposal, we now discuss the existence of extremal solutions and related results.

Theorem 1.3.1 *Assume that*

(i) $g \in C[J \times R, R]$, $H \in C[J \times J \times R, R]$, $H(t, s, u)$ *is nondecreasing in u for each (t, s) and $\int_{s}^{t} |H(\sigma, s, u(s))|\,d\sigma \leq N$ for $t_0 \leq s \leq t \leq t_0 + a$, $u \in \Omega^0$, where $J = [t_0, t_0 + a]$ and $\Omega^0 = \{u \in C[J, R]: |u(t) - u_0| \leq b\}$.*

Then there exist maximal and minimal solutions for the scalar IVP

(1.3.1) $$u' = g(t, u) + \int_{t_0}^{t} H(t, s, u(s))\,ds, \quad u(t_0) = u_0,$$

on $[t_0, t_0 + \alpha]$ for some $0 < \alpha < a$.

Proof We shall prove the existence of the maximal solution only, since the case of the minimal solution is very similar.

Let $0 < \epsilon \leq \frac{1}{2}b$ and consider the IVP

(1.3.2) $\quad u'(t) = g(t, u(t)) + \epsilon + \int_{t_0}^{t} H(t, s, u(s))\, ds, \quad u(t_0) = u_0 + \epsilon.$

Observing that $g_\epsilon(t, x) = g(t, x) + \epsilon$ is continuous on $D_\epsilon = \{(t, x): t \in J$ and $|u - (u_0 + \epsilon)| \leq \frac{1}{2}b\}$ and $D_\epsilon \subset D$, we get $|g_\epsilon(t, x)| \leq M + \frac{1}{2}b$ on D_ϵ. Consequently, we deduce from Theorem 1.1.1 that the initial value problem (1.3.2) has a solution $u(t, \epsilon)$ on the interval $[t_0, t_0 + \alpha]$ where $\alpha = \min(a, b/2M + 2N + b)$.

For $0 < \epsilon_2 < \epsilon_1 \leq \epsilon$, we have

$$u(t_0, \epsilon_2) < u(t_0, \epsilon_1),$$

$$u'(t, \epsilon_2) \leq g(t, u(t, \epsilon_2)) + \epsilon_2 + \int_{t_0}^{t} H(t, s, u(s, \epsilon_2))\, ds,$$

$$u'(t, \epsilon_2) > g(t, u(t, \epsilon_1)) + \epsilon_2 + \int_{t_0}^{t} H(t, s, u(s, \epsilon_1))\, ds, \quad t \in [t_0, t_0 + \alpha].$$

Hence by Theorem 1.2.1, it follows that $u(t, \epsilon_2) < u(t, \epsilon_1)$, $t \in [t_0, t_0 + \alpha]$. Since the family of functions $\{u(t, \epsilon)\}$ is equicontinuous and uniformly bounded on $[t_0, t_0 + \alpha]$, by the Ascoli – Arzelà theorem, there exists a decreasing sequence $\{\epsilon_n\}$ such that $\epsilon_n \to 0$ as $n \to \infty$, and the uniform limit $\gamma(t) = \lim_{t \to \infty} u(t, \epsilon_n)$ exists on $[t_0, t_0 + \alpha]$. Clearly $\gamma(t_0) = x_0$. Uniform continuity of g and H implies that $g(t, u(t, \epsilon_n))$ tends uniformly to $g(t, \gamma(t))$ and $H(\sigma, s, u(s, \epsilon_n))$ tends uniformly to $H(\sigma, s, \gamma(s))$ as $n \to \infty$, respectively, and thus term-by-term integration is applicable to

$$u(t, \epsilon_n) = u_0 + \int_{t_0}^{t} \left[g(s, u(s, \epsilon_n)) + \int_{s}^{t} H(\sigma, s, u(s, \epsilon_n))\, d\sigma \right] ds + \epsilon_n,$$

which in turn shows that the limit $\gamma(t)$ is a solution of (1.3.1) on $[t_0, t_0 + \alpha]$. We shall now show that $\gamma(t)$ is the desired maximal solution of (1.3.1) on $[t_0, t_0 + \alpha]$, satisfying $u(t) \leq \gamma(t)$ on $[t_0, t_0 + \alpha]$ for every solution $u(t)$ of (1.3.1).

Let $u(t)$ be any solution of (1.3.1) existing on $[t_0, t_0 + \alpha]$; then

$$u(t_0) = u_0 < u_0 + \epsilon = u(t_0, \epsilon),$$

$$u'(t) < g(t, u(t)) + \epsilon + \int_{t_0}^{t} H(t, s, u(s)) \, ds,$$

$$u'(t, \epsilon) \geq g(t, u(t, \epsilon)) + \epsilon + \int_{t_0}^{t} H(t, s, u(s, \epsilon)) \, ds,$$

for $t \in [t_0, t_0 + \alpha]$ and $\epsilon \leq \frac{1}{2} b$.

By Theorem 1.2.1, we obtain $u(t) < u(t, \epsilon)$ for $t \in [t_0, t_0 + \alpha]$. The uniqueness of the maximal solution shows that $u(t, \epsilon)$ tends uniformly to $\gamma(t)$ on $[t_0, t_0 + \alpha]$ as $\epsilon \to 0$ which completes the proof of the theorem. □

We next discuss a convergence theorem relating to maximal solutions since we shall need such a result later.

Lemma 1.3.1 *Let the hypothesis (i) of Theorem 1.3.1 hold. Suppose that the largest interval of existence of the maximal solution $\gamma(t)$ of (1.3.1) is $[t_0, t_0 + a)$. Then there is an $\epsilon_0 > 0$ such that for $0 < \epsilon < \epsilon_0$, the maximal solution $\gamma(t, \epsilon)$ of (1.3.2) exists over $J_1 = [t_0, t_1] \subset [t_0, t_0 + a)$, and $\lim_{\epsilon \to 0} \gamma(t, \epsilon) = \gamma(t)$ uniformly on J_1.*

Proof Let Ω_0 be an open bounded set such that $\bar{\Omega}_0 \subset R_+ \times R$ and $(t, \gamma(t)) \in \Omega_0$ for $t \in J_1$. We can choose a $b > 0$ such that for $t \in J_1$, the rectangle $\Omega_t^\epsilon : [t, t+b], |u(t) - (\gamma(t) + \epsilon)| \leq b$ is included in Ω_0 for $\epsilon \leq \frac{1}{2}b$. Let $|g(t, u)| \leq M$ on Ω_0. Then it is evident that $|g(t, u) + \epsilon| \leq M + \frac{1}{2}b$. Consider $\Omega_{t_0}^\epsilon$. By Theorem 1.3.1, the maximal solution $\gamma(t, \epsilon)$ of (1.3.2) exists on $[t_0, t_0 + \eta]$ where $\eta = \min[b, 2b/2M + 2N + b]$ and η does not depend on ϵ. Furthermore, proceeding as in Theorem 1.3.1, we can conclude, in view of the uniqueness of the maximal solution $\gamma(t)$ of (1.3.1) that $\lim_{\epsilon \to 0} \gamma(t, \epsilon) = \gamma(t)$ uniformly on $[t_0, t_0 + \eta]$ which implies $\lim_{\epsilon \to 0} \gamma(t_0 + \eta, \epsilon) = \gamma(t_0, \eta)$. Consequently, there is an $\epsilon_1 < \frac{1}{2}b$ such that for $0 < \epsilon \leq \epsilon_1$, we have $\gamma(t_0 + \eta, \epsilon) \leq \gamma(t_0, \eta) + \epsilon$. We can now repeat the foregoing argument with respect to the rectangle $\Omega_{t_0 + \eta}^\epsilon$, $\epsilon < \epsilon_1$ such that for $\epsilon < \epsilon_2$ the maximal solution $\hat{\gamma}(t, \epsilon)$ of

$$u' = g(t, u) + \epsilon + \int_{t_0}^{t} H(t, s, u(s)) \, ds, \quad u(t_0) = u_0, \quad u(t_0 + \eta) = \gamma(t_0 + \eta) + \epsilon$$

exists on $[t_0+\eta, t_0+2\eta]$ and $\lim_{\epsilon \to 0} \hat{\gamma}(t,\epsilon) = \gamma(t)$ uniformly on $[t_0+\eta, t_0+2\eta]$. For $\epsilon < \epsilon_2$, we can extend the function $\gamma(t,\epsilon)$ by defining $\gamma(t,\epsilon) = \hat{\gamma}(t,\epsilon)$, $t \in [t_0+\eta, t_0+2\eta]$. It is clear that $\gamma(t,\epsilon)$ is the maximal solution of (1.3.2) on $[t_0, t_0+2\eta]$ and $\lim_{\epsilon \to 0} \gamma(t,\epsilon) = \gamma(t)$ uniformly on $[t_0, t_0+2\eta]$.

By induction, it can be shown that there is an $\epsilon_0 = \epsilon_n$ such that $[t_0,t_1] \subset [t_0, t_1+n\eta]$, that the maximal solution of (1.3.2) exists on $[t_0, t_0+n\eta]$ for $0 < \epsilon < \epsilon_0$ and that $\lim_{\epsilon \to 0} \gamma(t,\epsilon) = \gamma(t)$ uniformly on $[t_0, t_0+n\eta]$. Hence the lemma is proved. □

The next result is concerned with the successive approximation of the comparison equation.

Lemma 1.3.2 *Assume that*

(i) $g \in C[J \times [0, 2b], R]$, $g(t,0) \equiv 0$, *and* $g(t,u)$ *is nondecreasing in* u *for each* $t \in J$ *where* $J = [t_0, t_0+a]$,

(ii) $u(t) \equiv 0$ *is the unique solution of* $u' = g(t,u) + \int_{t_0}^{t} H(t,s,u(s))ds$, $u(t_0) = 0$ *on* $[t_0, t_0+a]$,

(iii) $H \in C[J \times J \times [0, 2b], R]$, $H(t,s,0) \equiv 0$ *and* $H(t,s,u)$ *is nondecreasing in* u *for each* $(t,s) \in J \times J$ *and* $\int_{s}^{t} |H(\sigma, s, u(s))| \, d\sigma \leq N$ *for* $t_0 \leq s \leq t \leq t_0+a$, $u \in \Omega^0 = \{u \in C[J, R]: |u(t)| \leq 2b\}$.

Then the successive approximations

(1.3.3)
$$u_0(t) = (M+N)(t-t_0)$$
$$u_{n+1}(t) = \int_{t_0}^{t}\left[g(s,u_n(s)) + \int_{s}^{t} H(\sigma, s, u_n(s))d\sigma\right]ds$$

are well defined, $0 \leq u_{n+1}(t) \leq u_n(t)$ *on* $[t_0, t_0+a]$, *and*

(1.3.4)
$$\lim_{n \to \infty} u_n(t) \equiv 0 \text{ uniformly on } [t_0, t_0+a].$$

Moreover, for every $n \geq 1$, *the maximal solution* $\gamma_n(t)$ *of*

(1.3.5)
$$u' = g(t,u) + pg(t, u_{n-1}(t)) + \int_{t_0}^{t} H(t,s,u(s))ds$$
$$+ p\int_{t_0}^{t} H(t,s,u_{n-1}(s))ds,$$

$$u_n(t_0) = 0, \quad p > 0$$

exists on J and $\lim_{n \to \infty} \gamma_n(t) \equiv 0$ uniformly on J.

Proof An easy induction proves (1.3.3). Since by $|g(t,u)| \leq M$ and (ii), $|u'_n| \leq (M+N)$, by Ascoli–Arzelà theorem we can conclude that $\lim_{n \to \infty} u_n(t) = u(t)$ uniformly on J. It is clear that $u(t)$ satisfies $u' = g(t,u) + \int_{t_0}^{t} H(t,s,u(s))\,ds$ and $u(t_0) = 0$. By (ii), it follows that $u(t) \equiv 0$ and (1.3.4) is proved. Given ϵ, there is an $n \geq n(\epsilon)$ such that $|pg(t, u_{n-1}(t))| < \frac{1}{2}\epsilon$ and $|p\int_{t_0}^{t} H(t,s,u_{n-1}(s))ds| < \frac{1}{2}\epsilon$ because of $g(t,0) \equiv 0$, $H(t,s,0) \equiv 0$ and (1.3.4). Now an argument similar to that of Lemma 1.3.1 proves (1.3.5). Hence the proof is complete. □

1.4 Comparison Results

An important method in the theory of integro-differential equations is concerned with estimating a function satisfying an integro-differential inequality by the extremal solutions of the corresponding integro-differential equation. One of the results of this type is the following comparison theorem.

Theorem 1.4.1 Assume that $g \in C[R_+^2, R]$, $H \in C[R_+^3, R]$, $H(t,s,u)$ is nondecreasing in u for each (t,s) and for $t \geq t_0$,

$$(1.4.1) \qquad D_- m(t) \leq g(t, m(t)) + \int_{t_0}^{t} H(t,s,m(s))\,ds,$$

where $m \in C[R_+, R]$ and $D_- m(t) = \liminf_{h \to 0^-} h^{-1}[m(t+h) - m(t)]$. Suppose that $\gamma(t)$ is the maximal solution of

$$(1.4.2) \qquad u'(t) = g(t, u(t)) + \int_{t_0}^{t} H(t,s,u(s))\,ds, \quad u(t_0) = u_0 \geq 0,$$

existing on $[t_0, \infty)$. Then

$$(1.4.3) \qquad m(t) \leq \gamma(t), \quad t \geq t_0,$$

provided $m(t_0) \leq u_0$.

Proof Let $t_0 < \tau$. By Theorem 1.3.1, the maximal solution $\gamma(t,\epsilon)$ of

$$(1.4.4) \qquad u'(t) = g(t, u(t)) + \epsilon + \int_{t_0}^{t} H(t, s, u(s)) \, ds, \qquad u(t_0) = u_0 + \epsilon$$

exists on $[t_0, \tau]$ for all $\epsilon > 0$ sufficiently small, and

$$(1.4.5) \qquad \gamma(t) = \lim_{\epsilon \to 0} \gamma(t, \epsilon) \text{ uniformly on } [t_0, \tau].$$

Using (1.4.4) and (1.4.1) and applying the inequality, we find that $m(t) < \gamma(t,\epsilon)$, $t \in [t_0, \tau]$. The last inequality together with (1.4.5) proves the assertion (1.4.3) of the theorem. □

Next we give an alternative proof for Theorem 1.4.1, which does not depend on Theorem 1.3.1.

Proof of Theorem 1.4.1 Define the functions

$$(1.4.6) \qquad F(t, u) = g(t, p(t, u)), \quad G(t, s, u) = H(t, s, p(s, u)),$$

where $p(t, u) = \max[m(t), u]$ and let $u(t)$ be any solution of

$$u'(t) = F(t, u(t)) + \int_{t_0}^{t} G(t, s, u(s)) \, ds, \qquad u(t_0) \geq m(t_0).$$

We shall show that $m(t) \leq u(t)$ for $t \geq t_0$. If this is not true, suppose that for some $t^* > t_0$, $m(t^*) > u(t^*)$. Then there exists a $t_1 > t_0$ such that

$$(1.4.7) \qquad u(t_1) \leq m(t_1) \quad \text{and} \quad u'(t_1) < D_- m(t_1).$$

By definition of $p(t, u)$, it is clear that $m(t) \leq p(t, u(t))$ and $m(t_1) = p(t_1, u(t_1))$. Hence, using the monotonic character of H, we get

$$\begin{aligned} D_- m(t_1) &\leq g(t_1, m(t_1)) + \int_{t_0}^{t_1} H(t_1, s, m(s)) \, ds \\ &\leq g(t_1, p(t_1, u(t_1))) + \int_{t_0}^{t_1} H(t_1, s, p(s, u(s))) \, ds \\ &= F(t_1, u(t_1)) + \int_{t_0}^{t_1} G(t_1, s, p(s, u(s))) \, ds = u'(t_1), \end{aligned}$$

which is a contradiction to (1.4.7). Thus $m(t) \leq u(t)$, which implies that $u(t)$ is also a solution of (1.4.2). Since $\gamma(t)$ is the maximal solution of (1.4.2), it follows that $m(t) \leq \gamma(t)$ for $t \geq t_0$ and the proof is complete. □

Corollary 1.4.1 Let (1.4.1) hold with $g(t,u) = a(t)u$ and $H(t,s,u) = h(t,s)u$. Then $m(t) \leq R(t,t_0)u(t_0)$, $t \geq t_0$, where $R(t,s)$ is the solution of

$$\frac{\partial R}{\partial s}(t,s) + R(t,s)a(s) + \int_s^t R(t,\sigma)h(\sigma,s)\,d\sigma = 0, \quad R(t,t) = I$$

on the interval $t_0 \leq s \leq t$.

For the proof, it is sufficient to observe that $\gamma(t) = R(t,t_0)u(t_0)$ is the solution of (1.4.2). □

Although the foregoing comparison result plays an important role in the study of qualitative theory of integro-differential systems, it is not fruitful in many situations, because finding solutions of integro-differential equations is more difficult. Thus it would be more useful in applications if it were possible to reduce the study of scalar integro-differential inequalities to that of differential inequalities. We take this approach and prove below some comparison results.

We need the following lemma before we proceed further.

Lemma 1.4.1 Let $g_0, g \in C[R_+^2, R]$ satisfy

(1.4.8) $$g_0(t,u) \leq g(t,u), \quad (t,u) \in R_+^2.$$

Then the right maximal solution $\gamma(t, t_0, u_0)$ of

(1.4.9) $$u' = g(t,u), \quad u(t_0) = u_0 \geq 0$$

and the left maximal solution $\eta(t, T, v_0)$ of

(1.4.10) $$u' = g_0(t,u), \quad u(T) = v_0 \geq 0$$

satisfy the relation

(1.4.11) $$\gamma(t,t_0,u_0) \leq \eta(t,T,v_0), \quad t \in [t_0, T],$$

whenever $\gamma(T, t_0, u_0) \leq v_0$.

Proof It is known that $\lim_{\epsilon \to 0} u(t,\epsilon) = \gamma(t,t_0,u_0)$ and $\lim_{\epsilon \to 0} v(t,\epsilon) = \eta(t,T,v_0)$ where $u(t,\epsilon)$ is any solution of $u' = g(t,u) + \epsilon$, $u(t_0) = u_0 + \epsilon$, existing to the right of t_0, $v(t,\epsilon)$ is any solution of $v' = g_0(t,v) - \epsilon$, $v(T) = v_0$, existing to the left of T, and $\epsilon > 0$ is sufficiently small. Note that (1.4.11) follows if we first establish the inequality $u(t,\epsilon) < v(t,\epsilon)$, $t_0 \leq t < T$. Since $g_0 \leq g$ and $\gamma(T,t_0,u_0) \leq v_0$, it is easy to see that for a sufficiently small $\delta > 0$, we have $u(t,\epsilon) < v(t,\epsilon)$, $T - \delta \leq t < T$, and, in particular, $u(T-\delta,\epsilon) < v(T-\delta,\epsilon)$. We claim that $u(t,\epsilon) < v(t,\epsilon)$, $t_0 \leq t < T - \delta$. If this is not true, there exists a $t^* \in [t_0, T-\delta)$ such that

$$u(t,\epsilon) < v(t,\epsilon), \quad t^* < t \leq T - \delta, \quad \text{and} \quad u(t^*,\epsilon) = v(t^*,\epsilon).$$

This leads to the contradiction

$$g(t^*, u(t^*,\epsilon)) + \epsilon = u'(t^*,\epsilon) \leq v'(t^*,\epsilon) = g_0(t^*, v(t^*,\epsilon)) - \epsilon.$$

Hence $u(t,\epsilon) < v(t,\epsilon)$, $t_0 \leq t \leq T - \delta$ and the proof of the lemma is complete. $\square$

We are now in a position to prove the following comparison result, which plays an important role in the study of integro-differential inequalities.

Theorem 1.4.2 *Let* $m \in C[R_+, R_+]$, $g \in C[R_+^2, R]$, $H \in C[R_+^3, R]$ *and*

$$(1.4.12) \qquad D_- m(t) \leq g(t, m(t)) + \int_{t_0}^t H(t, s, m(t))\, ds, \quad t \in I_0,$$

where $I_0 = \{t \geq t_0 : m(s) \leq \eta(s, t, m(t)), t_0 \leq s \leq t\}$, $\eta(t, T, v_0)$ *being the left maximal solution of* (1.4.10) *existing on* $[t_0, T]$. *Assume that*

$$(1.4.13) \qquad g_0(t, u) \leq F(t, u; t_0),$$

where

$$(1.4.14) \qquad F(t, u; t_0) = g(t, u) + \int_{t_0}^t H(t, s, u)\, ds$$

and $\gamma(t)$ *is the maximal solution of*

$$(1.4.15) \qquad u' = F(t, u; t_0), \quad u(t_0) = u_0$$

existing on $[t_0, \infty)$. *Then*

$$(1.4.16) \qquad m(t_0) \leq u_0 \quad \text{implies} \quad m(t) \leq \gamma(t), \quad t \geq t_0.$$

Proof Since it is known that $\lim_{\epsilon \to 0} u(t,\epsilon) = \gamma(t)$, where $u(t,\epsilon)$ is a solution of $u' = F(t,u;t_0) + \epsilon$, $u(t_0) = u_0 + \epsilon$, for $\epsilon > 0$ sufficiently small, on any compact set $[t_0, T] \subset [t_0, \infty)$, it is enough to prove that $m(t) < u(t,\epsilon)$, $t_0 \leq t \leq T$. If this is not true then there exists a $t^* \in (t_0, T]$ such that $m(s) < u(s,\epsilon)$, $t_0 \leq s < t^*$, and $m(t^*) = u(t^*, \epsilon)$. This implies that

$$(1.4.17) \qquad D_- m(t^* \geq u'(t^*,\epsilon) = F(t^*, u(t^*,\epsilon); t_0) + \epsilon.$$

Consider now the left maximal solution $\eta(s, t^*, m(t^*))$, $t_0 \leq s \leq t^*$, of $u' = g_0(t,u)$, $\eta(t^*) = m(t^*)$. By Lemma 1.4.1, $\gamma(s, t_0, u_0) \leq \eta(s, t^*, m(t^*))$, $t_0 \leq s \leq t^*$. Since $\gamma(t^*, t_0, u_0) = \lim_{\epsilon \to 0} u(t^*, \epsilon) = m(t^*) = \eta(t^*, t^*, m(t^*))$ and $m(s) \leq u(s, \epsilon)$, $t_0 \leq s \leq t^*$, it follows that

$$m(s) \leq \gamma(s, t_0, u_0) \leq \eta(s, t^*, m(t^*)), \quad t_0 \leq s \leq t^*.$$

This inequality implies that $t^* \in I_0$, and as a result (1.4.10) yields

$$D_- m(t^*) \leq F(t^*, m(t^*); t_0)$$

which contradicts (1.4.17). Thus, $m(t) \leq \gamma(t)$ for $t \geq t_0$ and the proof is complete. $\square$

In Theorem 1.4.2, we could choose $g_0 = g$ if $H \geq 0$.

We shall now use the comparison Theorem 1.4.2 to study the integro-differential inequality

$$(1.4.18) \qquad D_- m(t) \leq g(t, m(t)) + \int_{t_0}^{t} H(t, s, m(s)) \, ds,$$

where $m \in C[R_+, R_+]$. The advantage of this approach will be clear when we consider some special cases.

Theorem 1.4.3 Let (1.4.18) hold with $g \in C[R_+^2, R]$ and $H \in C[R_+^3, R]$, and let $H(t, s, u)$ be nondecreasing in u for each (t,s) and for $t \geq t_0$. Suppose that $g_0 \in C[R_+^2, R]$ be such that $g_0 \leq F$, where

$$F(t, u, t_0) = g(t, u) + \int_{t_0}^{t} K(t, s, u) \, ds,$$

and $K(t, s, u) = H(t, s, \eta(s, t, u))$, where $\eta(s, T, v_0)$ is the left maximal solution of (1.4.10) existing on $t_0 \leq t \leq T$. Then $m(t) \leq \gamma(t)$, $t \geq t_0$, whenever $m(t_0) \leq u_0$, where $\gamma(t)$ is the maximal solution of (1.4.15) existing on $[t_0, \infty)$.

Proof We set $I_0 = \{t \geq t_0 : m(s) \leq \eta(s,t,m(t)), t_0 \leq s \leq t\}$. Since H is nondecreasing in u, (1.4.18) yields the inequality

$$D_- m(t) \leq g(t, m(t)) + \int_{t_0}^{t} H(t, s, \eta(s, t, m(t))) \, ds, \quad t \in I_0.$$

Hence, by Theorem 1.4.2, we have the stated result. □

Some of the interesting special cases of Theorem 1.4.3 are given below as corollaries.

Corollary 1.4.2 *Consider the special case when $g \geq 0$ and $H \geq 0$ in Theorem 1.4.3. Then $m(t) \leq \gamma(t)$, $t \geq t_0$, where $\gamma(t)$ is the maximal solution of*

$$u'(t) = g(t, u(t)) + \int_{t_0}^{t} H(t, s, u(t)) \, ds, \quad u(t_0) = u_0,$$

existing on $[t_0, \infty)$, provided $m(t_0) \leq u_0$. If, in particular, $g(t, u) = a(t)u$ and $H(t, s, u) = H(t, s)u$, with $a(t) \geq 0$ and $H(t, s) \geq 0$, then

$$m(t) \leq m(t_0) \exp\left[\int_{t_0}^{t} A(s, t_0) \, ds\right], \quad t \geq t_0,$$

where $A(t, t_0) = a(t) + \int_{t_0}^{t} H(t, s) \, ds$.

Corollary 1.4.3 *Consider the special case with $g = g_0$ and $H \geq 0$ in Theorem 1.4.3. Then $m(t) \leq \gamma(t)$, $t \geq t_0$, where $\gamma(t)$ is the maximal solution of (1.4.15) existing on $[t_0, \infty)$. If, in particular, $g = g_0 = -\alpha u$, $\alpha > 0$, and $H(t, s, u) = H(t, s)u$, with $H(t, s) \geq 0$, then*

$$m(t) \leq m(t_0) \exp\left[\int_{t_0}^{t} B(s, t_0) \, ds\right], \quad t \geq t_0,$$

where $B(t, t_0) = -\alpha + \int_{t_0}^{t} H(t, s) \exp[\alpha(t - s)] \, ds$.

Corollary 1.4.4 *Consider the special case with $g = g_0 = -C(u)$ and $H \geq 0$, where $C(u)$ is continuous, nondecreasing in u and $C(0) = 0$. Then $m(t) \leq \gamma(t)$, $t \geq t_0$, where $\gamma(t)$ is the maximal solution of*

$$u'(t) = -C(u(t)) + \int_{t_0}^{t} H(t, s, J^{-1}(J(u(t)) - (t - s))) \, ds, \quad u(t_0) = u_0$$

existing on $[t_0, \infty)$, provided $m(t_0) \leq u_0$. Here $J(u) = \int_{u_0}^{u} ds/C(s)$ and J^{-1} is the inverse of J.

Finally, we give another useful corollary.

Corollary 1.4.5 Consider the integral inequality

$$m(t) \leq h(t) + \int_{t_0}^{t} K(t,s) m(s) \, ds, \quad t \geq t_0$$

where $m \in C[R_+, R_+]$, $h \in C[R_+, R_+]$, $K \in C[R_+^2, R_+]$ and $K_t(t,s)$ exists, is continuous and nonnegative. Then

$$m(t) \leq h(t) + \int_{t_0}^{t} \sigma(s, t_0) \exp\left[\int_{s}^{t} d(\xi, t_0) \, d\xi\right] ds, \quad t \geq t_0,$$

where $d(t, t_0) = K(t,t) + \int_{t_0}^{t} K_t(t,s) \, ds$ and $\sigma(t, t_0) = K(t,t) h(t) + \int_{t_0}^{t} K_t(t,s) h(s) \, ds$.

1.5 Convergence of Successive Approximations

It is well known that one cannot always obtain a solution of (1.1.1) as a limit of the sequence of successive approximations. However, if the comparison functions are non-decreasing, we can show that successive approximations do converge to the unique solution. This is the content of the following result.

Theorem 1.5.1 Assume that

(A1) $f \in C[J \times R^n, R^n]$, $K \in C[J \times J \times R^n, R^n]$ and $\int_{s}^{t} |K(\sigma, s, x(s))| \, d\sigma \leq N$ for $t_0 \leq s \leq t \leq t_0 + a$, $x \in \Omega = \{\phi \in C[J, R^n]: \phi(t_0) = x_0$ and $|\phi(t) - x_0| \leq b\}$;

(A2) $|f(t,x) - f(t,y)| \leq g(t, |x-y|)$ and $|K(t,s,x) - K(t,s,y)| \leq H(t,s, |x-y|)$, where $g \in C[J \times [0, 2b], R_+]$, $H \in C[J \times J \times [0, 2b], R_+]$, $g(t, 0) \equiv 0$, $H(t, s, 0) \equiv 0$, $g(t, u)$ and $H(t,s,u)$ are nondecreasing in u for each $(t,s) \in J \times J$ and $\int_{s}^{t} H(\sigma, s, u(s)) \, d\sigma \leq N_0$ for $t_0 \leq s \leq t \leq t_0 + a$, $u \in \Omega^0 = \{u \in C[J, R_+]: |u(t)| \leq 2b\}$;

(A3) the comparison equation (1.4.2) admits only the trivial solution.

Then the successive approximations defined by

$$(1.5.1) \quad x_{n+1}(t) = x_0 + \int_{t_0}^{t} \left[f(s, x_n(s)) + \int_{s}^{t} K(\sigma, s, x_n(s)) \, d\sigma\right] ds$$

exist on $t_0 \leq t \leq t_0 + \alpha$, where $\alpha = \min(a, b/M + N)$, as continuous functions and converge uniformly on this interval to the solution of $x(t)$ of (1.1.1).

Proof It is easy to see, by induction, that successive approximations (1.5.1) are defined and continuous on $t_0 \leq t \leq t_0 + \alpha$ and satisfy $|x_k(t) - x_0| \leq b$ for $k = 0, 1, 2, \ldots$. Now

$$|x_1(t) - x_0| \leq \int_{t_0}^{t} \left[|f(s, x_0)| + \int_{s}^{t} |K(\sigma, s, x_0)| d\sigma \right] ds$$

$$\leq (M + N)(t - t_0) = u_0(t),$$

where $u_k(t)$ are the successive approximations given in Lemma 1.3.2.

Assume that $|x_k(t) - x_{k-1}(t)| \leq u_{k-1}(t)$ for given k. Since

$$|x_{k+1}(t) - x_k(t)|$$

$$\leq \int_{t_0}^{t} \left[|f(s, x_k(s)) - f(s, x_{k-1}(s))| + \int_{s}^{t} |K(\sigma, s, x_k(s)) - K(\sigma, s, x_{k-1}(s))| d\sigma \right] ds,$$

using monotonicity of g and H and assumption $(A2)$, we have

$$|x_{k+1}(t) - x_k(t)| \leq \int_{t_0}^{t} \left[g(s, u_{k-1}(s)) + \int_{s}^{t} H(\sigma, s, u_{k-1}(s)) d\sigma \right] ds = u_k(t).$$

Thus, by induction, the inequality $|x_{n+1}(t) - x_n(t)| \leq u_n(t)$, $t_0 \leq t \leq t_0 + a$, is true for all n. Also,

$$|x'_{n+1}(t) - x'_n(t)| \leq |f(t, x_n(t)) - f(t, x_{n-1}(t))|$$

$$+ \int_{t_0}^{t} |K(\sigma, s, x_n(s)) - K(\sigma, s, x_{n-1}(s))| ds$$

$$\leq g(t, |x_n(t) - x_{n-1}(t)|) + \int_{t_0}^{t} H(\sigma, s, |x_n(s) - x_{n-1}(s)|) ds$$

$$\leq g(t, u_{n-1}(t)) + \int_{t_0}^{t} H(\sigma, s, u_{n-1}(s)) ds.$$

Let $1 \leq n \leq m$. Then we can easily obtain

$$|x'_n(t) - x'_m(t)|$$
$$\leq |x'_n(t) - x'_{n+1}(t)| + |x'_m(t) - x'_{m+1}(t)|$$
$$+ |x'_{n+1}(t) - x'_{m+1}(t)|$$
$$\leq g(t, u_{n-1}(t)) + \int_{t_0}^{t} H(\sigma, s, u_{n-1}(s)) ds$$
$$+ g(t, u_{m-1}(t)) + \int_{t_0}^{t} H(\sigma, s, u_{m-1}(s)) ds$$
$$+ g(t, |x_n(t) - x_m(t)|) + \int_{t_0}^{t} H(\sigma, s, |x_n(s) - x_m(s)|) ds.$$

Since by Lemma 1.3.2, $u_{n+1}(t) \leq u_n(t)$ for all n, it follows that

$$D^+ |x_n(t) - x_m(t)| \leq g(t, |x_n(t) - x_m(t)|) + \int_{t_0}^{t} H(\sigma, s, |x_n(s) - x_m(s)|) ds$$
$$+ 2g(t, u_{n-1}(t)) + 2\int_{t_0}^{t} H(\sigma, s, u_{n-1}(s)) ds,$$

because of the monotonicity of $g(t, u)$ and $H(t, s, u)$ in u.

An application of the comparison Theorem 1.4.1 yields that

$$|x_n(t) - x_m(t)| \leq \gamma_n(t), \quad t_0 \leq t \leq t_0 + \alpha$$

where $\gamma_n(t)$ is the maximal solution of

$$y' = g(t, y) + \int_{t_0}^{t} H(\sigma, s, y(s)) ds + 2g(t, u_{n-1}(t))$$
$$+ 2\int_{t_0}^{t} H(\sigma, s, u_{n-1}(s)) ds, \quad y_n(t_0) = 0, \quad \text{for each } n.$$

By Lemma 1.3.2, $\gamma_n(t) \to 0$ uniformly on $t_0 \leq t \leq t_0 + \alpha$ as $n \to \infty$. This implies that $x_n(t)$ converges uniformly to $x(t)$ on $t_0 \leq t \leq t_0 + \alpha$ as $n \to \infty$, and it is easy to show that $x(t)$ is a solution of (1.1.1) by standard argument.

Now we shall show that solution is unique. Suppose that $x(t)$ and $y(t)$ are two solutions of (1.1.1). Then, setting $m(t) = |x(t) - y(t)|$, we get, using ($A2$) and ($A3$),

$$D_- m(t) \leq g(t, m(t)) + \int_{t_0}^{t} H(t, s, m(s))\, ds, \quad m(t_0) = 0.$$

By the comparison Theorem 1.4.1, it now follows that

$$m(t) \leq \gamma(t, t_0, 0), \quad t_0 \leq t \leq t_0 + a,$$

where $\gamma(t, t_0, 0)$ is the maximal solution of (1.4.2). But by hypothesis $\gamma(t, t_0, 0) \equiv 0$ and hence $x(t) \equiv y(t)$ for $t_0 \leq t \leq t_0 + a$. The proof of the theorem is complete. $\square$

1.6 Continuous Dependence

We shall consider the problem of continuity of solutions of (1.1.1) with respect to initial values (t_0, x_0).

We need the following result.

Lemma 1.6.1 *Assume that*

($A1$) $f \in C[J \times R^n, R^n]$ *and let* $G(t, \gamma) = \max\limits_{|x - x_0| \leq \gamma} |f(t, x)|$;

($A2$) $K \in C[J \times J \times R^n, R^n]$ *and let* $H(t, s, \gamma) = \max\limits_{|x - x_0| \leq \gamma} |K(t, s, x)|$;

($A3$) $\gamma^*(t, t_0, 0)$ *is the maximal solution of*

$$u'(t) = G(t, u(t)) + \int_{t_0}^{t} H(t, s, u(s))\, ds, \quad u(t_0) = 0.$$

If $x(t) = x(t, t_0, x_0)$ is any solution (1.1.1) then

$$|x(t, t_0, x_0) - x_0| < \gamma^*(t, t_0, 0), \quad t \geq t_0.$$

Proof Define $v(t) = |x(t) - x_0|$. Then

$$D^+ v(t) \leq |x'(t)| = \left| f(t, x(t)) + \int_{t_0}^{t} K(t, s, x(s))\, ds \right|$$

$$\leq \max_{|x - x_0| \leq v(t)} |f(t, x)| + \int_{t_0}^{t} \max_{|x - x_0| < v(s)} |K(t, s, x(s))|\, ds$$

$$= G(t, v(t)) + \int_{t_0}^{t} H(t, s, v(s)) \, ds.$$

This implies by Theorem 1.4.1 that

$$v(t) = |x(t, t_0, x_0) - x_0| \leq \gamma^*(t, t_0, 0), \qquad t \geq t_0,$$

and this proves the lemma. □

Now we are in a position to prove the continuous dependence of solutions $x(t, t_0, x_0)$ of (1.1.1) with respect to initial values (t_0, x_0).

Theorem 1.6.1 *Let the hypothesis of Theorem 1.5.1 be satisfied. Suppose that the solutions $u(t, t_0, u_0)$ of (1.4.2) through every point (t_0, u_0) are continuous with respect to initial conditions (t_0, u_0). Then the solutions $x(t, t_0, x_0)$ of (1.1.1) are unique and continuous with respect to the initial values (t_0, x_0).*

Proof Since the uniqueness follows from Theorem 1.5.1, we have to prove the continuity part only. To that end, let $x(t, t_0, x_0)$ and $y(t, t_0, y_0)$ be the solutions of (1.1.1) through (t_0, x_0) and (t_0, y_0) respectively. Define $v(t) = |x(t, t_0, x_0) - y(t, t_0, y_0)|$; then condition (A2) implies the integro-differential inequality

$$D^+ v(t) \leq g(t, v(t)) + \int_{t_0}^{t} H(t, s, v(s)) \, ds.$$

By Theorem 1.4.1, we obtain $v(t) \leq \gamma(t, t_0, |x_0 - y_0|)$, $t \geq t_0$, where $\gamma(t, t_0, |x_0 - y_0|)$ is the maximal solution of (1.4.2), such that $u(t_0) = |x_0 - y_0|$. Since the solution $u(t, t_0, u_0)$ of (1.4.2) is assumed to be continuous with respect to the initial values, it follows that $\lim_{x_0 \to y_0} \gamma(t, t_0, |x_0 - y_0|) = \gamma(t, t_0, 0)$. By hypothesis, $\gamma(t, t_0, 0) \equiv 0$. This in view of the fact the definition of $v(t)$ yields $\lim_{x_0 \to y_0} x(t, t_0, x_0) = y(t, t_0, x_0)$, which shows the continuity of $x(t, t_0, x_0)$ with respect x_0.

We shall next prove continuity with respect to initial time t_0.

If $x(t, t_0, x_0)$ and $y(t, t_1, x_0)$, $t_1 > t_0$, are the solutions of (1.1.1) through (t_0, x_0) and (t_1, x_0) respectively then, as before, we obtain the inequality

$$D^+ v(t) \leq g(t, v(t)) + \int_{t_0}^{t} H(t, s, v(s)) \, ds,$$

where $v(t) = |x(t,t_0,x_0) - y(t,t_1,x_0)|$. Also, $v(t_1) = |x(t_1,t_0,x_0) - x_0|$. Hence, by Lemma 1.6.1, $v(t_1) \leq \gamma^*(t_1,t_0,0)$, and consequently $v(t) \leq \tilde{\gamma}(t)$, $t > t_1$, where $\tilde{\gamma}(t) = \tilde{\gamma}(t,t_1,\gamma^*(t_1,t_0,0))$ is the maximal solution of (1.4.2) through $(t_1,\gamma^*(t_1,t_0,0))$. Since $\gamma^*(t_0,t_0,0) = 0$, we have

$$\lim_{t_1 \to t_0} \tilde{\gamma}(t,t_1,\gamma^*(t_1,t_0,0)) = \tilde{\gamma}(t,t_0,0),$$

and by hypothesis $\tilde{\gamma}(t,t_0,0)$ is identically zero, thus proving the continuity of $x(t,t_0,x_0)$ with respect to t_0. The proof of the theorem is complete. □

1.7 Linear Variation of Parameters

Consider the linear integro-differential equation

$$(1.7.1) \qquad x'(t) = A(t)x(t) + \int_{t_0}^t K(t,s)x(s)\,ds + F(t), \quad x(t_0) = x_0,$$

where $A(t)$ and $K(t,s)$ are $n \times n$ continuous matrices for $t \in R_+$ and $(t,s) \in R_+ \times R_+$ respectively and $F \in C[R_+, R^n]$. Define, for $t_0 \leq s \leq t < \infty$,

$$(1.7.2) \qquad \psi(t,s) = A(t) + \int_s^t K(t,\sigma)\,d\sigma,$$

and

$$(1.7.3) \qquad R(t,s) = I + \int_s^t R(t,\sigma)\psi(\sigma,s)\,d\sigma,$$

where I is the identity matrix and $K(t,s) = \psi(t,s) = R(t,s) = 0$ if $s > t \geq t_0$. Then we have the following result.

Theorem 1.7.1 *Assume that $A(t)$ and $K(t,s)$ are continuous $n \times n$ matrices for $t \in R_+$, $(t,s) \in R_+ \times R_+$ and $F \in C[R_+, R^n]$. Then the solution $x(t)$ of (1.7.1) satisfies*

$$(1.7.4) \qquad x(t) = R(t,t_0)x_0 + \int_{t_0}^t R(t,s)F(s)\,ds, \quad x(t_0) = x_0$$

where $R(t,s)$ is the unique solution of

$$(1.7.5) \quad \frac{\partial R(t,s)}{\partial s} + R(t,s)A(s) + \int_s^t R(t,\sigma)K(\sigma,s)\,d\sigma = 0, \quad R(t,t) = I.$$

Proof Since $\psi(t,s)$ is continuous in (t,s), the existence of $R(t,s)$ on $t_0 \le s \le t$ is trivial. Also, it is clear that $\partial R(t,s)/\partial s$ exists and satisfies (1.7.5). Let $x(t)$ be the solution of (1.7.1) for $t \ge t_0$.

Setting $p(s) = R(t,s)x(s)$, we have

$$\begin{aligned}
p'(s) &= \frac{\partial R(t,s)}{\partial s}x(s) + R(t,s)x'(s) \\
&= \frac{\partial R(t,s)}{\partial s}x(s) + R(t,s)\left[A(s)x(s) + \int_{t_0}^s K(s,u)x(u)\,du + F(s)\right].
\end{aligned}$$

Integrating between t_0 and t, we get

$$p(t) - p(t_0) = \int_{t_0}^t \left[\frac{\partial R(t,s)}{\partial s}x(s) + R(t,s)A(s)x(s) + R(t,s)F(s)\right]ds$$

$$+ \int_{t_0}^t R(t,s)\left[\int_{t_0}^s K(s,u)x(u)\,du\right]ds.$$

Using Fubini's theorem, we obtain

$$x(t) - R(t,t_0)x_0 = \int_{t_0}^t \left[\frac{\partial R}{\partial s} + R(t,s)A(s) + \int_s^t R(t,u)K(u,s)\,du\right]x(s)\,ds$$

$$+ \int_{t_0}^t R(t,s)F(s)\,ds.$$

This together with (1.7.5) yields (1.7.4).

Conversely, suppose that $y(t)$ is a solution of (1.7.4) with $y(t_0) = x_0$ existing for $t_0 \le t < \infty$. Then from the identity

$$\int_{t_0}^t R(t,s)y'(s)\,ds = R(t,t)y(t) - R(t,t_0)x_0 - \int_{t_0}^t \frac{\partial R(t,s)}{\partial s}y(s)\,ds,$$

it follows that

$$\int_{t_0}^{t} R(t,s)y'(s)\,ds = \int_{t_0}^{t} R(t,s)F(s)\,ds - \int_{t_0}^{t} \frac{\partial R(t,s)}{\partial s} y(s)\,ds.$$

From (1.7.5) and Fubini's theorem, we get

$$\int_{t_0}^{t} R(t,s)\left[y'(s) - A(s)y(s) - \int_{t_0}^{s} K(s,u)y(u)\,du - F(s)\right]ds = 0.$$

Since $R(t,s)$ is a nonzero continuous function for $t_0 \leq s \leq t < \infty$, it is clear that

$$y'(s) - A(s)y(s) - \int_{t_0}^{s} K(s,u)y(u)\,du - F(s) = 0.$$

This implies that $y(t)$ solves (1.7.1), and hence the proof is complete. □

Remark 1.7.1 The formula (1.7.4) that expresses the solution $x(t)$ of (1.7.1) in terms of the differentiable resolvent $R(t,s)$ and the source function $F(t)$ is called the linear variation of parameters formula for the integro-differential equation (1.7.1).

Remark 1.7.2 One can also sometimes utilize the classical variation of parameters formula for linear differential systems to obtain an integral equation for the solutions of (1.7.1). For this purpose, let $Y(t)$ be a fundamental matrix solution of

$$y' = A(t)y,$$

so that any solution of (1.7.1) with the initial function ϕ on $[t_0, \tau]$ is given by

$$x(t,\tau,\phi) = Y(t)Y^{-1}(\tau)\phi(\tau) + \int_{\tau}^{t} Y(t)Y^{-1}(s)\left[\int_{t_0}^{s} K(s,u)x(u)\,du + F(s)\right]ds.$$

Given a linear integro-differential equation, it is interesting and important to develop a method of finding an equivalent linear differential system. The next result is in this direction.

Theorem 1.7.2 Assume that there exists an $n \times n$ continuous matrix function $L(t,s)$ on R^2_+ such that $L_s(t,s)$ exists, is continuous and satisfies

$$(1.7.6) \qquad K(t,s) + L_s(t,s) + L(t,s)A(s) + \int_s^t L(t,u)K(u,s)\,du = 0$$

where $A(t)$ and $K(t,s)$ are continuous $n \times n$ matrices on R_+ and R_+^2 respectively. Then, the initial value problem for the linear integro-differential system

$$(1.7.7) \qquad u'(t) = A(t)u(t) + \int_{t_0}^t K(t,s)u(s)\,ds + F(t), \qquad u(t_0) = x_0,$$

where $F \in C[R_+, R^n]$, is equivalent to the initial value problem for the linear differential system

$$(1.7.8) \qquad v'(t) = B(t)v(t) + L(t,t_0)x_0 + H(t), \qquad v(t_0) = x_0,$$

where $B(t) = A(t) - L(t,t)$ and $H(t) = F(t) + \int_{t_0}^t L(t,s)F(s)\,ds$.

Proof Let $u(t)$ be any solution of (1.7.7) existing on $[t_0, \infty)$. Set $p(s) = L(t,s)u(s)$ so that $p'(s) = L_s(t,s)u(s) + L(t,s)u'(s)$. Substituting for $u'(s)$ in (1.7.7) and integrating, we get

$$L(t,t)u(t) - L(t,t_0)x_0 = \int_{t_0}^t [L_s(t,s) + L(t,s)A(s)]u(s)\,ds.$$

Since by Fubini's theorem,

$$\int_{t_0}^t L(t,s)\left[\int_{t_0}^s K(s,\theta)u(\theta)\,d\theta\right]ds = \int_{t_0}^t \left[\int_s^t L(t,u)K(u,s)\,du\right]u(s)\,ds,$$

it follows using (1.7.6) and (1.7.7) that

$L(t,t)u(t) - L(t,t_0)x_0$

$$= \int_{t_0}^t \left[L_s(t,s) + L(t,s)A(s) + \int_s^t L(t,u)K(u,s)\,du\right]u(s)\,ds + H(t) - F(t)$$

$$= -\int_{t_0}^t K(t,s)u(s)\,ds + H(t) - F(t)$$

$$= -u'(t) + A(t)u(t) + H(t).$$

Hence $u(t)$ satisfies (1.7.8).

Let $v(t)$ be any solution of (1.7.8) existing on $[t_0,\infty)$. Then, defining

$$z(t) = v'(t) - A(t)v(t) - \int_{t_0}^{t} K(t,s)v(s)\,ds - F(t),$$

we shall show that $z(t) \equiv 0$, which proves that $v(t)$ satisfies (1.7.7). Now, substituting for $v'(t)$ from (1.7.8) and using (1.7.6) together with Fubini's theorem, we get

$$z(t) = -\left[L(t,t)v(t) - L(t,t_0)x_0 - \int_{t_0}^{t} L_s(t,s)v(s)\,ds\right]$$

$$+ \int_{t_0}^{t} L(t,s)\left[A(s)v(s) + F(s) + \int_{t_0}^{s} K(s,u)v(u)\,du\right]ds.$$

Since $(d/ds)[L(t,s)v(s)] = L_s(t,s)v(s) + L(t,s)v'(s)$, we have, by integration,

$$L(t,t)v(t) - L(t,t_0)x_0 = \int_{t_0}^{t} [L_s(t,x)v(s) + L(t,s)v'(s)]\,ds.$$

It therefore follows, using (1.7.7), that

$$z(t) = \int_{t_0}^{t} L(t,s)\left[-v'(s) + A(s)v(s) + \int_{t_0}^{s} K(s,u)v(u)\,du + F(s)\right]ds$$

$$= -\int_{t_0}^{t} L(t,s)z(s)\,ds,$$

which implies $z(t) \equiv 0$ because of the uniqueness of solutions of Volterra linear integral equations. The proof is therefore complete. $\square$

In some situations, it may be convenient to reduce (1.7.7) to

(1.7.9) $\qquad v' = B(t)v(t) + \int_{t_0}^{t} \Phi(t,s)v(s)\,ds + H(t), \quad v(t_0) = x_0,$

where

$$B(t) = A(t) - L(t,t), \quad H(t) = F(t) + L(t,t_0)x_0 + \int_{t_0}^{t} \Phi(t,s)F(s)\,ds.$$

The following result, whose proof is similar to that of Theorem 1.7.2, is a modification of this case.

Theorem 1.7.3 *If in Theorem 1.7.2, the relation (1.7.6) is replaced by*

$$(1.7.10) \quad \Phi(t,s) = K(t,s) + L_s(t,s) + L(t,s)A(s) + \int_s^t L(t,u)K(u,s)\,du,$$

where $\Phi(t,s)$ is an $n \times n$ continuous matrix on R_+^2, then (1.7.7) is equivalent to (1.7.9).

When A is an $n \times n$ constant matrix and K is of convolution type, that is, $K(t,s) = K(t-s)$, the linear integro-differential system (1.7.1) with $t_0 = 0$ reduces to

$$(1.7.11) \quad x' = Ax(t) + \int_0^t K(t-s)x(s)\,ds + F(t), \quad x(0) = x_0.$$

The variation of parameters formula can, in this special case, be derived using the Laplace transform.

Let $Z(t)$ be a differential resolvent satisfying the adjoint equation

$$(1.7.12) \quad Z'(t) = AZ(t) + \int_0^t K(t-s)Z(s)\,ds, \quad Z(0) = I.$$

Then we have the following result.

Theorem 1.7.4 *Suppose that $K, F \in L^1(R_+)$, and $Z(t)$ satisfies (1.7.12). Assume further that*

$$(1.7.13) \quad \det[sI - A - \hat{K}(s)] \neq 0 \quad \text{for} \quad \operatorname{Re} s > 0,$$

where $\hat{K}(s)$ is the Laplace transform of $K(t)$. Then the solution $x(t)$ of (1.7.11) is of the form

$$(1.7.14) \quad x(t) = Z(t)x_0 + \int_0^t Z(t-s)F(s)\,ds.$$

Proof We have from (1.7.11)

$$x(t) = x_0 + \int_0^t F(s)\,ds + \int_0^t \left[A + \int_s^t K(u-s)\,du \right] x(s)\,ds,$$

and hence

$$|x(t)| \leq |x_0| + \int_0^t |F(s)|\,ds + \int_0^t \left[|A| + \int_s^t |K(u-s)|\,du \right] |x(s)|\,ds.$$

Since $K, F \in L^1(R_+)$,

$$\int_0^\infty |F(s)|\,ds \leq M_1 \quad \text{and} \quad \int_0^\infty |K(s)|\,ds \leq M_2.$$

Consequently, setting $M = \max[M_1, |A| + M_2]$, we get

$$|x(t)| \leq |x_0| + M + M \int_0^t |x(s)|\,ds,$$

which yields by Gronwall's inequality the estimate

$$|x(t)| \leq [|x_0| + M] e^{Mt}, \quad t \geq 0.$$

Thus $x(t)$ is of exponential order and since $Z \in L^1(R_+)$ (this follows from (1.7.13), see Grossman and Miller [1]), the Laplace transforms of $x(t)$ and $Z(t)$ exist. Hence taking the Laplace transform on both sides of (1.7.12), we obtain

$$s\mathcal{L}[Z] - Z(0) = A\mathcal{L}[Z] + \mathcal{L}[K]\mathcal{L}[Z],$$

which because of (1.7.13) gives

$$\mathcal{L}[Z] = [sI - A - \hat{K}(s)]^{-1}.$$

Similarly, taking the Laplace transform of (1.7.11), we get

$$s\mathcal{L}[x] - x_0 = A\mathcal{L}[x] + \mathcal{L}[K]\mathcal{L}[x] + \mathcal{L}[F].$$

It therefore follows that

$$\mathcal{L}[x] = [sI - A - \hat{K}(s)]^{-1}(x_0 + \mathcal{L}[F])$$

$$= \mathcal{L}[Z] x_0 + \mathcal{L}[Z]\mathcal{L}[F]$$

$$= \mathcal{L}\left[Zx_0 + \int_0^t Z(t-s)F(s)\,ds\right].$$

This together with the continuity of x, Z and F gives (1.7.14), and the proof is complete. $\square$

1.8 Nonlinear Variation of Parameters

Our aim, in this section, is to develop the nonlinear variation of parameters formula for integro-differential systems. For this purpose, we shall first investigate the problem of differentiability of solutions of nonlinear integro-differential equations and obtain a relation between the derivatives with respect to initial values. We then prove the nonlinear variation of parameters formula for solutions of perturbed integro-differential equations.

Let us begin by recalling the following well known result.

Theorem 1.8.1 *Let $f \in C[R_+ \times D, R^n]$, where D is an open, convex set in R^n, and let f_x exist and be continuous on $R_+ \times D$. Then*

$$f(t,x) - f(t,y) = \left[\int_0^1 f_x(t, sx + (1-s)y)\,ds\right](x-y).$$

Consider now the IVP for the nonlinear integro-differential equation

$$(1.8.1) \qquad x'(t) = f(t, x(t)) + \int_{t_0}^t g(t, s, x(s))\,ds, \quad x(t_0) = x_0.$$

We shall first discuss the problem of continuity and differentiability of solutions $x(t, t_0, x_0)$ of (1.8.1) with respect to (t_0, x_0).

Theorem 1.8.2 *Assume that $f \in C[R_+ \times R^n, R^n]$, $g \in C[R_+ \times R_+ \times R^n, R^n]$ and that f_x and g_x exist and are continuous on $R_+ \times R^n$ and $R_+ \times R_+ \times R^n$ respectively. Let $x(t, t_0, x_0)$ be the unique solution of (1.8.1) existing on some interval $t_0 \leq t \leq a < \infty$ and set $J = [t_0, t_0 + T]$, $t_0 + T < a$. Define $H(t, t_0, x_0) = f_x(t, x(t, t_0, x_0))$ and $G(t, s, t_0, x_0) = g_x(t, s, x(s, t_0, x_0))$. Then*

(i) $\Phi(t, t_0, x_0) = (\partial x/\partial x_0)(t, t_0, x_0)$ *exists and is the solution of*

(1.8.2) $$y'(t) = H(t, t_0, x_0) y(t) + \int_{t_0}^{t} G(t, s; t_0, x_0) y(s) \, ds,$$

such that $\Phi(t_0, t_0, x_0) = I$;

(ii) $\psi(t, t_0, x_0) = (\partial x/\partial t_0)(t, t_0, x_0)$ exists and is the solution of

(1.8.3) $$z'(t) = H(t, t_0, x_0) z(t) + \int_{t_0}^{t} G(t, s; t_0, x_0) z(s) \, ds - g(t, t_0, x_0),$$

such that $\psi(t_0, t_0, x_0) = -f(t_0, x_0)$;

(iii) the functions $\Phi(t, t_0, x_0)$ and $\psi(t, t_0, x_0)$ satisfy the relation

(1.8.4) $$\psi(t, t_0, x_0) + \Phi(t, t_0, x_0) f(t_0, x_0) + \int_{t_0}^{t} R(t, \sigma; t_0, x_0) g(\sigma, t_0, x_0) \, d\sigma = 0,$$

where $R(t, s; t_0, x_0)$ is the solution of the IVP

(1.8.5) $$\frac{\partial R}{\partial s}(t, s; t_0, x_0) + R(t, s; t_0, x_0) H(s, t_0, x_0)$$
$$+ \int_{s}^{t} R(t, \sigma; t_0, x_0) G(\sigma, s; t_0, x_0) \, d\sigma = 0,$$

$R(t, t; t_0, x_0) = I$ on the interval $t_0 \leq s \leq t$ and $R(t, t_0; t_0, x_0) = \Phi(t, t_0, x_0)$.

Proof Under the assumptions on f and g, it is clear that solutions $x(t, t_0, x_0)$ exist, are unique, and are continuous in (t, t_0, x_0) on some interval. Consequently, the functions H and g are continuous in (t, t_0, x_0) and (t, s, t_0, x_0) respectively, and therefore the solutions of the linear IVPs (1.8.2) and (1.8.3) exist and are unique on the same interval for which $x(t, t_0, x_0)$ is defined. Let us first prove the conclusion (i).

Let $e_k = (e_k^1, \ldots, e_k^n)$ be the vector such that $e_k^j = 0$ if $j \neq k$ and $e_k^k = 1$. Then for small h, $x(t, h) = x(t, t_0, x_0 + e_k h)$ is defined on J and $\lim_{h \to 0} x(t, h) = x(t, t_0, x_0)$ uniformly on J. Setting $x(t) = x(t, t_0, x_0)$, it follows by Theorem 1.8.1 that

$$[x(t, h) - x(t)]' = \int_0^1 f_x[t, sx(t, h) + (1-s)x(t)] \, ds [x(t, h) - x(t)]$$

$$+ \int_{t_0}^{t} \int_{0}^{1} g_x[t, s, \sigma x(s, h) + (1-\sigma)x(s)] d\sigma [x(s, h) - x(s)] ds.$$

If $x_h(t) = (x(t, h) - x(t))/h$, $h \neq 0$, we see that the existence of $\partial x(t, t_0, x_0)/\partial x_0$ is equivalent to the existence of the limit of $x_h(t)$ as $h \to 0$. Since $x(t_0, h) = x_0 + e_k h$, $x_h(t_0) = e_k$, it is clear that $x_h(t)$ is the solution of the IVP

(1.8.6) $\quad y'(t) = H(t, t_0, x_0, h) y(t) + \int_{t_0}^{t} G(t, s, t_0, x_0, h) y(s) \, ds, \quad y(t_0) = e_k,$

where

$$H(t, t_0, x_0, h) = \int_{0}^{1} f_x[t, sx(t, h) + (1-s)x(t)] \, ds,$$

and

$$G(t, s, t_0, x_0, h) = \int_{0}^{1} g_x[t, s, \sigma x(s, h) + (1-\sigma)x(s)] \, d\sigma.$$

Since $\lim_{h \to 0} x(t, h) = x(t)$ uniformly on J, continuity of f_x and g_x implies that

$$\lim_{h \to 0} H(t, t_0, x_0, h) = H(t, t_0, x_0)$$

and

$$\lim_{h \to 0} G(t, s, t_0, x_0, h) = G(t, s, t_0, x_0)$$

uniformly on J and $J \times J$ respectively. Consider (1.8.6) and as a family of IVPs depending on a parameter, where H and G are continuous, h being small. Since the solutions of (1.8.6) are unique, the general solution of (1.8.6) is a continuous function of h for fixed (t, t_0, x_0). In particular, $\lim_{h \to 0} x_h(t)$ exists and is the solution of (1.8.2) on J. This implies that $(\partial x/\partial x_0)(t_0, t_0, x_0) = I$. Furthermore, it is easy to see from the IVP of linear integro-differential equation (1.8.2) that $(\partial x/\partial x_0)(t, t_0, x_0)$ is also continuous with respect to its arguments.

To prove (ii), defining $x(t, h) = x(t, t_0 + h, x_0)$, we have, by Theorem 1.8.1,

$$[x(t, h) - x(t)]' = \int_{0}^{1} f_x[t, sx(t, h) + (1-s)x(t)] \, ds [x(t, h) - x(t)]$$

$$+ \int_{t_0+h}^{t} \int_0^1 g_x[t,s,\sigma x(s,h)+(1-\sigma)x(s)]d\sigma[x(s,h)-x(s)]ds$$

$$- \int_{t_0}^{t_0+h} g(t,s,x(s))ds.$$

Setting as before $x_h(t) = (x(t,h) - x(t))/h$, $h \neq 0$, it is obvious that $x_h(t)$ is the solution of the IVP

$$y'(t) = H(t,t_0,x_0,h)y(t) + \int_{t_0+h}^{t} G(t,s,t_0,x_0,h)y(s)ds - L(t,t_0,x_0,h),$$

such that $y(t_0+h) = a(h)$, where

$$L(t,t_0,x_0,h) = \frac{1}{h} \int_{t_0}^{t_0+h} g(t,s,x(s))ds$$

and

$$a(h) = -\frac{1}{h} \int_{t_0}^{t_0+h} f(s,x(s))ds - \frac{1}{h} \int_{t_0}^{t_0+h} \int_{t_0}^{s} g(s,\sigma,x(\sigma))d\sigma \, ds.$$

Noting that $\lim_{h \to 0} L(t,t_0,x_0,h) = g(t,t_0,x_0)$, $\lim_{h \to 0} a(h) = -f(t_0,x_0)$ and using an argument similar to the proof of (i), we see that $(\partial x/\partial t_0)(t,t_0,x_0)$ exists, is continuous in its arguments, and is the solution of (1.8.3).

Finally, to prove (iii), we observe that $\Phi(t,t_0,x_0)$ and $\psi(t,t_0,x_0)$ are the solutions of $IVPs$ (1.8.2) and (1.8.3) respectively. Consequently by Theorem 1.7.1, we get immediately the desired relation (1.8.4). The proof of the theorem is therefore complete. □

We are now in a position to prove the nonlinear variation of parameters formula for solutions $y(t,t_0,x_0)$ of the perturbed IVP for the integro-differential equation

(1.8.7) $\quad y'(t) = f(t,y(t)) + \int_{t_0}^{t} g(t,s,y(s))ds + F(t,y(t),(Sy)(t)), \quad y(t_0) = x_0.$

BASIC THEORY   35

Theorem 1.8.3 *Suppose that the hypotheses of Theorem 1.8.2 hold. Assume further that $F \in C[R_+ \times R^n \times R^n, R^n]$ and $S(y)(t) = \int_{t_0}^{t} K(t,s,y(s))\,ds$ with $K \in C[R_+ \times R_+ \times R^n, R^n]$. If $x(t,t_0,x_0)$ is the solution of (1.8.1) existing on J, any solution $y(t,t_0,x_0)$ of (1.8.7) existing on J satisfies the integral equation*

$$(1.8.8) \quad y(t,t_0,x_0) = x(t,t_0,x_0) + \int_{t_0}^{t} \Phi(t,s,y(s))F(s,y(s),(Sy)(s))\,ds$$

$$+ \int_{t_0}^{t}\int_{s}^{t} [\Phi(t,\sigma,y(\sigma)) - R(t,\sigma;s,y(s))]g(\sigma,s,y(s))\,d\sigma\,ds$$

for $t_0 \geq t_0$, where $y(t) = y(t,t_0,x_0)$ and $R(t,s;t_0,x_0)$ is the solution of the IVP (1.8.5).

Proof Setting $p(s) = x(t,s,y(s))$, where $y(s) = y(s,t_0,x_0)$, we have

$$p'(s) = \psi(t,s,y(s)) + \Phi(t,s,y(s))y'(s).$$

Substituting for $y'(s)$, integrating from t to t_0, and using Fubini's theorem, we get

$$p(t) - p(t_0) - \int_{t_0}^{t} \Phi(t,s,y(s))F(s,y(s),(Sy)(s))\,ds$$

$$= \int_{t_0}^{t}\left[\psi(t,s,y(s)) + \Phi(t,s,y(s))f(s,y(s)) + \int_{s}^{t}\Phi(t,\sigma,y(\sigma))g(\sigma,s,y(s))\,d\sigma\right]ds.$$

Now the relation (1.8.4) together with the fact $x(t,t,y(t)) = y(t,t_0,x_0)$ yields (1.8.8), completing the proof. □

Remark (i) Let $f(t,x) = A(t)x$ and $g(t,s,x) = B(t,s)x$ in (1.8.1), where A, B are $n \times n$ continuous matrices. In this special case,

$$x(t,t_0,x_0) = R(t,t_0)x_0 \quad \text{and} \quad \Phi(t,t_0,x_0) = \frac{\partial x}{\partial x_0}(t,t_0,x_0) = R(t,t_0),$$

where $R(t,s)$ is the solution of (1.8.5) such that $R(t,t) = I$. Consequently, it is easy to see that the relation (1.8.8) reduces to the linear variation of parameters formula

$$y(t, t_0, x_0) = R(t, t_0)x_0 + \int_{t_0}^{t} R(t,s)F(s, y(s), (Sy)s)\, ds, \qquad t \geq t_0,$$

since $\Phi(t, \sigma, y(\sigma)) = R(t, \sigma; s, y(s))$, being independent of y and s.

(ii) If $g(t, s, x) \equiv 0$, the relation (1.8.8) reduces to the usual Alekseev formula. However, when $g \neq 0$ or is not linear in x, $\Phi(t, t_0, x_0) \neq R(t, s; t_0, x_0)$, and hence the relation (1.8.8) cannot have a simpler form that corresponds to the Alekseev formula as one would like to expect.

1.9 Monotone Iterative Technique

In recent years, the monotone iterative technique coupled with lower and upper solutions has been employed to obtain the existence of extremal solutions (which are limits of monotone sequences) for several nonlinear problems. In this section, we shall discuss the existence of solutions of periodic boundary value problem for nonlinear integro-differential equations.

Consider the periodic boundary value problem $(PBVP)$

(1.9.1) $$u'(t) = f(t, u(t), (Tu)(t)), \qquad u(0) = u(2\pi)$$

where $f \in C[[0, 2\pi] \times R \times R, R]$, $Tu(t) = \int_0^t K(t,s)u(s)ds$ and $K \in C[[0, 2\pi] \times [0, 2\pi], R_+]$.

We need the following comparison result in our subsequent discussion.

Lemma 1.9.1 *Let $m \in C^1([0, 2\pi], R)$ be such that*

(1.9.2) $$m' \leq -Mm - NTm, \qquad m(0) \leq m(2\pi),$$

where $M > 0$, $N \geq 0$. Then, $m(t) \leq 0$ for $0 \leq t \leq 2\pi$ provided one of the following conditions hold:

(a) $2Nk_0\pi(e^{2M\pi} - 1) \leq M;$

or

(b) $2\pi[M + 2\pi Nk_0] \leq 1$, where $0 \leq k_0 = \max K(t, s)$ for $(t, s) \in [0, 2\pi] \times [0, 2\pi]$ and $K(t, s) \geq 0$.

Proof Suppose that (a) holds. Set $v(t) = m(t)e^{Mt}$ so that the inequality (1.9.2) reduces to

$$\text{(1.9.3)} \qquad v' \leq -N \int_0^t K^*(t,s) v(s)\, ds,$$

where $K^*(t,s) = K(t,s) e^{M(t-s)}$. It is then enough to prove $v(t) \leq 0$ for $t \in [0, 2\pi]$. If this is not true then we have the following cases:

(A) $v(t) \geq 0$ for $t \in [0, 2\pi]$ and $v(t) \not\equiv 0$;

(B) there exist $t_1, t_2 \in [0, 2\pi]$ such that $v(t_1) > 0$ and $v(t_2) < 0$.

In case (A), we have $v(0) \leq v(2\pi)$, and also from (1.9.3) that $v'(t) \leq 0$ on $[0, 2\pi]$. Since $v(0) \leq v(2\pi)$ and $v(t)$ is nonincreasing, $v(t) \equiv C \geq 0$. Hence $v(t) = m(t) e^{Mt}$ implies $m(t) = C e^{-Mt}$. This is impossible unless $C = 0$, since $m(0) \leq m(2\pi)$, and we get a contradiction to (A).

In case (B), we have two situations: (i) $v(2\pi) \geq 0$ and (ii) $v(2\pi) < 0$. When $v(2\pi) \geq 0$, it is clear that $v(0) \leq v(2\pi)$. Suppose that $v(t_2) = -\lambda$ where $\min_{0 \leq t \leq 2\pi} v(t) = -\lambda$, $\lambda > 0$. Then it is clear that $t_2 \in [0, 2\pi)$ and, using the mean value theorem on $[t_2, 2\pi]$, we get

$$v'(t_0) = \frac{v(2\pi) + \lambda}{2\pi - t_0} > \frac{\lambda}{2\pi},$$

for some $t_0 \in (t_2, 2\pi)$. On the other hand, it follows from (1.9.3) that

$$v'(t_0) \leq -N \int_0^{t_0} K^*(t_0, s) v(s)\, ds \leq \frac{k_0 N (e^{2M\pi} - 1) \lambda}{M},$$

since

$$\int_0^{t_0} K^*(t_0, s)\, ds \leq \frac{k_0}{M}\left(e^{M t_0} - 1 \right) \leq \frac{k_0}{M} \left(e^{2M\pi} - 1 \right).$$

This is a contradiction to the assumption (a).

When $v(2\pi) < 0$, we also have $v(0) < 0$ and there exists a $t^* \in (0, 2\pi)$ such that $v(t^*) = 0$ and $v(t) < 0$ for $t \in (t^*, 2\pi]$. It is clear that $\min_{0 \leq t \leq t^*} v(t) < 0$. Let $-\lambda = \min_{0 \leq t \leq t^*} v(t) = v(t_2)$, where $t_2 \in [0, t^*)$ and $\lambda > 0$. Observing that $v(0) \leq v(2\pi)$, we can repeat the argument employed in cases above in the interval $[t_2, t^*]$ and obtain a contradiction as before. The proof of the lemma is complete for the case when (a) holds.

If the condition (b) holds, we proceed with the proof directly from (1.9.2) instead of transforming it into (1.9.3). Again, we shall have the two cases (A) and (B) relative to $m(t)$. When (A) holds, since $m(0) \leq m(2\pi)$, it is clear that $m(t_0) = \max_{0 \leq t \leq 2\pi} m(t) > 0$ and $t_0 \in (0, 2\pi]$. Hence

$$0 \leq m'(t_0) \leq -Mm(t_0) - N \int_0^{t_0} K(t_0, s) m(s) \, ds \leq -Mm(t_0) < 0,$$

which is a contradiction.

If case (B) holds, we argue exactly as before and find that condition (b) leads to a contradiction. The proof of the lemma is therefore complete. □

The following counterexample shows that our lemma is sharp in the sense that it is almost impossible to relax the restrictions on M and N.

Let $p(t) = at - b$, with $a, b > 0$ and choose $b = 2\pi\delta a$, where $\sqrt{\tfrac{1}{2}}\delta < 1$. Clearly, $p(0) \leq p(2\pi)$; we set

$$q(t) = -Mp(t) - N \int_0^t p(s) \, ds - p'(t)$$

and note that, in this case, $K(t, s) \equiv 1$. Taking $M = 2\pi\delta N > 0$ and noting that $Nb - Ma = 0$ and $Mb - a = [(2\pi\delta)^2 N - 1]a$, we find that

$$q(t) = -\tfrac{1}{2}Nat^2 + [(2\pi\delta)^2 N - 1]a.$$

From the condition (b) of Lemma 1.9.1, we see that

$$N \leq \frac{1}{4\pi^2(\delta + 1)}.$$

On the other hand, if we choose

$$N \geq \frac{1}{2\pi^2(2\delta^2 - 1)},$$

the conclusion of the lemma does not hold for $p(t)$. Now, letting $\delta \to 1$, we have the following estimates:

$$N \leq \frac{1}{4\pi^2(\delta + 1)} \to \frac{1}{8\pi^2}, \quad N \geq \frac{1}{2\pi^2(2\delta^2 - 1)} \to \frac{1}{2\pi^2}.$$

In view of this, it is clear that the Lemma 1.9.1 is valid whenever $N \in \Omega_1 = \{x \in R_+ : x < 1/8\pi^2\}$, but is violated whenever $N \in \Omega_2 = \{x \in R_+ : x > 1/2\pi^2\}$. We only have the gap $\Omega_3 = [1/8\pi^2, 1/2\pi^2]$ for which we are not sure. In this sense our lemma is sharp and the best possible.

BASIC THEORY

Let us now consider the $PBVP$ (1.9.1). Assume that $\alpha, \beta \in C^1([0, 2\pi], R)$ are lower and upper solutions relative to (1.9.1) and for convenience, we list the following hypotheses that α, β and f must satisfy:

(H_0) $\alpha, \beta \in C^1([0, 2\pi], R)$ such that $\alpha(t) \leq \beta(t)$,

$$\alpha' \leq f(t, \alpha, T\alpha), \quad \alpha(0) \leq \alpha(2\pi)$$

and

$$\beta' \geq f(t, \beta, T\beta), \quad \beta(0) \geq \beta(2\pi);$$

(H_1) whenever $\alpha(t) \leq \overline{u} \leq u \leq \beta(t)$ and $\alpha(t) \leq \overline{\phi}(t) \leq \phi(t) \leq \beta(t)$,

$$f(t, u, T\phi) - f(t, \overline{u}, T\overline{\phi}) \geq -M(u - \overline{u}) - NT(\phi - \overline{\phi}), \quad t \in [0, 2\pi],$$

where M and N are positive constants satisfying $2Nk_0\pi e^{2M\pi} < M$, k_0 being $\max K(t, s)$ on $[0, 2\pi] \times [0, 2\pi]$.

We shall now prove the following main result.

Theorem 1.9.1 *Assume that* (H_0) *and* (H_1) *hold. Then, there exist monotone sequences* $\{\alpha_n(t)\}$, $\{\beta_n(t)\}$ *with* $\alpha_0 = \alpha$, $\beta_0 = \beta$ *such that* $\lim_{n \to \infty} \alpha_n(t) = \rho(t)$ *and* $\lim_{n \to \infty} \beta_n(t) = r(t)$ *uniformly on* $[0, 2\pi]$, *and* ρ *and* r *are minimal and maximal solutions of* $PBVP$ (1.9.1) *respectively, satisfying* $\alpha(t) \leq \rho(t) \leq r(t) \leq \beta(t)$ *on* $[0, 2\pi]$.

Proof For any $\eta \in C([0, 2\pi], R)$ such that $\alpha \leq \eta \leq \beta$, consider the $PVBP$ for linear integro-differential equation

(1.9.4) $$u' + Mu = -NTu + \sigma(t), \quad u(0) = u(2\pi)$$

where $\sigma(t) = f(t, \eta(t), [T\eta](t)) + M\eta(t) + N[T\eta](t)$. Using the method of variation of parameters and the boundary condition $u(0) = u(2\pi)$, we get

$$u(t) = e^{-Mt} \left\{ \frac{1}{e^{2M\pi} - 1} \int_0^{2\pi} \left[\sigma(s) - N \int_0^s K(s, \xi) u(\xi) d\xi \right] e^{Ms} ds \right\}$$

$$+ e^{-Mt} \int_0^t \left[\sigma(s) - N \int_0^s K(s, \xi) u(\xi) d\xi \right] e^{Ms} ds$$

$$\equiv Su, \quad \text{say.}$$

Now, an application of Schauder's fixed point theorem shows, in view of the condition $2Nk_0\pi e^{2M\pi} < M$, that there exists a solution $u(t)$ for the $PBVP$ (1.9.4).

The uniqueness of solutions of $PBVP$ (1.9.4) follows from Lemma 1.9.1. In fact, if u and v are two distinct solutions of (1.9.4) then setting $p = u - v$ gives

$$p' = -Mp - NTp, \quad p(0) = p(2\pi).$$

Hence Lemma 1.9.1 ensures that $u \equiv v$.

We define a mapping A by $A\eta = u$, where for any η such that $\alpha \leq \eta \leq \beta$, u is the unique solution of (1.9.4). We shall show that

(i) $\alpha \leq A\alpha$ and $\beta \geq A\beta$;
(ii) A is monotone nondecreasing on the sector $[\alpha, \beta]$ where $[\alpha, \beta] = [u \in C[[0, 2\pi], R]: \alpha(t) \leq u(t) \leq \beta(t)]$.

To prove (i), we set $p = \alpha - \alpha_1$, where $\alpha_1 = A\alpha$. We then have

$$p' \leq -Mp - NTp, \quad p(0) \leq p(2\pi),$$

and by Lemma 1.9.1, it follows that $\alpha \leq A\alpha$. One shows similarly that $\beta \geq A\beta$.

To prove (ii), let $\eta_1, \eta_2 \in [\alpha, \beta]$ such that $\eta_1 \leq \eta_2$. Let $A\eta_1 = u_1$ and $A\eta_2 = u_2$. Then, setting $p = u_1 - u_2$, it follows using (H_1) that

$$p' \leq -Mp - NTp, \quad p(0) = p(2\pi),$$

which implies, by Lemma 1.9.1, that $A\eta_1 \leq A\eta_2$, proving (ii).

It is therefore easy to see that we can define the sequences $\{\alpha_n\}$ and $\{\beta_n\}$ with $\alpha_0 = \alpha$ and $\beta_0 = \beta$ such that

$$\alpha_{n+1} = A\alpha_n, \quad \beta_{n+1} = A\beta_n,$$

and conclude that on $[0, 2\pi]$, we have

$$\alpha_0 \leq \alpha_1 \leq \ldots \leq \alpha_n \leq \beta_n \leq \ldots \leq \beta_1 \leq \beta_0.$$

It then follows, using standard arguments, that $\lim_{n\to\infty} \alpha_n(t) = \rho(t)$ and $\lim_{n\to\infty} \beta_n(t) = r(t)$ uniformly on $[0, 2\pi]$, and that ρ and r are solutions of the $PBVP$ (1.9.1).

To show that ρ and r are extremal solutions of (1.9.1), we let u be any solution of (1.9.1) such that $u \in [\alpha, \beta]$ and suppose that for some $k > 0$, $\alpha_{k-1} \le u \le \beta_{k-1}$ on $[0, 2\pi]$. Then, writing $p = \alpha_k - u$, we obtain $p' \le -Mp - NTp$, $p(0) = p(2\pi)$, which implies, by Lemma 1.9.1, that $\alpha_k \le u$ on $[0, 2\pi]$. Arguing similarly, one can conclude that $\alpha_k \le u \le \beta_k$ on $[0, 2\pi]$. By induction, it therefore follows that $\alpha_n \le u \le \beta_n$ on $[0, 2\pi]$ for all n, and hence $\rho \le u \le r$ on $[0, 2\pi]$, proving that ρ and r are extremal solutions of $PBVP$ (1.9.1). The proof of the theorem is complete. □

Remark 1.9.1 Since the existence of solutions of the $PBVP$ (1.9.1) requires the condition $2Nk_0\pi e^{2M\pi} \le M$, which implies condition (a) of Lemma 1.9.1, we have preferred the assumption $2Nk_0\pi e^{2M\pi} < M$ in Theorem 1.9.1. We do not know whether this assumption is compatible with condition (b) of Lemma 1.9.1.

One of the basic aims in the study of certain higher-order boundary value problems for differential equations is to reduce them to boundary value problems for lower-order integro-differential equations and then employ standard fixed point theorems. We shall consider a mixed boundary value problem of second order and, as an application of our main result Theorem 1.9.1, obtain its extremal solutions.

Consider the problem

$$(1.9.5) \qquad x'' = f(t, x', x), \qquad x(0) = a, \qquad x'(0) = x'(2\pi),$$

where $f \in C([0, 2\pi] \times R \times R, R)$, and we shall assume that the following hypotheses hold:

(H_0') $\hat{\alpha}, \hat{\beta} \in C^2([0, 2\pi], R), \hat{\alpha}(0) > \hat{\beta}(0)$ and

$$\hat{\alpha}'' \le f(t, \hat{\alpha}', \hat{\alpha}), \quad \hat{\alpha}'(0) \le \hat{\alpha}'(2\pi),$$
$$\hat{\beta}'' \ge f(t, \hat{\beta}', \hat{\beta}), \quad \hat{\beta}'(0) \ge \hat{\beta}'(2\pi);$$

(H_1') $f(t, x, y)$ is nonincreasing in x and y and

$$f(t, x, Tx) - f(t, y, Ty) \ge -M(x-y) - NT(x-y),$$

whenever $x \geq y$, and $M > 0$ and $N \geq 0$ are constants such that

$$2N\pi e^{2M\pi} < M, \quad \text{where } Tu = a + \int_0^t u(s)\,ds.$$

Choose $\epsilon, \delta_1, \delta_2 > 0$ such that

$$\hat{\alpha}' - \epsilon \leq \hat{\beta}' + \epsilon, \quad \hat{\alpha}(0) - \delta_1 = a, \quad \hat{\beta}(0) + \delta_2 = a,$$

where $\hat{\beta}(0) \leq a \leq \hat{\alpha}(0)$. Letting

$$\alpha_1(t) = \hat{\alpha}(t) - \epsilon t - \delta_1, \quad \beta_1(t) = \hat{\beta}(t) + \epsilon t + \delta_2,$$

we have $\alpha_1(0) = \beta_1(0) = a$. Also, setting

$$\alpha^*(t) = \alpha_1'(t), \quad \beta^*(t) = \beta_1'(t),$$

we find that $\alpha^*(t) \leq \beta^*(t)$ on $[0, 2\pi]$. Using the monotonicity of f, we can now show that

$$\alpha^{*\prime}(t) \leq f(\alpha^*, T\alpha^*), \quad \alpha^*(0) \leq \alpha^*(2\pi)$$
$$\beta^{*\prime}(t) \geq f(\beta^*, T\beta^*), \quad \beta^*(0) \geq \beta^*(2\pi),$$

where

$$T\alpha^* = a + \int_0^t \alpha^*(s)\,ds, \quad T\beta^* = a + \int_0^t \beta^*(s)\,ds.$$

Hence, by Theorem 1.9.1, we obtain

$$\alpha^*(t) \leq \rho(t) \leq r(t) \leq \beta^*(t), \quad t \in [0, 2\pi],$$

where ρ and r are extremal solutions of (1.9.1). Consequently, we can get two solutions $x(t)$ and $y(t)$ given by

$$x(t) = a + \int_0^t \rho(s)\,ds, \quad y(t) = a + \int_0^t r(s)\,ds$$

as extremal solutions of (1.9.5).

1.10 Interval Analytic Method

Interval analytic methods, in which the monotone inclusion property is inherent, provide simultaneous iterative bounds for solutions of several nonlinear problems. In this section, we apply this technique to first-order integro-differential equations.

Definition 1.10.1 If U is an interval, we shall denote its end points by $\underline{u}$ and $\bar{u}$; thus $U = [\underline{u}, \bar{u}]$.

Definition 1.10.2 If $U_1 = [\underline{u}_1, \bar{u}_1]$ and $U_2 = [\underline{u}_2, \bar{u}_2]$ are intervals then $U_1 \subseteq U_2$ if and only if $\underline{u}_2 \leq \underline{u}_1 \leq \bar{u}_1 \leq \bar{u}_2$.

Definition 1.10.3 If $U_1 = [\underline{u}_1, \bar{u}_1]$, the interval integral is simply

$$\int_a^b U(t)\,dt = \left[\int_a^b \underline{u}(t)\,dt, \int_a^b \bar{u}(t)\,dt\right].$$

Definition 1.10.4 If $U_1(t) \subseteq U_2(t)$ for all t in $[a,b]$ then this implies

$$\int_a^b U_1(t)\,dt \subseteq \int_a^b U_2(t)\,dt.$$

Lemma 1.10.1 Every nested sequence of intervals is convergent and has the limit $U = \bigcap_{k=1}^{\infty} U_k$.

Proof $\{\underline{u}_k\}$ is a nondecreasing monotone sequence of real numbers that is bounded above by $\bar{u}_1$ and hence has a limit $\underline{u}$. Similarly, $\{\bar{u}_k\}$ is a nonincreasing monotone sequence of real numbers that is bounded below by $\underline{u}_1$ and hence it has a limit $\bar{u}$. Since $\underline{u}_k \subset \bar{u}_k$ for every k, we have $\underline{u} \subset \bar{u}$. Therefore $\{u_k\}$ is convergent and its limit $u = [\underline{u}, \bar{u}]$.

Definition 1.10.5 Let $p: C[I, R] \to C[I, R]$ be an integral operator, where I is an interval. The corresponding interval integral operator P, which is inclusive monotonic, is defined by $PY = [py : y \in Y]$, where Y is an interval function.

Theorem 1.10.1 Suppose that there exists an interval function U_0 such that $PU_0 \subseteq U_0$. Then the successive iterates given by $U_{n+1} = PU_n$ form a decreasing sequence of interval-valued functions and $\lim_{n \to \infty} U_n(t) = U(t)$ exists as an interval function on I. Moreover if U is any solution of $pu = u$ then $u(t) \in U(t)$ on I.

Proof P is inclusive monotonic. Since $U_1 = P(U_0)$ by hypothesis, we get $P(U_0) = U_1 \subseteq U_0$. Hence we have by induction, $U_{n+1} \subseteq U_n$ for all $n = 0, 1, \ldots$. Hence the existence of u follows from Lemma 1.10.1. If u is any solution of $pu = u$ that is in U_0 then we have $pu = u \in P(u_0) = U_1$ by our definition PY. Therefore, it follows by induction that $u \in U_n$ for all $n = 0, 1, \ldots$. Hence by continuity $u \in U$.

Consider the IVP

(1.10.1) $$u' = f(t, u, Tu), \quad u(0) = u_0$$

where $f \in C[I \times R \times R, R]$, $(Tu)(t) = \int_0^t K(t,s) u(s) \, ds$,

$$K \in C[I \times I, R^+], \quad K(t,s) \not\equiv 0, \quad I = [0, \bar{T}],$$

with the following assumptions:

(C0) $\alpha, \beta \in C^1[I, R]$, $\alpha(t) \leq \beta(t)$ and

$$\alpha' \leq f(t, \alpha, T\alpha),$$

$$\beta' \geq f(t, \beta, T\beta) \quad \text{on } I;$$

(C1) $f(t, u, T\phi) - f(t, v, T\psi) \geq -M(u-v) - NT(\phi - \psi)$ whenever $\alpha(t) \leq v \leq u \leq \beta(t)$, $\alpha(t) \leq \psi(t) \leq \phi(t) \leq \beta(t)$, $t \in I$, with $M, N > 0$ and $\bar{T}[M + Nk_0 \bar{T}] < 1$, where $0 \leq K(t,s) \leq k_0$ on $I \times I$;

(C2) $\alpha, \beta \in C^1[[0, \bar{T}], R]$, $\alpha(t) \leq \beta(t)$ and

$$\alpha' \leq f(t, \sigma, T\psi) - NT(\alpha - \psi) - M(\alpha - \sigma),$$

$$\beta' \geq f(t, \sigma, T\psi) - NT(\beta - \psi) - M(\beta - \sigma)$$

for all σ, ψ such that $\alpha(t) \leq \sigma \leq \beta(t)$, $\alpha(t) \leq \psi(t) \leq \beta(t)$ and $\alpha(0) \leq u_0 \leq \beta(0)$.

It is easy to see that when (C0) and (C1) hold, (C2) is satisfied.

Lemma 1.10.2 Let $m \in C^1[I, R]$ and $I = [0, \bar{T}]$ be such that

$$m' \leq -Mm - NTm,$$

where $M, N \geq 0$, $M + N > 0$

$$(Tm)t = \int_0^t K(t,s)m(s)\,ds; \quad K \in C[I \times I, R^+]$$

and $K(t,s) \not\equiv 0$. Suppose further that $\overline{T}(M + Nk_0\overline{T}) \leq 1$, where $0 \leq K(t,s) \leq k_0$ on $I \times I$. Then either $m(0) \leq 0$ or $m(0) \leq m(\overline{T})$ implies that $m(t) \leq 0$ on I.

Proof This proof is the same as for Lemma 1.9.1, with minor modifications. □

Theorem 1.10.2 *Assume (C0), (C1) or (C2) holds. Then there exists a decreasing sequence $\{U_n(t)\}$ of interval functions with $U_0(t) = [\alpha(t), \beta(t)]$ such that $\lim_{n \to \infty} U_n(t) = U(t)$ exists and is an interval function on I. If $u(t)$ is any solution of (1.10.1) in the sector*

$$\{[\alpha, \beta] = u \in C^2[I, R]: \alpha \leq u \leq \beta]\}$$

then $u(t) \in U(t)$ on I. If (C0) and (C1) hold then $U(t) = [\rho(t), r(t)]$, where ρ and r are extremal solutions of (1.10.1) relative to $[\alpha, \beta]$. If further, f satisfies the condition

$$f(t, u_1, Tu_1) - f(t, u_2, Tu_2) \leq -M(u_1 - u_2) - NT(u_1 - u_2)$$

relative to the sector then $U(t)$ is the unique solution of (1.10.1).

Proof Let us suppose (C0) and (C1) hold. Consider the integral operator $p: C^1[I, R] \to C^1[I, R]$ defined by

(1.10.2) $$pu = G(t, 0)u_0 + \int_0^t G(t, s)F(s, u(s), Tu(s))\,ds$$

where $G(t, s)$ is the unique solution of

$$\frac{\partial}{\partial s}G(t, s) = -G(t, s)A(s) - \int_s^t G(t, v)B(v, s)\,dv,$$

with $G(t, t) = I$, where $A(t) = -M$, $B(t, s) = -NK(t, s)$ and $F(t, u, Tu) = f(t, u, Tu) + Mu + NTu$. Because of (C2), F is increasing in u with respect to both the variable u and Tu whenever $\alpha(t) \leq u(t) \leq \beta(t)$.

Now for any interval-valued function $Y \subset [\alpha, \beta]$, we let PY be the corresponding interval integral operator defined by $PY = \{py : y \in Y\}$ which is inclusive monotonic.

Since pu is increasing, we have

(1.10.3) $$PY = [p\underline{y}, p\bar{y}] \quad \text{if } Y = [\underline{y}, \bar{y}].$$

Because of $(C0)$ and the fact that F is increasing, we get

$$\alpha \leq p\alpha \leq p\beta \leq \beta;$$

$$P[\alpha, \beta] = [p\alpha, p\beta] \subseteq [\alpha, \beta].$$

Hence by Theorem 1.10.1, we have, by setting $u_0 = [\alpha, \beta]$, a decreasing sequence $\{U_n\}$ of interval functions such that $\lim_{n \to \infty} U_n(t) = U(t)$ exists and is an interval function on I. Furthermore, if $u(t)$ is any solution of (1.10.1) such that $u \in [\alpha, \beta]$, then $u(t) \in U(t)$ on I.

Suppose that ρ and r are minimal and maximal solutions of (1.10.1) in the sector $[\alpha, \beta]$. Then, observing that any solution u of $pu = u$ is also a solution of (1.10.1) and conversely, if $U(t) = [\underline{u}(t), \bar{u}(t)]$, we have $[\rho, r] = [\underline{u}, \bar{u}]$. Since PY is continuous, we also have $U = PU$. Hence it follows by (1.10.3) that $p\underline{u} = \underline{u}$ and $p\bar{u} = \bar{u}$, which implies that $\underline{u}' = f(t, \underline{u}, T\underline{u})$ and $\bar{u}' = f(t, \bar{u}, T\bar{u})$.

Consequently, $[\underline{u}, \bar{u}] \subseteq [\rho, r]$ on I, which implies that $U = [\rho, r]$.

Now, suppose that $(C2)$ holds. Consider the same operator p defined by (1.10.2). Since F is no longer increasing, the relation (1.10.3) need not hold. Now in order to apply Theorem 1.10.1, we need to show that $P[\alpha, \beta] \subseteq [\alpha, \beta]$. For this, let us show that for any $u \in [\alpha, \beta]$, $pu \in [\alpha, \beta]$, which implies that $P[\alpha, \beta] \subseteq [\alpha, \beta]$.

By (1.10.2) and $(C2)$, we easily get

$$\alpha - pu \leq \int_0^t G(t,s)[F(s, \sigma(s), T\psi(s)) - F(s, u(s), Tu(s))]\,ds \quad \text{for all } \sigma,$$

where $F(s, \sigma(s), T\psi(s)) = f(t, \sigma, T\psi) + NT\psi + M\sigma$. Choosing $\sigma = \psi = u$, we get $\alpha \leq pu$. Similarly, we can prove that $pu \leq \beta$.

BASIC THEORY

Suppose further that f satisfies $f(t, u_1, Tu_1) - f(t, u_2, Tu_2) \leq -M(u_1 - u_2) - NT(u_1 - u_2)$,

$$\alpha(t) \leq u_1 \leq u_2 \leq \beta(t), \quad t \in I.$$

Then by setting $m = \bar{u} - \underline{u}$, we easily see that $m' \leq -Mm - NTm$, $m(0) \leq 0$, which implies by Lemma 1.10.1 that $u(t) = \underline{u}(t) = \bar{u}(t)$ is the unique solution of (1.10.1) such that $\alpha(t) \leq u(t) \leq \beta(t)$ on I. This completes the proof. $\square$

Let us now consider the periodic boundary value problem

(1.10.4) $\qquad u' = f(t, u, Tu), \quad u(0) = u(\bar{T})$

where $f \in [C[0, \bar{T}] \times R \times R, R]$ and

$$(Tu)(t) = \int_0^t k(t, s) u(s)\, ds, \quad k \in C[[0, \bar{T}] \times [0, \bar{T}], R^+].$$

In the assumptions $(C0) - (C2)$, let us replace the initial conditions by $u(0) = u(\bar{T})$, $\alpha(0) \leq \alpha(\bar{T})$ and $\beta(0) \geq \beta(\bar{T})$.

With these modifications, we now define p as

(1.10.5) $\qquad pu = \dfrac{G(t,0)}{[1 - G(\bar{T},0)]} \displaystyle\int_0^{\bar{T}} G(t,s) F(s, u(s), Tu(s))\, ds$

$$+ \int_0^t G(t,s) F(s, u(s), Tu(s))\, ds$$

with $G(\bar{T}, 0) \neq 1$, $G(t, s)$ being the same function as in (1.10.2).

We can now prove Theorem 1.10.2 relative to the periodic boundary value problem (1.10.4).

For uniqueness, we need f to satisfy the condition

$$f(t, u_1, Tu_1) - f(t, u_2, Tu_2) \leq -M(u_1 - u_2) - NT(u_1 - u_2)$$

whenever $\alpha(t) \leq u_1 \leq u_2 \leq \beta(t)$, $t \in I$.

We only indicate the changes required in the proof of Theorem 1.10.2 for the corresponding result relative to the periodic boundary value problem (1.10.4) to hold.

In this case, define the integral operator p as in (1.10.5). Then the definition of p and the condition $(C0)$ gives $\alpha \leq p\alpha$ and $\beta \geq p\beta$.

Since F is increasing, it follows that $P[\alpha, \beta] \subseteq [\alpha, \beta]$. The rest of the proof is similar with suitable modification.

If f satisfies (1.10.4); by setting $m = \bar{u} - \underline{u}$, we easily get

$$m' \leq -Mm - NTm, \quad m(0) \leq m(\bar{T})$$

which implies by Lemma 1.10.1 that $u = \bar{u} = \underline{u}$ is the unique solution of the periodic boundary value problem (1.10.4) such that $u(t) \in [\alpha(t), \beta(t)]$ on I. The proof is complete. □

1.11 Notes and Comments

Section 1.1 contains the basic existence theorems which are proved by the application of standard fixed point methods. The material covered in Sections 1.2 and 1.3 is due to Lakshmikantham and Leela [1]. Similar results for integral equations are found in Corduneanu [3] and Miller [3]. Theorems 1.4.1 and 1.4.2 are adapted from the work of Lakshmikantham, Leela and Rama Mohana Rao [1]. The results of Sections 1.5 and 1.6 are formulated on similar lines as for ordinary differential equations (see Lakshmikantham and Leela [1] and Rama Mohana Rao [1]). Theorem 1.7.1 is due to Grossman and Miller [2]. Theorems 1.7.2 and 1.7.3 are taken from Rama Mohana Rao and Srinivas [2, 3]. Theorem 1.7.4 is found in Miller [5].

The nonlinear variation of parameters discussed in Section 1.8 is taken from Hu, Lakshmikantham and Rama Mohana Rao [1]. See also Beesack [1], Bernfeld and Lord [1] and Brauer [2]. The basic result for ordinary differential equations is found in Alekseev [1]. The results of Section 1.9 are due to Hu and Lakshmikantham [1]. For further results on boundary value problems in this direction, see Aftabizadeh and Leela [1]. See also Leela and Oguzstoneli [1] for integro-differential equations and Faheem and Rama Mohana Rao [1 − 3] for functional differential equations. For basic material on this topic, see Bernfeld and Lakshmikantham [1]. The material covered in Section 1.10 is taken from Sivasundaram [1]. Similar results for integral equations are found in Sivasundaram [2].

2 LINEAR ANALYSIS

2.0 Introduction

In this chapter, we study the properties of solutions of linear and weakly nonlinear integro-differential systems. We begin with the basic properties of solutions of general linear systems in Section 2.1. Section 2.2 discusses the various stability and boundedness results for linear convolution integro-differential systems using the method of variation of parameters. Stability properties of general linear systems are investigated in Section 2.3 in the same spirit as in Section 2.2, while the method of reduction, which transforms a given general linear systems to an equivalent linear differential system, is employed in Section 2.4. Stability in variation of nonlinear integro-differential system is considered in Section 2.5 utilizing linearization technique, while Section 2.6 is devoted to Lipschitz stability considerations. Asymptotic equivalence of solutions of integro-differential systems is discussed in detail in Section 2.7. In Section 2.8, we investigate the problem of finding sufficient conditions under which the ultimate behavior of solutions of integro-differential equations is determined by the corresponding ordinary differential system.

Difference equations of Volterra type form the content of Section 2.9, while Section 2.10 is devoted to the study of impulsive integro-differential equations. In both Section 2.9 and 2.10, we obtain corresponding stability properties of solutions developing the necessary theory of inequalities. Finally in Section 2.11, we discuss the existence of periodic solutions of linear integro-differential equations and their limiting equations. Examples are given throughout to illustrate the results discussed.

2.1 Basic Properties of Linear Systems

We consider the linear integro-differential system

$$(2.1.1) \qquad x'(t) = A(t)x(t) + \int_0^t K(t,s)x(s)\,ds,$$

and its perturbation

$$(2.1.2) \qquad x'(t) = A(t)x(t) + \int_0^t K(t,s)x(s)\,ds + F(t)$$

where $A(t)$ is an $n \times n$ continuous matrix on R_+, $K(t,s)$ is an $n \times n$ continuous matrix for $0 \le s \le t < \infty$ and $F \in C[R_+, R^n]$.

As in the case of ordinary differential equations, solutions of system of linear integro-differential equations can be linearly superimposed to obtain new solutions.

Theorem 2.1.1 *Let $A(t)$ and $K(t,s)$ be $n \times n$ continuous matrix functions for $0 \le t < \infty$ and $0 \le s \le t < \infty$ respectively, and let $F(t)$ be an n-vector continuous for $0 \le t < \infty$.*

(i) *If $x_1(t)$ and $x_2(t)$ are two solutions of (2.1.1), and c_1 and c_2 are real numbers, then $c_1 x_1(t) + c_2 x_2(t)$ is also a solution of (2.1.1).*

(ii) *If $x_1(t)$ is a solution of (2.1.2) with $F(t) = f_1(t)$ and $x_2(t)$ is a solution of (2.1.2) with $F(t) = f_2(t)$, and c_1 and c_2 are real numbers then $c_1 x_1(t) + c_2 x_2(t)$ is a solution of (2.1.2) with $F(t) = c_1 f_1(t) + c_2 f_2(t)$.*

(iii) *If $x_1(t), x_2(t), \ldots, x_n(t)$ are n – linearly independent solutions of (2.1.1) then every solution $x(t)$ of (2.1.1) can be expressed as a linear combination*

$$x(t) = \sum_{i=1}^{n} c_i x_i(t),$$

where c_i $(i = 1, 2, \ldots, n)$ are real constants. Moreover, $x(t)$ never vanishes unless $c_1 = c_2 = \ldots = c_n = 0$.

(iv) If $x_1(t), x_2(t), \ldots, x_n(t)$ are n - linearly independent solutions of (2.1.1) and $x_p(t)$ is any particular solution of (2.1.2) then every solution $x(t)$ of (2.1.2) can be expressed as $x(t) = x_p(t) + \sum_{i=1}^{n} c_i x_i(t)$, where $c_1, c_2, \ldots, c_n$ are real constants.

Proof The proofs of (i) and (ii) follow by direct substitution in (2.1.1) and (2.1.2) respectively. For the proof of (iii), let $x(t)$ be any solution of (2.1.1) defined for all $t \in [0, \infty)$ such that $x(0) = x_0$, where $x_0 = (x_{10}, x_{20}, \ldots, x_{n0})$. Let $x_1(t), x_2(t), \ldots, x_n(t)$ be the set of linearly independent solutions of (2.1.1) existing for $t \in [0, \infty)$ and satisfying $x_i(0) = e_i$, where $e_i (i = 1, 2, \ldots, n)$ are n-dimensional unit vectors. Set $\phi(t) = \sum_{i=1}^{n} x_{i0} x_i(t)$. It is easy to verify that $\phi(t)$ is a solution of (2.1.1) and moreover

$$\phi(0) = \sum_{i=1}^{n} x_{i0} e_i = x_0.$$

Since the solutions of (2.1.1) are unique, we have $\phi(t) \equiv x(t)$ for $t \in [0, \infty)$. The second assertion of (iii) follows from the linear independence of the solutions $x_1(t), x_2(t), \ldots, x_n(t)$ of (2.1.1). To prove (iv), let $x(t)$ and $x_p(t)$ be any two solutions of (2.1.2) existing for $t \in [0, \infty)$. Then it is easy to verify that $y(t) = x(t) - x_p(t)$ is a solution of (2.1.1). Hence the desired result follows by (iii) and the proof is complete. □

2.2 Stability of Linear Convolution Systems

In this section, we shall consider the linear integro-differential system of convolution type given by

(2.2.1) $$x'(t) = Ax(t) + \int_0^t K(t-s)x(s)\, ds,$$

where A is an $n \times n$ matrix and $K(t)$ is an $n \times n$ continuous matrix on R_+. Let $x(t) = x(t, t_0, \phi)$ be any solution of (2.2.1) with the initial function $\phi(t)$ on $0 \leq t \leq t_0$, for some $t_0 > 0$. Then we have

(2.2.2) $$x'(t) = Ax(t) + \int_{t_0}^{t} K(t-s)x(s)\, ds + \int_0^{t_0} K(t-s)\phi(s)\, ds.$$

Now using the variation of parameter formula (1.7.14), we obtain

$$(2.2.3) \qquad x(t) = Z(t-t_0)\phi(t_0) + \int_{t_0}^{t} Z(t-s)\left[\int_{0}^{t_0} K(s-\sigma)\phi(\sigma)\,d\sigma\right]ds.$$

Replacing t by $t+t_0$, we get

$$(2.2.4) \qquad x(t+t_0) = Z(t)\phi(t_0) + \int_{0}^{t} Z(t-\sigma)\left[\int_{0}^{t_0} K(\sigma+t_0-u)\phi(u)\,du\right]d\sigma$$

for all $t \geq 0$. Let the function $p(t)$ defined by

$$(2.2.5) \qquad p(t) = \int_{0}^{\infty}\left|\int_{0}^{t} Z(t-\sigma)K(\sigma+u)\,d\sigma\right|du$$

exist and be finite for all $t \geq 0$. Before proceeding further, we need a suitable definition of stability.

Definition 2.2.1 The trivial solution of (2.2.1) is said to be stable if, given $\epsilon > 0$ and $t_0 \in R_+$, there exists a $\delta = \delta(t_0, \epsilon) > 0$ such that $|\phi|_{t_0} = \max_{0 \leq s \leq t_0} |\phi(s)| < \delta$ implies $|x(t)| < \epsilon$, $t \geq t_0$.

Based on this definition, other stability notions can be defined. We are now in a position to prove the following result.

Theorem 2.2.1 Suppose that $K \in L^1(R_+)$. Then the equilibrium solution $x \equiv 0$ of (2.2.1) is

(i) uniformly stable if and only if the two functions $Z(t)$ and $p(t)$ are uniformly bounded on R_+;

(ii) uniformly asymptotically stable if and only if it is uniformly stable and both $Z(t)$ and $p(t)$ tend to zero as $t \to \infty$.

Proof Suppose the equilibrium $x \equiv 0$ of (2.1.1) is uniformly stable. Then there exists a constant B such that for any (t_0, ϕ) with $t_0 \geq 0$ and $|\phi|_{t_0} \leq 1$, we have

$$|x(t+t_0, t_0, \phi)| \leq B \quad \text{for all } t \geq 0.$$

If $t_0 = 0$ then this together with (2.2.3) gives

$$|x(t, 0, \phi)| = |Z(t)\phi(0)| \leq B \quad \text{for all } t \geq 0$$

and for all $\phi(0)$ satisfying $|\phi(0)| \leq 1$. In particular, $|Z(t)| \leq B$ for all $t \geq 0$. Similarly, if $y(t) = x(t + t_0, t_0, \phi) - Z(t)\phi(t_0)$ then $|y(t)| \leq 2B$. By (2.2.4), we obtain

$$y(t) = \int_0^t Z(t-s)\left[\int_0^{t_0} K(\sigma + t_0 - u)\phi(u)\,du\right]d\sigma$$

$$= \int_0^t Z(t-\sigma)\left[\int_0^{t_0} K(\sigma + u)\phi(t_0 - u)\,du\right]d\sigma.$$

By changing the order of integration, we get

(2.2.6) $$y(t) = \int_0^{t_0}\left[\int_0^t Z(t-\sigma)K(\sigma + u)\,d\sigma\right]\phi(t_0 - u)\,du.$$

This means that for all $t, t_0 \geq 0$,

$$\int_0^{t_0}\left|\int_0^t Z(t-\sigma)K(\sigma + u)\,d\sigma\right|du \leq 2B,$$

and this implies that $p(t) \leq 2B$ for all $t \geq 0$.

Conversely, suppose $|Z(t)| \leq A$ and $p(t) \leq A$ for some fixed constant A. Then (2.2.4) gives

$$|x(t + t_0, t_0, \phi)| \leq |Z(t)\phi(t_0)| + \left|\int_0^{t_0}\left[\int_0^t Z(t-s)K(s+u)\,ds\right]\phi(t_0 - u)\,du\right|$$

$$\leq A|\phi(t_0)| + A|\phi|_{t_0}$$

$$\leq 2A|\phi|_{t_0}.$$

Hence $x \equiv 0$ of (2.1.1) is uniformly stable. This completes the proof of part (i). The proof of part (ii) is derived in a similar manner and the proof of the theorem is complete. □

We shall next present a set of statements that are equivalent to uniform asymptotic stability of $x \equiv 0$ of (2.1.1).

Theorem 2.2.2 *Suppose that $K(t) \in L^1(R_+)$. Then the following statements for (2.2.1) are equivalent:*

(a) *the equilibrium solution $x \equiv 0$ of (2.2.1) is uniformly asymptotically stable;*

(b) *the equilibrium solution $x \equiv 0$ of (2.2.1) is exponentially stable;*

(c) *there exist positive real numbers M and λ such that $|Z(t)| \leq Me^{-\lambda t}$ for all $t \geq 0$;*

(d) $Z(t) \in L^1(R_+)$;

(e) *the solutions of (2.2.1) are uniformly bounded and uniformly ultimately bounded.*

Proof Since the equation (2.2.1) is linear, the proof that (a) is equivalent to (e) is exactly similar to that for ordinary differential equations.

To show that (a) implies (b), define a linear operator $Y(t, t_0)$: $C[0, t_0] \to C[0, t]$ by

$$(Y(t, t_0)\phi)(t) = x(t, t_0, \phi).$$

Then the uniqueness of solutions of (2.2.1) yields the relation

$$Y(t, t_0) = Y(t, s)Y(s, t_0) \quad \text{for } t \geq s \geq t_0,$$

where $Y(t_0, t_0) = I$. It follows from uniform stability that for $\epsilon_0 > 0$, there exists a $\delta = \delta(\epsilon_0) > 0$ such that $t_0 \geq 0$, $|\phi|_{t_0} < \delta$ and $t \geq t_0$ imply that

$$|(Y(t, t_0)\phi)(t)| = |x(t, t_0, \phi)| < \epsilon_0.$$

Hence for all $t \geq t_0$, we have

$$|Y(t, t_0)| = \sup_{|\phi|_{t_0} \leq 1} |(Y(t, t_0)\phi)(t)| \leq \frac{\epsilon_0}{\delta}.$$

The remainder of the proof is similar to that for ordinary differential equations.

To prove that (b) implies (c), we observe from (2.2.3) that $Z(t)\phi(0) = x(t, 0, \phi(0))$ where $x(t, 0, \phi(0))$ is a solution of (2.2.1) and hence the proof is clear.

Next we show that (c) implies (a). From (2.2.3) and (c), we see that

$$|x(t, t_0, \phi)| \leq |Z(t - t_0)| \, |\phi(t_0)|$$

$$+ \int_{t_0}^{t} |Z(t-s)| \left[\int_0^{t_0} |K(s-u)| \, |\phi(u)| \, du \right] ds$$

(2.2.7)
$$\leq M e^{-\lambda(t-t_0)} |\phi|_{t_0}$$

$$+ |\phi|_{t_0} M e^{-\lambda(t-t_0)} \int_0^{t-t_0} e^{\lambda \tau} \left[\int_\tau^\infty |K(s)| \, ds \right] d\tau.$$

Since $K(t) \in L^1(R_+)$, it is clear that $\int_\tau^\infty |K(s)| \, ds \to 0$ as $\tau \to \infty$. Thus the second term on the right-hand side of (2.2.7) tends to zero as $t \to \infty$ independently of t_0. Therefore, from (2.2.7), we obtain $|x(t, t_0, \phi)| \to 0$ as $t \to \infty$, independently of t_0, and the proof is complete.

The proof that (c) implies (d) is trivial. We shall show that (d) implies (c). Suppose (d) holds. Then there exists a $\lambda > 0$ such that

$$\int_0^t |Z(s)| \, ds \leq \frac{1}{\lambda} \qquad \text{for all } t \geq 0.$$

Moreover, for all $t, s \geq 0$, it is clear from the uniqueness of solutions of (2.2.1) that $Z(t) = Z(t-s)Z(s)$. Hence

(2.2.8)
$$|Z(t)| \int_0^t \frac{ds}{|Z(s)|}$$

$$\leq \int_0^t |Z(t-s)| \, |Z(s)| \frac{ds}{|Z(s)|} \leq \frac{1}{\lambda} \qquad \text{for all } t \geq 0.$$

Let $H(t) = \lambda \int_0^t ds / |Z(s)|$. Then $H'(t) = \frac{\lambda}{|Z(t)|}$. It follows from (2.2.8) that $H'(t) \geq \lambda H(t)$. Therefore for some $\alpha > 0$, we obtain

$$\frac{1}{|Z(t)|} \geq H(t) \geq H(\alpha) e^{\lambda(t-\alpha)} \qquad \text{for all } t \geq \alpha > 0.$$

This implies that $|Z(t)| \leq e^{-\lambda(t-\alpha)}/H(\alpha)$ for all $t \geq \alpha > 0$. Let

$$M^* = \max_{t \in [0,\alpha]}\left[e^{\lambda t}|Z(t)|\right], \quad M = \max\left[M^*, \frac{e^{\lambda \alpha}}{H(\alpha)}\right].$$

Then we have $|Z(t)| \leq Me^{-\lambda t}$ for all $t \geq 0$. This completes the proof of the theorem. $\square$

Let $C = C(R)$ be the set of all continuous functions $\phi: R \to R^n$ such that for any $t \in R$, the seminorms $|\phi|_t = \sup[|\phi(s)|: -\infty < s \leq t]$ are finite. Let $K \in L^1(R_+)$. Consider the initial value problem

(2.2.9) $$x'(t) = Ax(t) + \int_{-\infty}^{t} K(t-s)x(s)\,ds$$

for $t \geq t_0 \geq 0$ with $x(t) = \phi(t)$ on $-\infty < t \leq t_0$, where A is an $n \times n$ constant matrix and $K(t)$ is an $n \times n$ matrix continuous on R.

Various stability properties of the equilibrium solution $x \equiv 0$ of (2.2.9) can be defined in the same way as the corresponding type of stability for (2.2.1). We note that the system (2.2.9) is autonomous in the sense that for any (t_0, ϕ), we have $x(t, t_0, \phi) = x(t - t_0, 0, \phi_{t_0})$, where ϕ_{t_0} is the translated function $\phi_{t_0}(t) = \phi(t + t_0)$. In particular, it follows that we need only consider the system (2.2.9) with initial time $t_0 = 0$. Moreover, it is clear that stability and uniform stability are equivalent for the system (2.2.9).

We shall next establish a relationship between the stability or boundedness of (2.2.9) and the stability of (2.2.1).

Theorem 2.2.3 *Suppose $K \in L^1(R_+)$. Then the following statements are equivalent:*

(a) *the equilibrium solution $x \equiv 0$ of (2.2.1) is uniformly stable;*

(b) *the equilibrium solution $x \equiv 0$ of (2.2.9) is (uniformly) stable;*

(c) *$Z(t)$ is bounded and for each $\phi \in C(R)$, the solution $x(t,0,\phi)$ of (2.2.9) is bounded on R^+.*

Proof Given initial values $(0, \phi)$, let $x(t, \phi) = x(t, 0, \phi)$ be the corresponding solution of (2.2.9). Then for any $t \geq 0$, we have

$$x'(t,\phi) = Ax(t,\phi) + \int_0^t K(t-s)x(s,\phi)\,ds + \int_{-\infty}^0 K(t-s)\phi(s)\,ds$$

$$= Ax(t,\phi) + \int_0^t K(t-s)x(s,\phi)\,ds + \int_t^\infty K(u)\phi(t-u)\,du.$$

Using the variation of parameters formula (1.7.14) it follows that

$$x(t,\phi) = Z(t)\phi(0) + \int_0^t Z(t-s)\left[\int_s^\infty K(u)\phi(s-u)\,du\right]ds$$

(2.2.10)
$$= Z(t)\phi(0) + \int_0^\infty \left[\int_0^t Z(t-s)K(s+u)\,ds\right]\phi(-u)\,du$$

and hence $|x(t,\phi)| \leq |Z(t)||\phi(0)| + p(t)|\phi|_0$, where $|\phi|_0 = \sup_{-\infty < s \leq 0} |\phi(s)|$ and $p(t)$ is defined by (2.2.5). Suppose that (a) holds. Then it follows by Theorem 2.2.1 that $Z(t)$ and $p(t)$ are bounded, and hence the equilibrium solution $x \equiv 0$ of (2.2.9) is stable which proves (b).

To prove that (b) implies (c), suppose that $x \equiv 0$ of (2.2.9) is stable. Then there exists a constant B such that for any $\phi \in C(R)$ with $|\phi|_0 \leq 1$, we have $|x(t,\phi)| \leq B$ for all $t \geq 0$. Given any unit vector x_0 and any $\epsilon > 0$, let $\phi(t) = 0$ if $t \leq -\epsilon$ and $\phi(t) = (t/\epsilon + 1)x_0$ if $t \geq -\epsilon$. Then $|\phi|_0 \leq 1$ and hence $|x(t,\phi)| \leq B$ for all $t \geq 0$. Further, it follows from (2.2.10) that

$$x(t,\phi) = Z(t)x_0 + \int_0^t Z(t-s)\left[\int_0^\epsilon K(s+u)\phi(-u)\,du\right]ds$$

and hence

$$|Z(t)x_0| \leq B + \int_0^t |Z(t-s)|\left[\int_s^{s+\epsilon} |B(\tau)|\,d\tau\right]ds$$

for all $t \geq 0$, all $\epsilon > 0$ and all unit vectors x_0. By letting $\epsilon \to 0^+$, we obtain $|Z(t)x_0| \leq B$ for all $t \geq 0$. Thus $|Z(t)| \leq B$ uniformly in t. This proves (c).

Suppose (c) holds. Given any $\phi \in C(R)$. Since $x(t,\phi)$ and $Z(t)$ are bounded on R^+, it follows from (2.2.10) that

$$T_t\phi = \int_0^t Z(t-s)\left[\int_s^\infty K(u)\phi(s-u)\,du\right]ds$$

$$= \int_0^\infty \left[\int_0^t Z(t-s)K(s+u)\,ds \right] \phi(-u)\,du$$

is uniformly bounded in t. For any fixed $t \geq 0$, the symbol T_t represents a bounded linear mapping of $C((-\infty, 0])$ into R^n with norm $|T_t| = p(t)$. By the principle of uniform boundedness, it is clear that $|T_t| = p(t)$ is uniformly bounded in $t \in R^+$. This together with the fact that $Z(t)$ is assumed bounded implies by part (i) of Theorem 2.2.1 that $x \equiv 0$ of (2.2.1) is uniformly stable, and hence the proof of the theorem is complete. □

2.3 Stability Criteria for General Linear Systems

Employing the variation of parameters formula given in Remark 1.7.2, we shall discuss the stability properties of the integro-differential system (2.1.1). For this purpose, we use the fundamental matrix solution $Y(t)$ of

(2.3.1) $$y' = A(t)y$$

and prove stability results for the system (2.1.1). We begin with the following result.

Theorem 2.3.1 *Assume that*

(2.3.2) $$|Y(t)Y^{-1}(s)| \leq K, \quad t \geq s \geq 0,$$

and

(2.3.3) $$\int_0^\infty \int_0^t |K(t,s)|\,ds\,dt \leq M.$$

Then the trivial solution of (2.1.1) is uniformly stable.

Proof For any $\epsilon > 0$, let $\delta(\epsilon) < \epsilon/Ke^{KM}$ and $\|\phi\|_{t_0} < \delta(\epsilon)$. Suppose that there exists $t_1 \geq t_0$ such that $|x(t_1)| = \epsilon$ and $|x(t)| < \epsilon$ on $[t_0, t_1)$. By the variation of parameters formula, we have

$$|x(t)| \leq |Y(t)Y^{-1}(t_0)|\,|\phi(t_0)|$$

$$+ \int_{t_0}^t |Y(t)Y^{-1}(s)| \int_0^s |K(s,u)|\,|x(u)|\,du\,ds$$

$$\leq K\delta(\epsilon) + K \int_{t_0}^{t} \int_{0}^{s} |K(s,u)| \, |x(u)| \, du \, ds \quad \text{on } [t_0, t_1].$$

Define $r(t) \equiv \max_{0 < s \leq t} |x(s)|$ to obtain

$$r(t) \leq K\delta(\epsilon) + K \int_{t_0}^{t} \left[\int_{0}^{s} |K(s,u)| \, du \right] r(s) \, ds.$$

By Gronwall's inequality, we then get

$$|x(t)| \leq r(t) \leq K\delta(\epsilon) \exp\left[K \int_{t_0}^{t} \int_{0}^{s} |K(s,u)| \, du \, ds \right]$$

$$\leq K e^{KM} \delta(\epsilon) < \epsilon \quad \text{on } [t_0, t_1].$$

Therefore $|x(t_1)| < \epsilon$, which is a contradiction. Thus the zero solution of (2.1.1) is uniformly stable, completing the proof. □

We shall next prove a result on asymptotic stability.

Theorem 2.3.2 *Assume that*

(2.3.4) $$\int_{0}^{t} |Y(t)Y^{-1}(s)| \, ds \leq L \quad \text{for } t \geq 0$$

and

(2.3.5) $$\sup_{t \geq 0} \int_{0}^{t} |K(t,s)| \, ds < \frac{1}{L}.$$

Furthermore, suppose also that

(2.3.6) $$\lim_{s \to \infty} \int_{0}^{t} |K(s,u)| \, du = 0 \quad \text{for all } t \geq 0.$$

Then the zero solution of (2.1.1) is asymptotically stable.

Proof We first show that stability of the zero solution. From (2.3.5), there exists a positive constants γ such that

(2.3.7) $$0 < \gamma < \frac{1}{L}, \quad \sup_{s \geq 0} \int_{0}^{s} |K(s,u)| \, du \leq \gamma.$$

From (2.3.4), there exists a positive constant N such that

(2.3.8) $$|Y(t)| \leq N \quad \text{for all } t \geq 0.$$

For any $\epsilon > 0$ and $t_0 \geq 0$, let $\delta = \delta(\epsilon, t_0) < \min\{(1-\gamma L)\epsilon/(N|Y^{-1}(t_0)|), \epsilon\}$.

Consider the solution of (2.1.1) such that $|\phi|_{t_0} < \delta$. Suppose that there exists $t_1 > t_0$ such that $|x(t_1)| = \epsilon$ and $|x(t)| < \epsilon$ on $[t_0, t_1)$. For all $t \in [t_0, t_1]$, we have

$$|x(t)| \leq |Y(t)Y^{-1}(t_0)||\phi(t_0)|$$
$$+ \int_{t_0}^{t} |Y(t)Y^{-1}(s)| \int_0^s |K(s,u)||x(u)|\, du\, ds$$
$$< N|Y^{-1}(t_0)|\delta + \epsilon \int_{t_0}^{t} |Y(t)Y^{-1}(s)| \int_0^s |K(s,u)|\, du\, ds$$
$$< (1-\gamma L)\epsilon + \gamma L \epsilon = \epsilon.$$

Therefore $|x(t_1)| < \epsilon$, which is a contradiction. Thus the zero solution of (2.1.1) is stable.

Next we shall show that the zero solution of (2.1.1) is attractive. From the stability, let $\epsilon = 1$; then there exists $\delta_0 = \delta(1, t_0) < 1$ such that $t_0 \geq 0$ and $|\phi|_{t_0} < \delta_0$ imply

(2.3.9) $$|x(t; t_0, \phi)| < 1 \quad \text{for all } t \geq 0.$$

Hereinafter we consider the solutions such that $|\phi|_{t_0} < \delta_0$. Among them suppose that there exist $\phi(t)$ and $x(t) = x(t; t_0, \phi)$ such that

(2.3.10) $$\limsup_{t \to \infty} |x(t)| = \mu > 0.$$

Since $\gamma L < 1$ by (2.3.7), there exists a constant θ such that $\gamma L < \theta < 1$. By (2.3.10) there exists $t_1 \geq t_0$ such that

(2.3.11) $$|x(u)| \leq \frac{\mu}{\theta} \quad \text{for all } u \geq t_1.$$

By (2.3.6), there exists $T > t_1$ such that

(2.3.12) $$\int_0^{t_1} |K(s,u)|\, du < \frac{(\theta - \gamma L)\mu}{2\theta L} \quad \text{for } s \geq T.$$

Then we have

$$|x(t)| \leq |Y(t)| |Y^{-1}(t_0)| \delta_0$$
$$+ \int_{t_0}^{t} |Y(t)Y^{-1}(s)| \int_0^s |K(s,u)| |x(u)| \, du \, ds$$
$$\leq |Y(t)| |Y^{-1}(t_0)| \delta_0 + |Y(t)| \int_{t_0}^{T} |Y^{-1}(s)| \int_0^s |K(s,u)| |x(u)| \, du \, ds$$
$$+ \int_T^t |Y(t)Y^{-1}(s)| \int_0^{t_1} |K(s,u)| |x(u)| \, du \, ds$$
$$+ \int_T^t |Y(t)Y^{-1}(s)| \int_{t_1}^s |K(s,u)| |x(u)| \, du \, ds.$$

From (2.3.4), (2.3.9) and (2.3.12) we obtain

$$\int_T^t |Y(t)Y^{-1}(s)| \int_0^{t_1} |K(s,u)| |x(u)| \, du \, ds \leq \frac{(\theta - \gamma L)\mu}{2\theta}.$$

From (2.3.4), (2.3.7) and (2.3.11), we have

$$\int_T^t |Y(t)Y^{-1}(s)| \int_{t_1}^s |K(s,u)| |x(u)| \, du \, ds \leq \frac{L\gamma\mu}{\theta}.$$

Thus we get

$$|x(t)| \leq |Y(t)| |Y^{-1}(t_0)| \delta_0$$
$$+ |Y(t)| \int_{t_0}^{T} |Y^{-1}(s)| \int_0^s |K(s,u)| |x(u)| \, du \, ds + \frac{(\theta + \gamma L)\mu}{2\theta}.$$

Since $Y(t) \to 0$ as $t \to \infty$ by (2.3.4) (see Coppel [1]), it follows that $\mu \leq (\theta + \gamma L)\mu/2\theta$ and thus $\mu < \mu$, which is impossible. Therefore the zero solution of (2.1.1) is attractive and the proof is complete. □

We give an example that satisfies the conditions of Theorem 2.3.2.

Example 2.3.1 Let $\alpha(t)$ be a real, continuously differentiable function, equal to 1 except in the intervals

$$J_n = [n - 2^{-4n}, n + 2^{-4n}] \quad (n = 1, 2, \ldots);$$

in J_n, $\alpha(t)$ lies between 1 and 2^{2n} and $\alpha(n) = 2^{2n}$. Consider the scalar equation

(2.3.13) $$x'(t) = -\left[1 + \frac{\alpha'(t)}{\alpha(t)}\right]x(t) + \frac{3}{8}\int_0^t \frac{\sin t}{1 + (t-s)^2} x(s)\,ds.$$

Clearly,

$$\int_0^t |Y(t)Y^{-1}(s)|\,ds = \frac{1}{\alpha(t)}\int_0^t \alpha(s)e^{s-t}\,ds \leq \frac{5}{3} \quad \text{for } t \geq 0,$$

and

$$\sup_{t \geq 0} \frac{3}{8}\int_0^t \frac{ds}{1 + (t-s)^2} = \frac{3\pi}{16} < \frac{3}{5}.$$

Hence the conditions of Theorem 2.3.2 are satisfied and therefore the zero solution of (2.3.13) is asymptotically stable.

Finally, we shall prove a result on exponential stability of (2.1.1).

Theorem 2.3.3 *Suppose that there exist positive constants $K \geq 1$, λ and μ such that*

(2.3.14) $$|Y(t)Y^{-1}(s)| \leq Ke^{-\lambda(t-s)} \quad \text{for } 0 \leq s \leq t < \infty$$

and

(2.3.15) $$\sup_{t \geq 0} \int_0^t e^{\mu(t-s)}|K(t,s)|\,ds < \frac{\lambda}{K}.$$

Then the zero solution of (2.1.1) is exponentially asymptotically stable.

Proof For all $t \geq t_0$ and $|\phi|_{t_0} < 1/K$, we have

(2.3.16) $$|x(t)| \leq Ke^{-\lambda(t-t_0)}|\phi(t_0)|$$

$$+ K\int_{t_0}^t e^{-\lambda(t-s)}\int_0^s |K(s,u)|\,|x(u)|\,du\,ds.$$

There exist positive constants $\nu < \mu$ and σ such that $\lambda = \nu + \sigma$ and $\sup_{t \geq 0} \int_0^t e^{\nu(t-s)} |K(t,s)| \, ds < \sigma/K$. Multiply both sides of (2.3.16) by $e^{\nu t}$ to obtain

$$e^{\nu t}|x(t)| \leq Ke^{\nu t_0}e^{-\sigma(t-t_0)}|\phi(t_0)|$$

$$+ K\int_{t_0}^t e^{\nu s}e^{-\sigma(t-s)}\int_0^s |K(s,u)|\,|x(u)|\,du\,ds$$

$$= Ke^{\nu t_0}e^{-\sigma(t-t_0)}|\phi(t_0)|$$

$$+ K\int_{t_0}^t e^{-\sigma(t-s)}\int_0^s e^{\nu(s-u)}|K(s,u)|e^{\nu u}|x(u)|\,du\,ds.$$

If we denote $\sup_{0 \leq s \leq t} e^{\nu s}|x(s)|$ by $r(t)$, it follows that

(2.3.17) $\quad e^{\nu t}|x(t)| \leq Ke^{\nu t_0}e^{-\sigma(t-t_0)}|\phi(t_0)|$

$$+ Kr(t)\int_{t_0}^t e^{-\sigma(t-s)}\int_0^s e^{\nu(s-u)}|K(s,u)|\,du\,ds$$

$$\leq Ke^{\nu t_0}e^{-\sigma(t-t_0)}|\phi(t_0)| + \sigma r(t)\int_{t_0}^t e^{-\sigma(t-s)}\,ds$$

$$\leq Ke^{\nu t_0}e^{-\sigma(t-t_0)}|\phi(t_0)| + (1 - e^{-\sigma(t-t_0)})r(t).$$

(I) If $e^{\nu s}|x(s)| \leq e^{\nu t}|x(t)|$ for any $s \in [0,t]$, we see that $r(t) = e^{\nu t}|x(t)|$. Then from (2.3.17), we get

$$r(t) \leq Ke^{\nu t_0}e^{-\sigma(t-t_0)}|\phi(t_0)| + (1 - e^{-\sigma(t-t_0)})r(t).$$

Thus $r(t) \leq K|\phi(t_0)|e^{\nu t_0}$ for $t \geq t_0$. Then $r(t) = e^{\nu t}|x(t)|$ implies $|x(t)| \leq K|\phi(t_0)|\,e^{-\nu(t-t_0)}$ for $t \geq t_0$.

(II) If there exists $s \in [0,t]$ such that $e^{\nu s}|x(s)| > e^{\nu t}|x(t)|$, we have the following two further cases.

$(II)(i)$ There exists $t_1 \in [t_0, t)$ such that $e^{\nu t_1}|x(t_1)| = r(t)$. Then from (2.3.17), we have

$$r(t_1) = e^{\nu t_1}|x(t_1)|$$
$$\leq K|\phi(t_0)|e^{\nu t_0}e^{-\sigma(t_1-t_0)} + (1 - e^{-\sigma(t_1-t_0)})r(t_1),$$

Thus $r(t_1) \leq K|\phi(t_0)|e^{\nu t_0}$ for $t_1 \geq t_0$. Then $e^{\nu t}|x(t)| < r(t_1)$ implies that $|x(t)| \leq K|\phi(t_0)|e^{-\nu(t-t_0)}$ for $t \geq t_0$.

$(II)(ii)$ There exists $t_2 \in [0, t_0)$ such that $e^{\nu t_2}|x(t_2)| = r(t)$. Then $e^{\nu t}|x(t)| < e^{\nu t_2}|x(t_2)| < e^{\nu t_0}|\phi|_{t_0}$, and we have $|x(t)| \leq |\phi|_{t_0}e^{-\nu(t-t_0)}$.

Thus from (I) and (II), the zero solution of (2.1.1) is exponentially stable. The proof is complete. □

2.4 Stability by Method of Reduction

Let us consider the integro-differential system

$$(2.4.1) \qquad x'(t) = A(t)x(t) + \int_0^t K(t,s)x(s)\,ds + F(t), \qquad x(0) = x_0$$

where $A(t)$ is an $n \times n$ matrix continuous for $0 \leq t < \infty$, $K(t,s)$ is an $n \times n$ matrix continuous for $0 \leq s \leq t < \infty$, and $F(t)$ is an n-vector continuous for $0 \leq t < \infty$.

As we have seen in Theorem 1.7.3, the system (2.4.1) is equivalent to

$$(2.4.2) \qquad y'(t) = B(t)y(t) + \int_0^t \Phi(t,s)y(s)\,ds + H(t), \qquad y(0) = x_0$$

where $B(t) = A(t) - L(t,t)$,

$$\Phi(t,s) = K(t,s) + \frac{\partial L(t,s)}{\partial s} + L(t,s)A(s) + \int_s^t L(t,u)K(u,s)\,du,$$

and

$$H(t) = F(t) + L(t,0)x_0 + \int_0^t L(t,s)F(s)\,ds,$$

$L(t,s)$ being an $n \times n$ continuously differentiable matrix function for

$0 \leq s \leq t < \infty.$

Let us begin with the following result.

Theorem 2.4.1 *Let $B(t)$ be an $n \times n$ continuous matrix that commutes with its integral, and let M and α be positive real numbers. Suppose that the inequality*

$$(2.4.3) \qquad \left| \exp\left(\int_s^t B(\tau) d\tau \right) \right| \leq M e^{-\alpha(t-s)}, \qquad 0 \leq s \leq t < \infty,$$

holds. Then every solution $y(t)$ of (2.4.2) with $y(0) = x_0$ satisfies

$$(2.4.4) \qquad |y(t)| \leq M |x_0| e^{-\alpha t} + M \int_0^t e^{-\alpha(t-\tau)} |H(\tau)| d\tau$$

$$+ M \int_0^t \left[\int_s^t e^{-\alpha(t-\tau)} |\Phi(\tau, s)| d\tau \right] |y(s)| ds.$$

Proof Multiplying both sides of (2.4.2) by $\exp\left[-\int_0^t B(\tau) d\tau \right]$ and rearranging the terms, we obtain

$$\frac{d}{dt}\left\{ \exp\left[-\int_0^t B(\tau) d\tau \right] y(t) \right\} = \exp\left[-\int_0^t B(\tau) d\tau \right] \left[H(t) + \int_0^t \Phi(t,s) y(s) ds \right].$$

Integrating from 0 to t, we get

$$\exp\left\{ \left[-\int_0^t B(\tau) d\tau \right] y(t) \right\} = x_0 + \int_0^t \exp\left[-\int_0^s B(\tau) d\tau \right] H(s) ds$$

$$+ \int_0^t \left\{ \exp\left[-\int_0^s B(\tau) d\tau \right] \right\} \left[\int_0^s \Phi(s,u) y(u) du \right] ds.$$

By changing the order of integration on the right-hand side in the above relation and using (2.4.3), we obtain (2.4.4). □

Remark 2.4.1 If B is a constant matrix then it commutes with its integral. Further, the condition (2.4.3) holds if, in addition, all the characteristic roots of B have negative real parts.

Theorem 2.4.2 Let $L(t,s)$ be a continuously differentiable $n \times n$ matrix function such that, for $0 \leq s \leq t < \infty$,

(i) the hypotheses of Theorem 2.4.1 hold;

(ii) $|L(t,s)| \leq L_0 e^{-\gamma(t-s)}$;

(iii) $\displaystyle\sup_{0 \leq s \leq t < \infty} \int_s^t e^{\alpha(\tau-s)} |\Phi(\tau,s)| \, d\tau \leq \alpha_0$, where L_0, γ and α_0 are positive real numbers and $\alpha < \gamma$.

Suppose further that

(iv) $F(t) \equiv 0$, where $F(t)$ is defined in (2.4.1).

If $\alpha - M\alpha_0 > 0$ then every solution $x(t)$ of (2.4.1) tends to zero exponentially as $t \to \infty$.

Proof In view of Theorem 1.7.3, it is enough to show that every solution $y(t)$ of (2.4.2) tends to zero exponentially as $t \to \infty$. From hypotheses (i), (ii) and (iv), the inequality (2.4.4) implies

$$e^{\alpha t} |y(t)| \leq M |x_0| + ML_0 |x_0| \int_0^t e^{-(\gamma-\alpha)\tau} d\tau$$

$$+ M \int_0^t \left[\int_s^t e^{\alpha \tau} |\Phi(\tau,s)| \, d\tau \right] |y(s)| \, ds.$$

Using hypothesis (iii), we get

$$e^{\alpha t} |y(t)| \leq M |x_0| + \frac{ML_0 |x_0|}{\gamma - \alpha} + \int_0^t M\alpha_0 e^{\alpha s} |y(s)| \, ds.$$

Application of Gronwall's inequality yields

$$|y(t)| \leq M |x_0| \left(1 + \frac{L_0}{\gamma - \alpha}\right) e^{-(\alpha - M\alpha_0)t}.$$

Thus in view of $\alpha - M\alpha_0 > 0$, the result follows. □

Corollary 2.4.1 In addition to assumptions (i), (ii) and (iv) of Theorem 2.4.2, suppose that the following conditions hold:

(a) $|K(t,s)| \leq K_0 e^{-\beta(t-s)}$, $0 \leq s \leq t < \infty$;

(b) $|\partial L(t,s)/\partial s| \leq N_0 e^{-\delta(t-s)}$, $0 \leq s \leq t < \infty$;

(c) $|A(t)| \leq A_0$, where A_0, N_0, K_0, β, δ are positive real numbers;

(d) $\gamma > \beta > \alpha$, $\delta > \alpha$ and $\alpha - Mp_0 > 0$, where
$$p_0 = \frac{K_0}{\beta - \alpha} + \frac{N_0}{\delta - \alpha} + \frac{L_0 A_0}{\gamma - \alpha} + \frac{K_0 L_0}{(\beta - \alpha)(\gamma - \beta)}.$$

Then every solution $x(t)$ of (2.4.1) tends to zero exponentially as $t \to \infty$.

Proof Following the proof of Theorem 2.4.2, we obtain

$$e^{\alpha t} |y(t)| \leq M |x_0| \left(1 + \frac{L_0}{\gamma - \alpha}\right) + M \int_0^t \left[\int_s^t e^{\alpha \tau} |\Phi(\tau, s)| \, d\tau\right] |y(s)| \, ds.$$

Using the definition of $\Phi(t, s)$ and conditions (ii) and $(a) - (d)$, we obtain from (2.4.5) that

$$e^{\alpha t} |y(t)| \leq M |x_0| \left(1 + \frac{L_0}{\gamma - \alpha}\right) + \int_0^t M p_0 e^{\alpha s} |y(s)| \, ds.$$

Since $\alpha - Mp_0 > 0$, application of Gronwall's inequality yields the desired result.

Remark 2.4.2 If $F(t) \equiv 0$ in (2.4.1) then Theorem 2.4.2 asserts that the equilibrium solution $x \equiv 0$ of (2.4.1) is exponentially asymptotically stable.

Remark 2.4.3 If $F(t)$ is not zero in Theorem 2.4.2, the solutions of (2.4.1) still tend to zero as $t \to \infty$, provided $F(t) \in L^1(R_+)$. This is an immediate consequence of variation of constants formula and Theorem 2.4.2.

Let us provide a simple example.

Example 2.4.1 In (2.4.1), let $A(t) = a_1 e^{-b_1 t} - a_2$, $K(t, s) = e^{-b_2(t-s)}$ and $F(t) \equiv 0$ where a_1, a_2, b_1, and b_2 are positive real numbers. Choose $L(t, s) = a_1 e^{-b_1 t}$. Then $M = 1$, $\alpha = a_2$, $A_0 = a_1 + a_2$, $L_0 = a_1$, $K_0 = 1$, $\beta = b_2$, $\gamma = \delta = b_1$ and $N_0 = 0$ in Corollary 2.4.1. Thus condition (d) of Corollary 2.4.1 holds if $a_1 = ma_2$, $b_1 = (4m^2 + 4m + 1)a_2$, $b_2 = (4 + a_2^2)/a_2$ and $a_2 \geq 2/m(4m + 3)^{1/2}$ where m $(1 < m < \infty)$ is an arbitrary real number. For example, if $m = 2$ and $a_2 = 1$ then $Mp_0 \simeq 0.53$ and $\alpha = 1$.

2.5 Stability in Variation

Consider the initial value problem for the integro-differential system

$$(2.5.1) \quad x'(t) = f(t, x(t)) + \int_{t_0}^{t} g(t, s, x(s))\, ds, \quad x(t_0) = x_0, \quad t_0 \geq 0$$

where $f \in C[R^+ \times R^n, R^n]$ and $g \in C[R^+ \times R^+ \times R^n, R^n]$.

We shall assume that $f_x = \partial f/\partial x$ and $g_x = \partial g/\partial x$ exist and are continuous, and that $f(t,0) \equiv 0$ and $g(t,s,0) \equiv 0$. Let $x(t) = x(t, t_0, x_0)$ be the unique solution of (2.5.1) existing for $t \geq t_0 \geq 0$. We shall now consider the variational systems

$$(2.5.2) \quad y'(t) = f_x(t, x(t))y(t) + \int_{t_0}^{t} g_x(t, s, x(s))y(s)\, ds$$

and

$$(2.5.3) \quad z'(t) = f_x(t, 0)z(t) + \int_{t_0}^{t} g_x(t, s, 0)z(s)\, ds.$$

We observe that under the assumptions on f and g, $\partial x(t, t_0, x_0)/\partial x_0$ exists and is the solution of (2.5.2) such that $\partial x(t_0, t_0, x_0)/\partial x_0 = I$, the identity matrix.

We need the following results (see Lakshmikantham and Leela [1]) before we proceed further.

Theorem 2.5.1 *Let $G \in C[R^+ \times R^+ \times R^+, R]$, $G(t, s, u)$ be monotone nondecreasing in u for each (t, s) and*

$$m(t) \leq m_0(t) + \int_{t_0}^{t} G(t, x, m(s))\, ds, \quad t \geq t_0,$$

where $m, m_0 \in C[R^+, R^+]$. Suppose that $r(t)$ is the maximal solution of the integral equation

$$r(t) = m_0(t) + \int_{t_0}^{t} G(t, s, r(s))\, ds \quad \text{existing on } [t_0, \infty).$$

Then $m(t) \leq r(t)$ for all $t \geq t_0$.

Theorem 2.5.2 *Let f and g in (2.5.1) satisfy the assumptions listed above. Then if $x(t,t_0,x_0)$ and $x(t,t_0,y_0)$ are any two solutions of (2.5.1) existing on $[t_0,\infty)$, we have*

$$x(t,t_0,x_0) - x(t,t_0,y_0) = \left[\int_0^1 \frac{\partial x}{\partial x_0}(t,t_0,sx_0 + (1-s)y_0)\,ds\right](x_0 - y_0).$$

We shall now begin with the following definition of stability in variation of the nonlinear system (2.5.1).

Definition 2.5.1 The equilibrium solution $x \equiv 0$ of (2.5.1) is said to possess stability properties in variation if the equilibrium solution $x \equiv 0$ of (2.5.3) possesses the corresponding stability properties.

Theorem 2.5.3 *Suppose the assumptions of Theorem 1.7.2 hold. Then the stability properties of the function*

$$(2.5.4) \qquad p(t,t_0,|x_0|) = |\psi(t,t_0)|\,|x_0| + \int_{t_0}^t |\psi(t,s)|\,|L(t,t_0)|\,|x_0|\,ds,$$

where $\psi(t,s)$ is the fundamental matrix solution of $v' = B(t)v$, with $\psi(t_0,t_0) = I$, imply the corresponding stability properties of (2.5.1) in variation.

Proof Setting $A(t) = f_x(t,0)$, $K(t,s) = g_x(t,s,0)$ and $F(t) \equiv 0$, we see, in view of Theorem 1.7.2, that it is enough to consider the initial value problem (1.7.8) with $H(t) \equiv 0$ in order to deduce information concerning the variational system (2.5.3). By the variation of parameters formula applied to

$$v' = B(t)v + L(t,t_0)x_0, \qquad v(t_0) = x_0,$$

we have

$$v(t) = \psi(t,t_0)x_0 + \int_{t_0}^t \psi(t,s)L(s,t_0)x_0\,ds.$$

It then follows that

$$(2.5.5) \qquad\qquad |v(t)| \leq p(t,t_0,|x_0|),$$

where $p(t,t_0,|x_0|)$ is given by (2.5.4). Hence the conclusion of the theorem follows immediately. $\square$

Remark 2.5.1 In order to discuss the stability properties of the equilibrium solution ($x \equiv 0$) of (2.5.1), it is enough to consider the stability properties of the equilibrium solution ($x \equiv 0$) of the variational system (2.5.2) in view of Theorem 2.5.2. To see this, we note that any solution $v(t)$ of (2.5.2) such that $v(t_0) = x_0$ is given by $v(t) = (\partial x(t, t_0, x_0)/\partial x_0)x_0$. Furthermore, if $x(t, t_0, x_0)$ is any solution of (2.5.1), then because of Theorem 2.5.2, we have

$$x(t, t_0, x_0) = \left[\int_0^1 \frac{\partial x(t, t_0, x_0 s)}{\partial x_0} ds\right] x_0.$$

Hence if $|v(t)| \leq \rho$, $t \geq t_0$, whenever $|x_0| < \delta(\rho)$, then we get $|x(t, t_0, x_0)| \leq \rho$.

In view of Remark 2.5.1, we shall therefore concentrate now on the system (2.5.2), which can be written as

(2.5.6) $\quad y'(t) = A(t)y(t) + \int_{t_0}^{t} K(t, s)y(s)\,ds + F(t, y(t)), \quad y(t_0) = x_0$

where

$$F(t, y(t)) = [f_x(t, x(t)) - f_x(t, 0)]y(t) + \int_{t_0}^{t} [g_x(t, s, x(s)) - g_x(t, s, 0)]y(s)\,ds.$$

Then, because of Theorem 1.7.2, it is sufficient to investigate the stability properties of the equilibrium solution ($x \equiv 0$) of

(2.5.7) $\quad v'(t) = B(t)v(t) + L(t, t_0)x_0 + H(t, v), \quad v(t_0) = x_0.$

where

(2.5.8) $\quad H(t, v) = F(t, v) + \int_{t_0}^{t} L(t, s)F(s, v(s))\,ds.$

We shall assume that

(2.5.9) $\quad |f_x(t, x) - f(t, 0)| \leq \lambda(t), \quad \lambda \in C[R^+, R^+]$

and

(2.5.10) $\quad |g_x(t, s, x) - g_x(t, s, 0)| \leq \eta(t, s), \quad \eta \in C[R^+ \times R^+, R^+]$

whenever $|x| < \rho$ for some $\rho > 0$.

Let us now utilize the functions.

(2.5.11)
$$\gamma(t,s) \geq \eta(t,s) + \lambda(s)|L(t,s)| + \int_s^t |L(t,\sigma)|\eta(\sigma,s)\,d\sigma$$

$$d(t,s) \geq |\psi(t,s)|\left[\lambda(s) + \int_s^t |\psi(t,\sigma)|\gamma(\sigma,s)\,d\sigma\right].$$

Consider the integral equation

(2.5.12)
$$r(t) = p(t,t_0,|x_0|) + \int_{t_0}^t d(t,s)r(s)\,ds$$

whose solution $r(t)$ is given by

$$r(t) = p(t,t_0,|x_0|) - \int_{t_0}^t r(t,s)p(s,t_0,|x_0|)\,ds$$

where $r(t,s)$ is the resolvent kernel satisfying

$$r(t,s) = -d(t,s) + \int_s^t r(t,\sigma)d(\sigma,s)\,d\sigma.$$

Now we can prove the following result.

Theorem 2.5.4 *Suppose that the hypotheses of Theorem 1.7.2 and (2.5.9) and (2.5.10) hold. Then any solution $v(t)$ of (2.5.7) satisfies the estimate*

(2.5.13)
$$|v(t)| \leq r(t) \quad \text{for all } t \geq t_0$$

provided $|x(t)| \leq \rho$ for all $t \geq t_0$, where $r(t) = r(t,t_0,|x_0|)$ is the solution of (2.5.12).

Proof Let $v(t)$ be any solution of (2.5.7). Then by the variation of constants formula for linear ordinary differential equations,

$$v(t) = \psi(t,t_0)x_0 + \int_{t_0}^t \psi(t,s)\bigl[L(s,t_0)x_0 + H(s,v(s))\bigr]ds.$$

Since

$$|F(t,v)| \leq |f_x(t,x(t)) - f_x(t,0)|\,|v(t)|$$
$$+ \int_{t_0}^{t} |g_x(t,s,x(s)) - g_x(t,s,0)|\,|v(s)|\,ds,$$

we get by using Fubini's theorem, (2.5.8), (2.5.9) and (2.5.10),

$$|H(t,v)| \leq \lambda(t)|v(t)| + \int_{t_0}^{t} r(t,s)|v(s)|\,ds.$$

Employing Fubini's theorem again, it follows that

$$\int_{t_0}^{t} |\psi(t,s) H(s,v(s))|\,ds \leq \int_{t_0}^{t} |\psi(t,s)| \left[\lambda(s) + \int_{s}^{t} |\psi(t,\sigma)|\,r(\sigma,s)\,d\sigma\right]|v(s)|\,ds.$$

Consequently, we obtain from (2.5.8), (2.5.4) and the definition of $d(t,s)$ that

$$|v(t)| \leq p(t,t_0,|x_0|) + \int_{t_0}^{t} d(t,s)|v(s)|\,ds, \quad t \geq t_0.$$

It therefore follows from Theorem 2.5.1 that

$$|v(t)| \leq r(t,t_0,|x_0|), \quad t \geq t_0$$

where $r(t,t_0\,|\,x_0|)$ is the solution of the linear integral equation (2.5.12). The proof is therefore complete.

Remark 2.5.2 Since (2.5.6) and (2.5.7) are equivalent by Theorem 1.7.2, and (2.5.6) is nothing but a restatement of (2.5.2), it is clear from the conclusion of Theorem 2.5.4 that the stability properties of (2.5.12) imply the corresponding stability properties of (2.5.1). One could also directly consider (2.5.1) instead of (2.5.2). For this purpose it is enough to note that (2.5.1) can be written as

$$x'(t) = A(t)x(t) + \int_{t_0}^{t} K(t,s)x(s)\,ds + \widehat{F}(t,x(t)), \quad x(t_0) = x_0$$

where

$$\widehat{F}(t,x(t)) = [f(t,x(t)) - f_x(t,0)]x(t) + \int_{t_0}^{t} [g(t,s,x(s)) - g_x(t,s,0)]x(s)\,ds$$

which also can be written, because of Theorem 1.8.1, as

$$\widehat{F}(t,x) = \left\{ \int_0^1 [f_x(t,\theta x) - f_x(t,0)]d\theta \right\} x$$

$$+ \left\{ \int_{t_0}^t \int_0^1 [(g_x(t,s,x(s)\theta) - g_x(t,s,0)]d\theta \right\} x(s)\,ds.$$

We therefore see that $\widehat{F}$ is of similar form to F, and hence it is easy to see that we could obtain the estimate (2.5.13) even in this case, namely

$$|x(t,t_0,x_0)| \leq r(t,t_0,|x_0|), \qquad t \geq t_0.$$

Finally, we shall provide some examples to illustrate these results.

Example 2.5.1 Consider the scalar equation

(2.5.14) $$x' = -5x - 2\int_0^t e^{-2(t-s)}\sin x(s)\,ds.$$

Here $f(t,x) = -5x$, $f_x(t,x) = -5$, $g(t,s,x) = -2e^{-2(t-s)}\sin x$ and $g_x(t,s,0) = -2e^{-2(t-s)}$. It is easy to check that $L(t,s) = -e^{-3(t-s)}$ satisfies (1.7.6), so that (1.7.8) becomes

$$v' = -4v - e^{-3t}x_0 + H(t,v), \qquad v(0) = x_0.$$

Since $f_x(t,x) - f_x(t,0) = 0$ and $g_x(t,s,x) - g_x(t,s,0) = 2e^{-2(t-s)}(1-\cos x)$, we see that $\lambda(t) \equiv 0$ and $\eta(t,s) = 4e^{-2(t-s)}$. Consequently, we can obtain $\gamma(t,s) = 8e^{-2(t-s)}$ and $d(t,s) = 4e^{-6(t-s)}$. We can then find $r(t,s) = -4e^{-2(t-s)}$, which yields

$$r(t) \leq |x_0|(e^{-3t} + 6e^{-2t} + e^{-4t}),$$

since $p(t) \leq |x_0|(e^{-3t} + e^{-4t})$. This shows that the equilibrium solution $x \equiv 0$ of (2.5.14) is asymptotically stable.

If, on the other hand, we treat $\int_0^t g(t,s,x(s))\,ds$ as a perturbation and $x' = f(t,x)$ as an unperturbed system in (2.5.1) then, using the variation of parameters formula for (2.5.14), we get

$$|x(t)| \leq |x_0|e^{-5t} + 2\int_0^t e^{-5(t-s)}\left[\int_0^s e^{-2(s-\sigma)}|\sin x(\sigma)|\,d\sigma\right]ds,$$

and hence $|x(t)| \leq |x_0|e^{-5t} + \frac{1}{5}$, $t \geq 0$. This shows only that the equilibrium solution $(x \equiv 0)$ of (2.5.14) is stable.

The next example does not involve f, and hence it is interesting in itself.

Example 2.5.2 Consider the scalar equation

(2.5.15) $$x' = -\frac{1}{2}\int_0^t e^{-7(t-s)}\left[24x(s) + x^2(s)\right]ds.$$

Here $f(t,x) \equiv 0$ and $g(t,s,x) = -\frac{1}{2}e^{-7(t-s)}(24x + x^2)$. Hence $g_x(t,s,0) = -12e^{-7(t-s)}$. Choose $L(t,s) = 4e^{-3(t-s)}$. Then (1.7.8) becomes

$$v' = -4v + 4e^{-3t}x_0 + H(t,v), \quad v(0) = x_0.$$

For $|x| \leq \rho = 1$, we see that $\eta(t,s) = e^{-7(t-s)}$, and hence we find $\gamma(t,s) = e^{-3(t-s)}$, $d(t,s) = e^{-7(t-s)}$ and $r(t,s) = -e^{6(t-s)}$. Since $p(t) \leq |x_0|(e^{-4t} + 4e^{-3t})$, we obtain $r(t) \leq |x_0|\left(\frac{3}{2}e^{-4t} + \frac{16}{3}e^{-3t}\right)$, $t \geq 0$. Hence the equilibrium solution $(x \equiv 0)$ of (2.5.15) is asymptotically stable.

Finally we consider an example where the basic system $(x' = f(t,x))$ is unstable.

Example 2.5.3 Consider the scalar equation

(2.5.16) $$x' = \frac{1}{2}x - \frac{3}{2}\int_0^t e^{-2(t-s)}\sin x(s)\,ds.$$

Here $f_x(t,0) = \frac{1}{2}$, $g_x(t,s,0) = -\frac{3}{2}e^{-2(t-s)}$ and $L(t,s) = e^{-(t-s)}$ satisfies (1.7.6). Hence (1.7.8) becomes

$$v' = -\frac{1}{2}v + e^{-t}x_0 + H(t,v), \quad v(0) = x_0.$$

Considering first the estimate on $v(t)$ with $H \equiv 0$, we see from (2.5.8) that $|v(t)| \leq p(t,t_0,|x_0|)$, $t \geq 0$, where $p(t,t_0,|x_0|) \leq 3|x_0|e^{-t/2}$. This shows that the equilibrium solution $(x \equiv 0)$ of (2.5.16) is asymptotically stable in variation. However, if $H \not\equiv 0$, proceeding as in Examples 2.5.2 and 2.5.3, we can compute $\gamma(t,s) = 3e^{-(t-s)}$, $d(t,s) = 6e^{-(t-s)}$, $r(t,s) = -6e^{5(t-s)}$ and

$r(t) \leq |x_0|\left(3e^{-t/2} + \frac{36}{11}e^{5t}\right)$. This does not offer any information about the equilibrium solution ($x \equiv 0$).

2.6 Lipschitz Stability

We shall continue to consider the systems $(2.5.1)-(2.5.3)$ in order to introduce Lipschitz stability. Let the fundamental matrix solutions of the variational systems $(2.5.2)$ and $(2.5.3)$ be respectively $\Phi(t,t_0,x_0) = \partial x(t,t_0,x_0)/\partial x_0$ and $\Phi(t,t_0,0)$, where $x(t) = x(t,t_0,x_0)$ is the solution of $(2.5.1)$. Let us now define Lipschitz stability.

Definition 2.6.1 The trivial solution $x \equiv 0$ of $(2.5.1)$ is said to be uniformly Lipschitz stable (ULS) if there exist constants $M \geq 1$ and $\delta > 0$ such that

$$|x(t,t_0,x_0)| \leq M|x_0| \quad \text{for } t \geq t_0$$

whenever $|x_0| < \delta$ and $t_0 \geq 0$.

Definition 2.6.2 The trivial solution $x \equiv 0$ of $(2.5.1)$ is said to be uniformly Lipschitz stable in variation $(ULSV)$ if there exists constants $M \geq 1$ and $\delta > 0$ such that

$$|\Phi(t,t_0,x_0)| \leq M \quad \text{for } t \geq t_0$$

whenever $|x_0| < \delta$ and $t_0 \geq 0$.

Let us begin with the following result.

Theorem 2.6.1 *For the system $(2.5.3)$, the following statements are equivalent:*
  (i) *the trivial solution of $(2.5.3)$ is globally uniformly Lipschitz stable in variation;*
  (ii) *the trivial solution of $(2.5.3)$ is uniformly Lipschitz stable in variation;*
  (iii) *the trivial solution of $(2.5.3)$ is globally uniformly Lipschitz stable;*
  (iv) *the trivial solution of $(2.5.3)$ is uniformly stable.*

Proof Apply Theorem 1.7.1 to the system $(2.5.3)$ with $A(t) = f_x(t,0)$ and $K(t,s) = g_x(t,s,0)$, and on observing that $\Phi(t,t_0,x_0) = R(t,t_0)$, where

$R(t,s)$ is the differentiable resolvent corresponding to $K(t,s)$, it follows that the solution $z(t,t_0,x_0) = \Phi(t,t_0,0)x_0$. The rest of the proof is essentially the same as for linear ordinary differential equations, and hence is omitted. $\square$

Remark 2.6.1 From Theorem 2.6.1, it is clear that the concept of Lipschitz stability is essentially a nonlinear phenomenon, and differs from the usual uniform stability (US) only in the case of nonlinear systems.

Theorem 2.6.2 *If the trivial solution of (2.5.1) is uniformly Lipschitz stable (ULS) then it is uniformly stable (US).*

Proof Assume that $x \equiv 0$ of (2.5.1) is ULS. Then there exist $M \geq 1$ and $\delta > 0$ such that

$$|x(t,t_0,x_0)| \leq M|x_0| \quad \text{for all } t \geq t_0$$

whenever $|x_0| < \delta$ and $t_0 \geq 0$. Now for a given $\epsilon > 0$, let $\delta = \min(\delta, \epsilon/2M)$. Then for $|x_0| < \delta$, it follows that

$$|x(t,t_0,x_0)| \leq M|x_0| < M\delta < M(\epsilon/2M) < \epsilon \quad \text{for all } t \geq t_0 \geq 0.$$

This implies that $x \equiv 0$ of (2.5.1) is US, and the proof is complete. $\square$

We remark here that the converse of Theorem 2.6.2 is false, as the following example shows.

Example 2.6.1 Consider the two dimensional system

(2.6.1)
$$\begin{cases} x_1'(t) = x_2(t) - \int_0^t x_2(t)x_2^3(s)\,ds, \\ \\ x_2'(t) = -x_1^3(t) + \int_0^t x_1^3(t)x_2^3(s)\,ds. \end{cases}$$

It is easy to see that the equilibrium solution $x_1 \equiv 0$, $x_2 \equiv 0$ of (2.6.1) is US. However, $(0,0)$ of the linearized system $x_1' = x_2$, $x_2' = 0$ is unstable.

Theorem 2.6.3 *If the equilibrium solution $x \equiv 0$ of (2.5.1) is uniformly Lipschitz stable (ULS) then the equilibrium solution $x \equiv 0$ of (2.5.3) is uniformly Lipschitz stable (ULS).*

LINEAR ANALYSIS

Proof By assumption, there exist $M \geq 1$ and $\delta > 0$ such that

$$|x(t, t_0, x_0)| \leq M |x_0| \quad \text{for all } t \geq t_0$$

whenever $|x_0| < \delta$ and $t_0 \geq 0$. Let $y_{0j} = (0, 0, \ldots 0, x_{0j}, 0, \ldots 0)$. Then $|y_{0j}| = h \leq \delta$. From the fact that $x(t, t_0, 0) \equiv 0$, it follows that

$$\left| \frac{\partial x(t, t_0, 0)}{\partial x_{0j}} \right| = \left| \lim_{h \to 0} \frac{x(t, t_0, y_{0j}) - x(t, t_0, 0)}{h} \right|$$

$$\leq \lim_{h \to 0} \frac{M |y_{0j}|}{h} = \lim_{h \to 0} \frac{Mh}{h} = M.$$

Thus $|\Phi(t, t_0, 0)| = |R(t, t_0)| \leq M$, and consequently $x \equiv 0$ of (2.5.3) is US. Hence the application of Theorem 2.6.1 yields the required result. □

It should be pointed out here that the converse of Theorem 2.6.3 is, in general, false, as may be seen from the following example.

Example 2.6.2 Consider the two dimensional system

(2.6.2)
$$\begin{cases} x_1'(t) = x_1^3(t) - \int_0^t e^{-2(t+s)} x_2(t) x_2^2(s) \, ds, \\ x_2'(t) = x_2^3(t) - \int_0^t e^{-2(t+s)} x_1(t) x_2^2(s) \, ds. \end{cases}$$

It is easy to verify that $(0, 0)$ of (2.6.2) is unstable. However, the linearized system $x_1' = 0$, $x_2' = 0$ is clearly ULS.

Theorem 2.6.4 *If the trivial solution of (2.5.1) is uniformly Lipschitz stable in variation (ULSV) then it is uniformly Lipschitz stable (ULS).*

Proof Let $x(t) = x(t, t_0, x_0)$ be any solution of (2.5.1). Then by Theorem 2.5.3, we have

(2.6.3)
$$x(t, t_0, x_0) = \left[\int_0^1 \Phi(t, t_0, s x_0) \, ds \right] x_0.$$

Since $x \equiv 0$ of (2.5.1) is $ULSV$, it follows from Definition 2.6.2 that $|\Phi(t, t_0, x_0)| \leq M$, $t \geq t_0$, whenever $|x_0| < \delta$ and $t_0 > 0$ for some $\delta > 0$ and $M \geq 1$.

This implies from (2.6.3) that $|x(t,t_0,x_0)| \leq M|x_0|$ for all $t \geq t_0$ whenever $|x_0| < \delta$ and $t_0 \geq 0$, and hence $x \equiv 0$ of (2.5.1) is ULS. Thus the proof is complete. □

Theorem 2.6.5 Assume that

(H2) (i) $|f_x(t,x) - f_x(t,0)| \leq \alpha(t)$

and

(ii) $|g_x(t,s,x) - g_x(t,s,0)| \leq \beta(t,s)$

whenever $|x| \leq \rho$ for some $\rho > 0$, where $\alpha \in C[R_+, R_+]$ and $\beta \in C[R_+ \times R_+, R_+]$. If $x \equiv 0$ of (2.5.3) is uniformly Lipschitz stable (ULS) and if

(2.6.4) $$\sup_{t \geq t_0} \int_{t_0}^{t} [\alpha(s) + \lambda(t,s)]ds \leq K$$

where $\lambda(t,s) = \int_s^t \beta(t,\sigma)d\sigma$, such that $MK < 1$, then $x \equiv 0$ of (2.5.1) is uniformly Lipschitz stable in variation (ULSV).

Proof Let $y(t) = y(t,t_0,y_0)$ be a solution of (2.5.2). If we write (2.5.2) in the form

$$y(t) = f_x(t,0)y(t) + \int_{t_0}^{t} g_x(t,s,0)y(s)ds$$
$$+ [f_x(t,x(t)) - f_x(t,0)]y(t)$$
$$+ \int_{t_0}^{t} [g_x(t,s,x(s)) - g_x(t,s,0)]y(s)ds,$$

then the application of Theorem 1.7.1 and the hypothesis (H2) give

(2.6.5) $$|y(t)| \leq |R(t,t_0)||y_0| + \int_{t_0}^{t} |R(t,s)|[\alpha(s) + \lambda(t,s)]|y(s)|ds.$$

Since $x \equiv 0$ of (2.5.3) is ULS, there exists $M \geq 1$ such that

(2.6.6) $|R(t,t_0)| \leq M$ for all $t \geq t_0 \geq 0$.

Therefore it follows from (2.6.4) − (2.6.6) that

$$|y(t)| \leq M|y_0| + MK \sup_{t_0 \leq s \leq t} |y(s)|.$$

Thus $\sup_{t_0 \leq s \leq t} |y(s)| \leq M|y_0| + MK \sup_{t_0 \leq s \leq t} |y(s)|.$

Since $MK < 1$, we have

$$\sup_{t_0 \leq s \leq t} |y(s)| \leq \frac{M}{(1-MK)} |y_0| = M_1 |y_0|,$$

where $M_1 \geq 1$, and therefore

$$|\Phi(t,t_0,x_0)| = \sup_{|y_0| \leq 1} |\Phi(t,t_0,x_0)y_0| \leq \sup_{|y_0| \leq 1} |y(t,t_0,y_0)| \leq M_1$$

whenever $|x_0| < \delta$. Hence $x \equiv 0$ of (2.5.1) is $ULSV$, completing the proof. □

Theorem 2.6.6 *Suppose that the assumptions of Theorem 1.7.2 and (H2) of Theorem 2.6.5 hold. Then the uniform Lipschitz stability (ULS) property of the function*

(2.6.7) $\qquad p(t,t_0,|x_0|) = |\psi(t,t_0)| \, |x_0| + \int_{t_0}^{t} |\psi(t,s)| \, |L(s,t_0)| \, |x_0| \, ds$

where $\psi(t,s)$ is a fundamental matrix solution of $v' = B(t)v$, implies that $x \equiv 0$ of (2.5.1) is uniformly Lipschitz stability in variation (ULSV).

Proof Let us first consider (2.5.3) with $f_x(t,0) = A(t)$ and $g_x(t,s,0) = K(t,s)$. Then, in view of Theorem 2.5.4, it is enough to investigate the equivalent system

(2.6.8) $\qquad v'(t) = B(t)v(t) + L(t,t_0)x_0, \quad v(t_0) = x_0.$

By variation of parameters, we have

$$v(t,t_0,x_0) = \psi(t,t_0)x_0 + \int_{t_0}^{t} \psi(t,s)L(s,t_0)x_0 \, ds,$$

and hence, from (2.6.7), we obtain $|v(t,t_0,x_0)| \leq p(t,t_0,|x_0|)$. Therefore, from the assumptions of the theorem, it follows that $|v(t,t_0,x_0)| \leq M|x_0|$ for all $t \geq t_0 \geq 0$ for some $M \geq 1$ and $|x_0| < \delta$. Hence $x \equiv 0$ of (2.6.8) is ULS, and so is that of (2.5.3). This, with the assertion of Theorem 2.6.5, implies that $x \equiv 0$ of (2.5.1) is $ULSV$, and hence the proof is complete. □

80   THEORY OF INTEGRO-DIFFERENTIAL EQUATIONS

2.7 Asymptotic Equivalence

Consider the operator equation

(2.7.1) $$Lx = f(x)$$

where L is a linear operator with domain $D(L)$ and range $R(L)$ contained in Banach spaces X and Y respectively and $f(x)$ is a nonlinear operator from X into Y. Let $N(L)$ denote the null space of L. We shall first obtain a correspondence between the solutions of (2.7.1) and $N(L)$, and then use this result to investigate sufficient conditions for boundedness and asymptotic equivalence of solutions of Volterra integro-differential equations.

We shall assume the following hypotheses on L and f.

(H1) L is closed on $D(L)$ and there exists a subspace $S \subseteq D(L)$ such that the restriction of L to S, denoted by L_S, is closed and one-to-one and has closed range.

Let $R(L_S)$ be the range of L_S and let $\Sigma(\rho) = \{x \in X : |x|_X \le \rho\}$. Concerning the nonlinear operator f, we assume, with no loss of generality, that $f(x) = G(x) + F$ where $G(0) = 0$ and $F \in Y$.

(H2) G maps $D(L)$ into $R(L_S)$ continuously in such a way that for some constants β and ρ, $0 \le \beta < \infty$, $0 < \rho < \infty$, $|G(x) - G(y)|_Y \le \beta |x - y|_X$ for all $x, y \in D(L) \cap \Sigma(\rho)$.

Remark 2.7.1 Under the hypothesis (H1), it follows by the closed graph theorem that L_S has a bounded inverse L_S^{-1}.

Theorem 2.7.1 Suppose that the hypotheses (H1) and (H2) hold. Assume further that $\beta |L_S^{-1}| < 1$. Then there exists a constant $\delta > 0$ such that for each $F \in R(L_S)$ satisfying $|F|_Y \le \delta\rho$, a one-ton-one bicontinuous mapping Q exists from the set $N(L) \cap \Sigma(\delta\rho)$ into the set of solutions of (2.7.1) contained in $D(L) \cap \Sigma(\rho)$.

Proof We shall first show that Q is well defined. Given $n \in N(L) \cap \Sigma(\delta\rho)$, define the operator

$$T: D(L) \to D(L)$$

by the relation $Tx = n + L_S^{-1} f(x)$. For $x \in \Sigma(\rho)$, it follows from (H2) with $y = 0$ that

$$|Tx|_X \leq \left[\delta(1 + |L_S^{-1}|) + |L_S^{-1}|\beta\right]\rho.$$

Thus if we choose $\delta < \left[1 - |L_S^{-1}|\beta\right]/(1 + |L_S^{-1}|)$ then it is clear that T maps $D(L) \cap \Sigma(\rho)$ into itself. Moreover, from hypothesis (H2), we have

$$|Tx - Ty|_Y \leq |L_S^{-1}|\beta|x - y|_X,$$

and hence T is a contraction on $D(L) \cap \Sigma(\rho)$. Choosing $x_1 \in D(L) \cap \Sigma(\rho)$ and setting

$$x_n = Tx_{n-1} \in D(L) \cap \Sigma(\rho), \qquad n \geq 2,$$

it follows from (H2) that

$$x_n \to x_0 \text{ in } X \quad \text{and} \quad f(x_n) \to f(x_0) \text{ in } Y.$$

Since L is closed on $D(L)$ and

$$Lx_n = LTx_{n-1} = LL_S^{-1} f(x_{n-1}) = f(x_{n-1}) \to f(x_0) \quad \text{in } Y,$$

it is clear that the point x_0 lies in $D(L) \cap \Sigma(\rho)$ and $Lx_0 = f(x_0)$. This implies that x_0 solves (2.7.1) and further we shall have a well defined function $Q: Q(n) = x_0$.

Suppose that $n_i \in N(L) \cap \Sigma(\delta\rho)$ for $i = 1, 2$ and $Q(n_1) = Q(n_2) = x$. That is

(2.7.2) $$x = n_i + L_S^{-1} f(x), \qquad i = 1, 2.$$

Then by subtracting, we find that $n_1 = n_2$, and hence Q is continuously one-to-one. Further, from (2.7.2), we have

$$\begin{aligned}|Q(n_1) - Q(n_2)|_X &= |x_1 - x_2|_X = |Tx_1 - Tx_2|_X \\ &\leq |n_1 - n_2|_X + |L_S^{-1}|\beta|x_1 - x_2|_X \\ &= |n_1 - n_2|_X + |L_S^{-1}|\beta|Q(n_1) - Q(n_2)|_X.\end{aligned}$$

This implies that

(2.7.3) $$|Q(n_1) - Q(n_2)|_X \leq \frac{|n_1 - n_2|_X}{1 - |L_S^{-1}|\beta}.$$

Similarly,

$$|Q^{-1}(x_1) - Q^{-1}(x_2)|_X = |n_1 - n_2|_X$$
$$= |x_1 - Tx_1 - x_2 + Tx_2|_X$$
(2.7.4)
$$\leq |x_1 - x_2|_X + |L_S^{-1}|\beta |x_1 - x_2|_X$$
$$\leq (1 + |L_S^{-1}|\beta)|x_1 - x_2|_X.$$

Thus the inequalities (2.7.3) and (2.7.4) respectively imply that Q and Q^{-1} are continuous, and this completes the proof. □

Remark 2.7.2 If $|F|_Y \leq \delta\rho/|L_S^{-1}|$ then there exists a constant $\rho^* > 0$ such that the range of Q contains all solutions of (2.7.1) contained in $D(L) \cap \Sigma(\rho^*)$. To see this, choose ρ^* so small that $|x - L_S^{-1}f(x)|_X \leq \delta\rho$ for $|x|_X \leq \rho^*$. This is possible by the way F is chosen, since $I - L_S^{-1}f$ is continuous. Let $x \in D(L) \cap \Sigma(\rho^*)$ be a solution of (2.7.1). Define $n = s - L_S^{-1}f(x)$, which, by the assumptions made, lies in $N(L) \cap \Sigma(\delta\rho)$. Thus there exists a unique solution $\hat{x} \in D(L) \cap \Sigma(\rho)$ of (2.7.1) such that $\hat{x} = Q(n)$. That is, $\hat{x} = n + L_S^{-1}f(\hat{x})$. Therefore $\hat{x} - x = L_S^{-1}[f(\hat{x}) - f(x)]$. This implies that $|\hat{x} - x|_X \leq |L_S^{-1}|\beta|\hat{x} - x|$. Since $|L_S^{-1}|\beta < 1$, it is clear that $\hat{x} = x$. Hence x is in the range of Q.

Consider now the perturbed system of integro-differential equations,

(2.7.5) $$x'(t) = A(t)x(t) + \int_0^t B(t,s)x(s)\,ds + F(t) + (Gx)(t)$$

and the corresponding unperturbed system

(2.7.6) $$y'(t) = A(t)y(t) + \int_0^t B(t,s)y(s)\,ds,$$

where $0 \leq t < \infty$, F and G are given vectors, and A and B are given $n \times n$ matrices. In general, the perturbations G are nonlinear functionals of the form

$$(Gx)(t) = G(t, x(t)) \quad \text{or} \quad \int_0^t K(t, s, x(s))\,ds$$

or

$$x(t) \int_0^t K(t, s, x(s))\,ds.$$

We shall now investigate sufficient conditions on the perturbations G and on the nonhomogeneous system

(2.7.7) $$z'(t) = A(t)z(t) + \int_0^t B(t,s)z(s)\,ds + F(t)$$

under which it is possible to assert (locally) that there is a one-to-one correspondence between the bounded solutions of (2.7.5) and those of (2.7.6).

Given any set $\Omega \subset R^n$, let $LL^1(\Omega)$ be the set of all measurable functions m such that the seminorms $\int_D |m(t)|\,dt$ are finite for all compact subsets D of Ω.

Let $BC = \{x(t) \in C[0,\infty): |x|_0 = \sup_{0 \le t \le \infty} |x(t)| < \infty\}$ and let L^p, $1 \le p < \infty$, be the Banach space of functions defined and measurable for $t \ge 0$, and for which

$$|x|_p = \left[\int_0^\infty |x(t)|^p\,dt\right]^{1/p} < \infty.$$

For convenience, we take L^∞ as BC and $|x|_\infty = |x|_0$.

By a solution of (2.7.5) or (2.7.6) or (2.7.7), we mean an absolutely continuous function satisfying the corresponding system for almost all $t \ge 0$. We now make the following hypotheses:

(H3) $A, F \in LL^1(R_+)$ and $B \in LL^1(R_+ \times R_+)$;

(H4) For each $F \in L^p$, $1 \le p < \infty$, there exists at least one bounded solution $z \in BC$ of (2.7.7).

Remark 2.7.3 Under the hypothesis (H3), the unique solution $z(t)$ of (2.7.7) with $z(0) = x_0$ is given by the variation of constants formula as

(2.7.8) $$z(t) = R(t,0)x_0 + \int_0^t R(t,s)F(s)\,ds$$

where $R(t,s)$ is the unique solution of the initial value problem

(2.7.8)* $$\frac{\partial R}{\partial t} = A(t)R(t,s) + \int_s^t B(t,u)R(u,s)\,du,$$

$$R(s,s) = I, \quad \text{for } 0 \le s \le t < \infty.$$

It is clear from (2.7.8) that sufficient conditions for $z \in BC$ are

(i) $R(t,0) \in BC$,

(ii) $R(t,s)$ satisfies

(2.7.9) $$\left[\int_0^t |R(t,s)|^p\, ds \right]^{1/p} \leq B \quad \text{for } 0 \leq t < \infty,\ p > 1,$$

(iii) $F \in L^q(R_+)$, $1/p + 1/q = 1$.

Here B is a positive constant. If $p = 1$ then (iii) can be replaced by $F \in BC$.

Remark 2.7.4 In order to apply Theorem 2.7.1 to the system of integro-differential equations, we make the following observations. Define X_1 to be those vectors in R^n that, as initial conditions $y(0)$, give rise to bounded solutions of (2.7.6). Then X_1 is clearly a linear subspace of R^n. Let X_2 be any space supplementary to X_1 such that $R^n = X_1 \oplus X_2$, and let P_i be the projection of R^n onto X_i.

Let $X = BC$ and $Y = L^p$, $1 \leq p < \infty$, in Theorem 2.7.1. Define the linear operator L by

$$Lx \equiv x' - A(t)x - \int_0^t B(t,s)x(s)\, ds$$

whose domain we take to the linear subspace $D^p(L) = \{x \in BC : x(t)$ is absolutely continuous for $t \geq 0$ and $Lx \in L^p\}$.

Let $S^p = \{x \in D^p(L) : x(0) \in X_2\}$. Thus $S^p \subseteq D^p(L)$. Under the hypothesis (H4), it is clear that the range R^p of L restricted to S^p is all of $Y = L^p$ and hence is closed. Therefore, for a given $F \in L^p$, there exists $z \in D^p(L)$ such that $Lx = F$, and if $y(t) \in BC$ is the unique solution of (2.7.6) satisfying $y(0) = P_1 z(0)$ then $x = z - y \in S^p$ and $Lx = F$. Moreover, L is one-to-one on the subspace S^p, since if $Lx_1 = Lx_2$ for $x_1, x_2 \in S^p$ then $L(x_1 - x_2) = 0$ and $x_1 - x_2 \in S^p$. This implies that $y = x_1 - x_2$ is a bounded solution of (2.7.6) with initial state in X_2. Since X_2 is supplementary to X_1, it must be that $y = 0$. To see that L is closed on $D^p(L)$ and S^p, let us suppose that $x_n \in D^p(L)$ and $F_n = Lx_n$ converge in BC and L^p respectively to $x^0 \in BC$ and $F^0 \in L^p$. Integrating $F_n = Lx_n$, we obtain

$$x_n(t) = x_n(0) + \int_0^t \left[A(s) + \int_s^t B(\sigma,s)\, d\sigma \right] x_n(s)\, ds + \int_0^t F_n(s)\, ds.$$

Using the hypothesis $(H3)$ and the dominated convergence theorem, we get

$$x^0(t) = x^0(0) + \int_0^t \left[A(s) + \int_s^t B(\sigma, s)\, d\sigma \right] x^0(s)\, ds + \int_0^t F^0(s)\, ds.$$

Hence $x^0(t)$ is absolutely continuous and solves $Lx^0 = F^0$. This proves that L is closed on $D^p(L)$. If, on the other hand, $x_n(t) \in S^p$ then, in addition to $x^0(t) \in D^p(L)$, it is obvious that $x_n(0) \in X_2$, which implies that $x^0(0) \in X_2$ and hence $x^0(t) \in S^p$. That is L is closed on S^p.

Theorem 2.7.2 *Suppose $(H3)$ and $(H4)$ hold. Further, suppose $(Gx)(t)$ in (2.7.5) maps BC into L^p, $1 \leq p < \infty$, such that*

(2.7.10) $$|(Gx)(t) - (Gy)(t)|_p \leq \beta |x - y|_0$$

holds for all $x, y \in BC$ satisfying $|x|_0, |y|_0 \leq \rho$. Then there exist positive constants a, b and ρ_0 with the following properties: if $\rho \leq \rho_0$ then

(i) *for every $F \in L^p$, $|F|_p \leq a$, there exists a one-to-one bicontinuous correspondence Q between the bounded solutions $y \in BC$ of (2.7.6) satisfying $|y(0)| \leq b$ and the bounded solutions $x \in BC$ of (2.7.5) satisfying $|x|_0 \leq \rho$, $|P_1 x(0)| \leq b$; and,*

(ii) *the correspondence Q is such that if $x = Qy$, then $P_1 x(0) = y(0)$.*

Proof For all the bounded solutions $y(t)$ of the linear homogeneous system (2.7.6), it is possible to assert that $|y|_0 \leq M |y(0)|$ for some constant $M > 0$. In view of the assumption (2.7.10) and Remark 2.7.4, it is clear that all the conditions of Theorem 2.7.1 are satisfied. Let ρ be the constant whose existence is guaranteed by Theorem 2.7.1 and take $\rho_0 = 1/(2 |L_S^{-1}|)$ where $S = S^p$, $a = \gamma \rho$ and $b = \gamma \rho / M$. Given a bounded solution $y(t)$ of (2.7.6) with $|y(0)| \leq b$ (hence $y(0) = X_1$), it follows that $|y|_0 \leq \gamma \rho$. Hence by Theorem 2.7.1, there exists a unique corresponding solution $x = Qz$ of (2.7.5) satisfying $|x_0| \leq \rho$. Moreover, Q is invertible and bicontinuous. As in the proof of Theorem 2.7.1, let $x = y + L_S^{-1}(f(x))(t)$. Hence $P_1 x(0) = P_1 y(0) = y(0)$, since y being bounded implies $y(0) \in X_1$ and since $L_S^{-1}(f(x))(t) \in S^p$ implies that $L_1^{-1}(fx)(0)$ lies in X_2. Finally, to show that Q is onto the set of solutions of (2.7.5), let $x(t)$ be a solution of (2.7.5) with $|x|_0 \leq \rho$ and $|P_1 x(0)| \leq b$. If we define $y = x - L_S^{-1}(f(x))(t)$ then we find that y is a bounded solution of (2.7.6) satisfying $|y(0)| = |P_1 y(0)| =$

$|P_1 x(0)| \le b$. Hence $|y|_0 \le \delta\rho$ and $\widehat{x} = Qy$ exists. But then $\widehat{x} = y + L_S^{-1}(f(x))(t)$, and hence

$$x - \widehat{x} = L_S^{-1}[f(x)(t) - f(\widehat{x})(t)].$$

In view of (2.7.10) this implies that

$$|x - \widehat{x}|_0 \le |L_S^{-1}| \rho_0 |x - \widehat{x}|_0.$$

Since $|L_S^{-1}| \rho_0 = \frac{1}{2}$, we conclude that $x = \widehat{x}$ and Q is onto. This completes the proof. □

We shall now discuss the asymptotic equivalence of the integro-differential equations (2.7.5) and (2.7.7).

Definition 2.7.1 The integro-differential systems (2.7.5) and (2.7.7) are said to be asymptotically equivalent if, for every solution $z \in BC$ of (2.7.7) ($x \in BC$ of (2.7.5)), there exists a solution $x \in BC$ of (2.7.5) ($z \in BC$ of (2.7.7)) such that $x(t) - z(t) \to 0$ as $t \to \infty$.

Theorem 2.7.3 Assume that

(i) (H3) holds;

(ii) $R(t,s)$ satisfies the condition (2.7.9) and for each fixed $T > 0$,

$$(2.7.11) \qquad \lim_{t \to \infty} \int_0^T |R(t,s)|^p \, ds = 0;$$

(iii) the perturbation $(G\phi)(t)$ in (2.7.5) is a continuous mapping of BC into $LL^1(R_+)$ and there exists a measurable function $\lambda(t) \ge 0$, $\lambda \in L^q[0,\infty)$, such that for all functions $\phi \in BC$,

$$(2.7.12) \qquad |(G\phi)(t)| \le (1 + |\phi|)\lambda(t) \quad \text{for } 0 \le t < \infty.$$

Then the integro-differential systems (2.7.5) and (2.7.7) are asymptotically equivalent on $[0,\infty)$.

Proof Let $z \in BC$ be a solution of (2.7.7). We first show that $x(t) \in BC$. Let $0 < \epsilon < 1$. Since $\lambda \in L^q[0,\infty)$, choose a number $T_1 > 0$ so large that

$$(2.7.13) \qquad \left[\int_{T_1}^t \lambda^q(s) \, ds \right]^{1/q} < \frac{\epsilon}{2B}, \quad T_1 < t < \infty, \quad 1 < q < \infty,$$

where $1/p + 1/q = 1$ and B is the constant in (2.7.9). Since $x(t)$ is continuous on $[0,\infty)$, there exists a constant $d = d(T_1) > 0$ such that $d = \sup\limits_{0 \geq t \leq T_1} |x(t)|$. Choose a number $\delta > d$ such that

(2.7.14) $\qquad |z|_0 + |1+d)B|\lambda|_q + \frac{\epsilon}{2} < \frac{1}{2}(1-\epsilon)\delta.$

We claim that $|x(t)| < \delta$ for $0 \leq t < \infty$. For, if not, there exists a $t_1 > T_1$ such that $|x(t)| < \delta$ for $0 \leq t < t_1$ and $|x(t_1)| = \delta$. From the variation of parameters formula, we have

(2.7.15) $\qquad x(t) = z(t) + \int_0^t R(t,s)(Gx)(s)\,ds,$

where $z(t)$ is given by (2.7.8). Thus from (2.7.12), (2.7.15) and the Hölder inequality, we obtain

$$\delta = |x(t_1)| \leq |z|_0 + \int_0^{T_1} |R(t,s)|\lambda(s)[1 + |x(s)|]\,ds$$

$$+ \int_{T_1}^{t_1} |R(t,s)|\lambda(s)[1 + |x(s)|]\,ds$$

$$\leq |z|_0 + (1+d)B|\lambda|_q + (1+\delta)B\left[\int_{T_1}^{t_1} \lambda^q(s)\,ds\right]^{1/q}.$$

Using (2.7.13) and (2.7.14), we get

$$\delta = |x(t_1)| < |z|_0 + (1+d)B|\lambda|_q + \tfrac{1}{2}(1+\delta)\epsilon$$
$$< \tfrac{1}{2}(1-\epsilon)\delta - \tfrac{1}{2}\epsilon + \tfrac{1}{2}(1+\delta)\epsilon = \tfrac{1}{2}\delta,$$

which is a contradiction and hence $|x(t)| < \delta$ for $0 \leq t < \infty$. We shall now show that $\lim\limits_{t\to\infty} |x(t) - z(t)| = 0$. Let $\sup\limits_{0 \geq t < \infty} |x(t)| = d$ and let $\epsilon > 0$ be given. Choose $T_2 > 0$ so large that

(2.7.16) $\qquad \left[\int_{T_2}^t \lambda^q(s)\,ds\right]^{1/q} = \dfrac{\epsilon}{2B(1+d)}, \quad t \geq T_2.$

By (2.7.11), choose $T_1 \geq T_2$ such that

$$\text{(2.7.17)} \qquad \left[\int_0^{T_2} |R(t,s)|^p\, ds\right]^{1/p} < \frac{\epsilon}{2(1+d)|\lambda|_q}, \quad t \geq T_1.$$

Then from (2.7.12), (2.715) and the Hölder inequality, it follows that

$$\text{(2.7.18)} \qquad \begin{aligned} |x(t) - z(t)| &\leq \int_0^{T_2} |R(t,s)|\lambda(s)[1 + |x(s)|]\, ds \\ &\quad + \int_{T_2}^t |R(t,s)|\lambda(s)(1 + |x(s)|)\, ds \\ &\leq (1+d)|\lambda|_q \left[\int_0^{T_2} |R(t,s)|^p\, ds\right]^{1/p} \\ &\quad + (1+d)B\left[\int_{T_2}^t \lambda^q(s)\, ds\right]^{1/q}. \end{aligned}$$

Hence (2.7.16) and (2.7.17) give

$$|x(t) - z(t)| < \tfrac{1}{2}\epsilon + \tfrac{1}{2}\epsilon = \epsilon \quad \text{for } t \geq T_2.$$

Since ϵ is arbitrary, we have $\lim_{t\to\infty} |x(t) - z(t)| = 0$. Finally, suppose that $u(t) \in BC$ is a solution of (2.7.5). We shall show that there exists a solution $v \in BC$ of (2.7.7) such that $\lim_{t\to\infty} |u(t) - v(t)| = 0$. To see this, let us define

$$\text{(2.7.19)} \qquad v(t) = u(t) - \int_0^t R(t,s)(Gu)(s)\, ds.$$

In view of (2.7.15), it is clear that $v(t)$ is a solution of (2.7.7). Using (2.7.9), (2.7.12), (2.7.19) and the Hölder inequality, we obtain

$$|v(t)| \leq |u|_0 + (1 + |u|_0)B|\lambda|_q < \infty.$$

Further, from the continuity of $R(t,s)$, it is clear that $v(t)$ is continuous and hence $v \in BC$. Let $\epsilon > 0$ be given. Define T_1 and T_2 as (2.7.16) and (2.7.17) with d replaced by $|u|_0$. Then from (2.7.18) and (2.7.19) with $x(t) = u(t)$ and $z(t) = v(t)$, we obtain $|u(t) - v(t)| < \tfrac{1}{2}\epsilon + \tfrac{1}{2}\epsilon = \epsilon$ for all $t \geq T_2$. Since ϵ is arbitrary, it follows that $\lim_{t\to\infty} |u(t) - v(t)| = 0$ and hence the proof is complete. □

Corollary 2.7.1 Let the hypothesis (i) and (ii) of Theorem 2.7.3 hold with $p=1$. Let $(G\phi)(t)$ be a continuous mapping of BC into BC such that (2.7.12) is satisfied for a function $\lambda \in BC$, $\lambda(t) \geq 0$ and $\lim_{t\to\infty} \lambda(t) = 0$. Then the integro-differential systems (2.7.5) and (2.7.7) are asymptotically equivalent on $[0, \infty)$.

Proof The proof is practically the same as the proof of Theorem 2.7.3. In the Hölder inequality, we take $p=1$ and $q=\infty$. Since $\lambda \in BC$ and $\lambda(t) \to 0$ as $t \to \infty$, the choices of T_1 and T_2 in (2.7.16) and (2.7.17) are respectively

$$|\lambda(t)| \leq \frac{\epsilon}{2B(1+d)}, \quad t \geq T_2$$

and

$$\int_0^{T_2} |R(t,s)|\,ds < \frac{\epsilon}{2(1+d)|\lambda|_0}, \quad t \geq T_1.$$

The rest of the proof follows exactly on the same lines as that of Theorem 2.7.3. □

Remark 2.7.4 The assumption (2.7.12) allows us a variety of perturbations G in (2.7.5) including, for example,

$$(Gx)(t) = \int_0^t C(t,s) x^\sigma(s)\,ds$$

where $C(t,s)$ is a given $n \times n$ matrix, and, of course, also $(Gx)(t) = H(t,x)$, where H is a given function which is either continuous in (t,x) or measurable in t for each fixed x and continuous in x for each fixed t, $0 \leq t < \infty$, $|x| < \infty$. In the latter case, the condition (2.7.12) takes the form

(2.7.20) $$|H(t,x)| \leq \lambda(t)(1 + |x|)$$

for $0 \leq t < \infty$, $|x| < \infty$. The hypothesis (2.7.20) includes the important special case

$$|H(t,x)| \leq \lambda(t)|x|^\sigma, \quad 0 \leq \sigma \leq 1.$$

If we consider $\sigma > 1$ then $|H(t,x)| = 0(|x|)$ as $|x| \to 0$. Then by the same argument as in Strauss [1], we can deduce the existence of a solution $x \in BC$ of (2.7.5) provided $|F|_0 \leq \delta$, where δ is a sufficiently small constant depending on σ and λ. It then follows from Theorem 2.7.3 that if

$|H(t,x)| \le \lambda(t)|x|^\sigma$, $\sigma > 1$, we have the asymptotic equivalence of (2.7.5) and (2.7.7) for those $F \in BC$ for which $|F|_0$ is sufficiently small.

The following result is somewhat similar to Theorem 2.7.2, which establishes the existence of a homeomorphism Q between the sets of M-bounded solution of (2.7.6) and its perturbed system

$$(2.7.21) \qquad x'(t) = A(t)x(t) + \int_0^t B(t,s)x(s)\,ds + G(t,x,Tx),$$

in which A and B are $n \times n$ matrices and $G: R_+ \times R^n \times C(R_+) \to R^n$ satisfies the Caratheodory conditions; that is, $G(t,u,v)$ is measurable in $t \in R_+$ for all $(u,v) \in R^n \times C(R_+)$ and continuous in u,v for all $t \in R_+$, $T: C(R_+) \to C(R_+)$ where

$$C(R_+) = \{u: u \in C[R_+, R^n]\}.$$

Under the hypotheses (H3) and (H4), it is easy to show that the unique bounded solution $z(t)$ of (2.7.7) with $z(0) \in X_2$ is given by

$$(2.7.22) \qquad z(t) = \int_0^t V(t,s)F(s)\,ds + \int_t^\infty W(t,s)F(s)\,ds,$$

where

$$V(t,s) = R(t,s) - R(t,0)P(s), \quad 0 \le s \le t,$$

and

$$W(t,s) = -R(t,0)P(s) \quad \text{for } 0 \le t \le s.$$

$R(t,s)$ being the unique solution of the initial value problem (2.7.8)*, and $z(0) = -\int_0^\infty P(s)F(s)\,ds$, $P(t)$ is an $n \times n$ matrix, $|P(t)| \in L^1(R_+)$, $1/p + 1/q = 1$ for $p \ne 1$ and $q = \infty$ for $p = 1$.

Let $M(t)$ be a continuous $n \times n$ matrix such that $M^{-1}(t)$ exists for all $t \ge 0$. We shall say that a function q is M-bounded on $[0,\infty)$, denoted by $q \in MB$, if and only if $\sup_{t \ge 0} |M^{-1}(t)q(t)| < \infty$. We shall also discuss, besides the existence of a homeomorphism Q between the sets of M-bounded solutions of (2.7.6) and (2.7.21), the M-asymptotic equivalence and (M,p)-integral equivalence $1 \le p < \infty$ between the solution $y \in MB$ of (2.7.6) and the solution $x \in MB$, $x(t) = Qy(t)$, of (2.7.21) in the sense of the following definitions.

Definition 2.7.2 The integro-differential systems (2.7.21) and (2.7.6) are said to be M-asymptotically equivalent if, for every solution $y \in MB$ of (2.7.6) ($x \in MB, x(t) = Qy(t)$ of (2.7.21)), there exists a solution $x \in MB$, $x(t) = Qy(t)$ of (2.7.21) ($y \in MB$ of (2.7.6)) such that $|M^{-1}(t)[Qy(t) - y(t)]| \to 0$ as $t \to \infty$.

Definition 2.7.3 The integro-differential systems (2.7.21) and (2.7.6) are said to be (M, p)-equivalent (with $p > 1$ an integer) if, for every solution $y \in MB$ of (2.7.6) ($x \in MB$, $x(t) = Qy(t)$ of (2.7.21)), there exists a solution $x \in MB$, $x(t) = Qy(t)$ of (2.7.21) ($y \in MB$ of (2.7.6)) such that

$$|M^{-1}(t)[Qy(t) - y(t)]| \in L^p[0, \infty).$$

Theorem 2.7.4 *Assume that*
(i) *hypotheses H(3) and H(4) hold;*
(ii) $\gamma: R_+ \times R_+ \to R_+$ *is a nondecreasing function with respect to each variable separately and such that*

$$\sup\left\{\frac{\gamma(u,v)}{\max(u,v)}: a \leq u, v \leq b, 0 < a \leq b\right\} < 1;$$

(iii) *there exists an $n \times n$ matrix $P(t)$ and constant $K > 0$ such that*

(2.7.23) $$\left[\int_0^t |M^{-1}(t)V(t,s)|^q ds\right]^{1/q} + \left[\int_t^\infty |M^{-1}(t)W(t,s)|^q ds\right]^{1/q} \leq K$$

for all $t \geq 0$, $p \neq 1$;
(iv) *there exists a continuous function $\lambda(t): L^p \to R_+$ and positive constants α and k_1 such that*

$$|G(t,u,v) - G(t,u_1,v_1)| \leq \lambda(t)\gamma(|M^{-1}(t)(u - u_1)|, |v - v_1|),$$

where $|v - v_1| \leq \alpha |M^{-1}(t)(u - u_1)|$, $0 < \alpha \leq 1$, $t \geq 0$, $\int_0^\infty \lambda^p(t) dt < \infty$ and $K\left[\int_0^\infty \lambda^p(t) dt\right]^{1/p} = k_1$, $p + q = pq$ $(p \geq 2)$ and $\int_0^\infty |G(s,0,0)|^p ds < \infty$.

Then there exists a homeomorphism Q from the set of M-bounded solutions of (2.7.6) into the M-bounded solutions of (2.7.21).

Proof In view of Remark 2.7.4, by the application of Theorem 2.7.1, the proof follows exactly on the same lines as that of Theorem 2.7.2. □

Theorem 2.7.5 *Suppose that all the assumptions of Theorem 2.7.4 hold, and that*

$$(2.7.24) \quad \lim_{t\to\infty}\left[\int_0^T |M^{-1}(t)V(t,s)|^q ds\right]^{1/q} = 0 \quad \text{for each fixed } T > 0.$$

Then the systems (2.7.21) and (2.7.6) are M-asymptotically equivalent on $[0,\infty)$.

Proof From the variation of parameters formula of the form (2.7.22), it follows that

$$(2.7.25) \quad M^{-1}(t)[Qy(t) - y(t)] = \int_0^t M^{-1}(t)V(t,s)G(s, Qy(s), TQy(s))\,ds$$

$$+ \int_t^\infty M^{-1}(t)W(t,s)G(s, Qy(s), TQy(s))\,ds.$$

Given any $\epsilon > 0$, in view of assumption (iii), choose $T = T(\epsilon) > 0$ so large that

$$(2.7.26) \quad \left[\int_T^\infty \lambda^p(s)\,ds\right]^{1/p} \leq \frac{\epsilon}{4K\rho_0}, \quad \left[\int_T^\infty |G(s,0,0)|^p\,ds\right]^{1/p} \leq \frac{\epsilon}{4K},$$

where ρ_0 is some positive constant. It follows from (2.7.25), for $t \geq T$, that

$$M^{-1}(t)[Qy(t) - y(t)] = \int_0^T M^{-1}(t)V(t,s)[G(s, Qy(s), TQy(s)) - G(s,0,0)]\,ds$$

$$+ \int_T^t M^{-1}(t)V(t,s)[G(s, Qy(s), TQy(s)) - G(s,0,0)]\,ds$$

$$+ \int_t^\infty M^{-1}(t)W(t,s)[G(s, Qy(s), TQy(s)) - G(s,0,0)]\,ds$$

$$+ \int_0^T M^{-1}(t)V(t,s)G(s,0,0)\,ds$$

$$+ \int_T^t M^{-1}(t)V(t,s)G(s,0,0)\,ds$$

$$+ \int_t^\infty M^{-1}(t)W(t,s)G(s,0,0)\,ds.$$

Thus by using assumptions (ii) and (iii), (2.7.23) and (2.7.26), and the Hölder inequality, we obtain

$$|M^{-1}(t)[Qy(t)-y(t)]| \leq \rho_0 \left[\int_0^T |M^{-1}(t)V(t,s)|^q ds\right]^{1/q} + \epsilon.$$

Since ϵ is arbitrary, it follows from (2.7.24) that

$$|M^{-1}(t)[Qy(t)-y(t)]| \to 0 \quad \text{as } t \to \infty.$$

This completes the proof. □

Theorem 2.7.6 *In addition to assumptions of Theorem 2.7.4, suppose that the following conditions hold:*

$$\int_0^\infty s\lambda^p(s)\,ds < \infty, \qquad \int_0^\infty s|G(s,0,0)|^p\,ds < \infty$$

and

$$\int_0^\infty \left[\int_0^t |M^{-1}(t)V(t,s)|^q ds\right]^{1/q} dt < \infty.$$

Then the integro-differential systems (2.7.21) and (2.7.6) are (M,p)-integral equivalent on $[0,\infty)$.

Proof The proof of this theorem is practically the same as the proof of Theorem 2.7.5. □

2.8 Ultimate Behavior of Solutions

Consider the integro-differential system

(2.8.1) $\qquad y'(t) + \int_0^t K(t,s)y(s)\,ds + f(y(t)) = g(t), \quad t \in R_+.$

The basic problem we shall discuss in this section is to find conditions on (2.8.1) so that the ultimate behavior of its solution is determined by the ordinary differential system

(2.8.2) $$z'(t) + f(z(t)) = 0.$$

Let us begin with the following definition of M_0 which will play an important role in our subsequent discussion:

$$M_0 = \left\{ x\colon R_+ \to R^n,\ x \in L_{loc}(R_+, R^n),\ \int_t^{t+1} |x(s)|\,ds \to 0 \text{ as } t \to \infty \right\}.$$

It is clear that $L^p(R_+, R^n) \subset M_0(R_+, R^n)$, $1 \leq p < \infty$. The Volterra integral operator A is defined by

(2.8.3) $$(Ay)(t) + \int_0^t K(t,s) y(s)\,ds, \quad t \in R_+.$$

Theorem 2.8.1 *Assume that*

(i) $f\colon R^n \to R^n$ *is a continuous map;*

(ii) $K\colon \Delta \to \mathcal{L}(R^n, R^n)$, *where* $\Delta = \{(t,s)\colon 0 \leq s \leq t\}$, *is a continuous map, and*

(2.8.4) $$\int_0^t |K(t,s)|\,ds \in M_0(R_+, R);$$

(iii) *for any $y(t)$ continuous and bounded on R_+, with values in R^n the inequality*

(2.8.5) $$\int_0^t \langle (Ay)(s) + f(y(s)), y(s) \rangle \, ds \geq 0$$

holds for $t \in R_+$;

(iv) $g \in L^1(R_+, R_n)$.

Then any solution of the system (2.8.1) exists on R_+, and is bounded there, and its limit agrees with that of a convenient solution of the ordinary differential system (2.8.2).

Proof Let $y = y(t)$ be a solution of (2.8.1) with $y(0) = y_0 \in R^n$, existing in the interval $[0, T)$, $T \leq \infty$, and on this interval we have the following relationship:

(2.8.6) $$\tfrac{1}{2}\tfrac{d}{dt}|y(t)|^2 + \langle (Ay)(t), y(t) \rangle + \langle f(y(t)), y(t) \rangle = \langle g(t), y(t) \rangle.$$

On integrating both sides of (2.8.6) from 0 to t, $t < T$, we obtain

(2.8.7)
$$\frac{1}{2}|y(t)|^2 + \int_0^t \langle (Ay)(s), y(s) \rangle \, ds + \int_0^t \langle f(y(s)), y(s) \rangle \, ds$$
$$= \frac{1}{2}|y_0|^2 + \int_0^t \langle g(s), y(s) \rangle \, ds.$$

Therefore it follows from (2.8.5) and (2.8.7) that

$$|y(t)|^2 \leq |y_0|^2 + 2\int_0^t |\langle g(s), y(s) \rangle| \, ds.$$

Then by using Schwartz's inequality, we obtain

(2.8.8)
$$Y^2(t) \leq |y_0|^2 + 2Y(t) \int_0^\infty |g(s)| \, ds$$

where $Y(t) = \sup[|y(s)|, 0 \leq s \leq t]$.

The inequality (2.8.8) further implies that

$$Y(t) \leq \left\{ |y_0|^2 + \left[\int_0^\infty |g(s)| \, ds \right]^2 \right\}^{1/2} + \int_0^\infty |g(s)| \, ds \quad \text{for all } t \in [0,T).$$

Taking into account $y(t) \leq Y(t)$ and the assumption (iii), we conclude that $y(t)$ is bounded on $[0, T)$. This implies that the solution $y(t)$ can be continued in the future. Hence any solution of (2.8.1) does exist on R_+ and remains bounded there. We shall now prove that the limit set of the solution $y(t)$ (the limit set is obviously nonempty) coincides with that of a convenient solution of (2.8.2). To this end, let $y(t)$ be an arbitrary solution of (2.8.1). Since any solution of (2.8.1) is bounded on R_+, the family of functions (from R_+ into R^n) $\{y(t+h), h \in R_+\}$ is uniformly bounded on R_+. On the other hand, it is clear from (2.8.1) that $y'(t) \in M_0(R_+, R^n) \oplus L^\infty(R_+, R^n)$ and $Ay \in M_0$ (see the condition (2.8.4)). This implies the uniform continuity of $y(t)$ on R_+, which has as consequence the compactness of the family $\{y(t+h), h \in R_+\}$ on every compact interval of R_+ (with respect to uniform convergence). Therefore there exists at least one sequence $\{\tau_m\} \subset R_+$, $\tau_m \to \infty$ as $m \to \infty$ such that

$$\lim_{m\to\infty} y(t+\tau_m) = z(t), \qquad t \in R_+.$$

We shall show now that $z(t)$ is a solution of (2.8.2). Let $h > 0$ be chosen arbitrarily and $t \geq 0$. Integrating (2.8.1) both sides from t to $t+h$, we obtain

$$(2.8.9) \qquad \int_t^{t+h} y'(s+\tau_m)\,ds \; + \int_t^{t+h} (Ay)(s+\tau_m)\,ds + \int_t^{t+h} f(y(s+\tau_m))\,ds$$

$$= \int_t^{t+h} g(s+\tau_m)\,ds.$$

It is clear from (2.8.4) and (2.8.5) that $Ay \in M_0$ and $g \in L^1 \subset M_0$, and hence as $m \to \infty$, (2.8.8) yields that

$$(2.8.10) \qquad z(t+h) - z(t) + \int_t^{t+h} f(z(s))\,ds = 0.$$

Since $h > 0$ is arbitrary, (2.8.10) implies (2.8.2), and therefore $z(t)$ is a solution of (2.8.2). Finally, to prove the last assertion of the theorem, let P and Q be respectively the limit sets of solutions $y(t)$ of (2.8.1) and $z(t)$ of (2.8.2). If $z(t_k) \to \hat{z} \in Q$ then by taking into account that

$$z(t_k) = \lim_{m\to\infty} y(t_k + \tau_m^*),$$

we obtain for some subsequence $\{\tau_{m_k}^*\} \subset \{\tau_m^*\}$ that

$$\hat{z} = \lim_{k\to\infty} y(t_k + \tau_{m_k}^*) \quad \text{with } \tau_k + \tau_{m_k}^* \to \infty \text{ as } k\to\infty.$$

This shows that $\hat{z} \in P$ and hence $Q \subseteq P$. Let us now assume that $\hat{y} \in P$. Then $y(t_k) \to \hat{y}$ for some sequence $\{t_k\}$, with $t_k \to \infty$ as $k\to\infty$. Denote by $z(t)$ the solution of (2.8.2) with $z(0) = \hat{y}$. It is easy to show that $y(t+t_k)$ converges to $z(t)$ as $k\to\infty$, uniformly on any compact interval of R_+. Now choose a subsequence $\{t_{k_m}\} \subset \{t_k\}$ such that $t_{k_m} > 2t_m$. Then $\bar{t}_m = t_{k_m} - t_m \to \infty$ as $m\to\infty$, and

$$|y(\bar{t}_m + t_m) - z(\bar{t}_m)| \;=\; |y(t_{k_m}) - z(t_{k_m} - t_m)| \to 0 \quad \text{as } m\to\infty.$$

Hence $\bar{y} = \lim_{m\to\infty} y(t_{k_m}) = \lim_{m\to\infty} z(t_{k_m} - t_m) \in Q$. This completes the proof of the theorem. □

Remark 2.8.1 The monotonicity condition (2.8.5) on the operator $A + f$ can be secured by many ways. For example, if A is a monotone operator and f satisfies the condition $\langle f(y), y \rangle \geq 0$ for any $y \in R^n$ then we have (2.8.5). Certainly, by choosing A in a more specialized class of operators, then there is a lot of freedom for f. For instance, if A is strictly monotone then f needs to satisfy to a milder condition of the type $\langle f(y), y \rangle > -m|y|^2$, $y \in R^n$, where m is a positive number.

We shall now consider the integro-differential system

(2.8.11) $$y'(t) + \int_0^t K(t,s) y'(s)\, ds + f(y(t)) = g(t), \quad t \in R_+$$

which provides a mathematical model for the description of certain electrical networks (see Moser [1]), with a view to establishing a similar result to that given in Theorem 2.8.1.

Theorem 2.8.2 *Assume that*

(i) $f(y) = \operatorname{grad} V(y)$, $y \in R^n$, *where* $V \in C^1(R^n, R)$ *is such that*

(2.8.12) $$\lim_{|y| \to \infty} V(y) = \infty;$$

(ii) $K(t,s)$ *is a continuous map from* Δ *into* $\mathcal{L}[R^n, R^n]$ *such that*

(2.8.13) $$\left[\int_0^t |K(t,s)|^2\, ds \right]^{1/2} \in M_0(R_+, R),$$

and for some real η, $\eta < 1$,

(2.8.14) $$\int_0^t \langle (Ay)(s) + \eta y(s), y(s) \rangle\, ds \geq 0, \quad t \in R_+$$

for any $y(t)$ *continuous on* R_+, *with values in* R^n;

(iii) $g \in L^2(R_+, R^n)$.

Then any solution of the system (2.8.11) is defined on R_+, *it is bounded there, and its limit set agrees with that of a convenient solution of the ordinary differential equation (2.8.2).*

Proof Let $y(t)$ be a solution of (2.8.11), satisfying the initial condition $y(0) = y_0 \in R^n$. It is defined for some interval $[0, T)$, $T < \infty$. We shall prove that $y(t)$ can be continued to the whole positive real axis. It follows from the assumption (i) that

$$(2.8.15) \qquad \int_0^t \langle f(y(s)), y'(s) \rangle \, ds = V(y(t)) - V(y_0).$$

Taking the scalar product of both sides of (2.8.11) with $y'(t)$, and integrating between 0 and t, $t < T$, we obtain from (2.8.15) that

$$(2.8.16) \qquad \int_0^t \langle y'(s) + (Ay')(s), y'(s) \rangle \, ds + V(y(t)) = V(y_0) + \int_0^t \langle g(s), y'(s) \rangle \, ds.$$

From (2.8.14) and (2.8.16), we get

$$(2.8.17) \qquad (1-\eta) \int_0^t |y'(s)|^2 \, ds + V(y(t)) \le V(y_0) + \int_0^t \langle g(s), y'(s) \rangle \, ds.$$

Let $\delta = 1 - \eta > 0$. From the fact that $(a\delta - b)^2 \ge 0$ it follows that

$$\int_0^t \langle g(s), y'(s) \rangle \, ds \le \frac{\delta}{2} \int_0^t |y'(s)|^2 \, ds + \frac{1}{2\delta} \int_0^t |g(s)|^2 \, ds,$$

and hence (2.8.17) gives

$$(2.8.18) \qquad \frac{\delta}{2} \int_0^t |y'(s)|^2 \, ds + V(y(t)) \le V(y_0) + \frac{1}{2\delta} \int_0^t |g(s)|^2 \, ds.$$

This implies, in view of the assumption (iii), that the right-hand side of (2.8.18) is finite (and independent of t), and therefore the left-hand side must remain bounded on R_+. This shows that a finite escape time for $y(t)$ cannot exist. Moreover, any solution $y(t)$ of (2.8.11) must be defined on the whole of R_+ and must remain bounded on that half axis, because of the condition (2.8.12) on $V(y)$, while the integral on the left-hand side of (2.8.18) must be bounded on R_+. Hence we can assert that $y(t) \in L^\infty(R_+, R^n)$ and $y'(t) \in L^2(R_+, R_n)$. Now by using (2.8.13) and (2.8.19), it is easy to show that

$$(Ay')(t) \in M_0(R_+, R^n),$$

and the uniform continuity of $y(t)$ on R_+. The rest of the proof follows on exactly the same lines as that of Theorem 2.8.1. □

Corollary 2.8.1 *Suppose all the assumptions of Theorem 2.8.2 hold. Let $V(y)$ have only a finite number of critical points (i.e. such that $f(y) = 0$).*

Then every solution of (2.8.11) approaches one of these points as $t\to\infty$.

Proof The proof follows from Theorem 2.8.2, taking also into account the fact for each solution of the system (2.8.2), $f(z(t))$ must tend to zero as $t\to\infty$. □

Remark 2.8.2 The condition (2.8.14) is, actually, a condition of positive definiteness for the operator $A + \eta I$ on the space $L^2_{\text{loc}}(R_+, R^n)$. Thus, in view of Remark 2.8.1, the condition (2.8.14) can easily be replaced by a somewhat weaker form.

2.9 Difference Equations

Consider the difference equation of Volterra type

$$(2.9.1) \qquad \Delta x(n) = x(n+1) - x(n) = A(n)x(n) + \sum_{s=x_0}^{n-1} K(n,s)x(s) + F(n)$$

with initial condition $x(n_0) = x_0$, where $A(n)$ and $K(n,s)$ are $d \times d$ matrices for each $n, s \in N$ and $F: N^+_{n_0} \to R^d$, $N^+_{n_0} = \{n_0, n_0+1, \ldots, n_0+k, \ldots\}$, $n_0, k \in N$.

We need the following results in our subsequent discussions. Let us first state a result corresponding to Fubini's theorem, which can be proved by induction.

Lemma 2.9.1 *Let $L(n,s)$ and $K(n,s)$ be $d \times d$ matrices defined for $s, n \geq n_0$ such that L and K are zero matrices for $n, s < n_0$. Then the relation*

$$\sum_{s=n_0}^{n-1} L(n,s+1) \sum_{\sigma=n_0}^{s-1} K(s,\sigma)x(\sigma) = \sum_{s=n_0}^{n-1} \sum_{\sigma=s+1}^{n-1} L(n,\sigma+1)K(\sigma,s)x(s)$$

holds, where $x: N^+_{n_0} \to R^d$.

Lemma 2.9.2 *Let $n \in N^+_{n_0}$, $\gamma \geq 0$, and let $g(n,\gamma)$ and $G(n,s,\gamma)$ be nonnegative functions, nondecreasing with respect to γ for fixed n and s. Suppose that*

$$\Delta u(n) \leq g(n, u(n)) + \sum_{s=n_0}^{n-1} G(n,s,u(s)), \quad u(n_0) = u_0 > 0$$

for any $u: N^+_{n_0} \to R_+$. Then $u(n) \leq \gamma(n)$ $n \geq n_0$, where $\gamma(n) = \gamma(n, n_0, u_0)$ is the solution of the difference equation

$$\Delta \gamma(n) = g(n, \gamma(n)) + \sum_{s=n_0}^{n-1} G(n, s, \gamma(n)), \qquad \gamma(n_0) = u_0.$$

Proof Set $v(n) = g(n, u(n)) + \sum_{s=n_0}^{n-1} G(n, s, u(s))$, so that $u(n) \leq \Delta^{-1} v(n) + w(n)$, where Δ^{-1} is the antidifference operator and $w(n)$ is an arbitrary function of period 1. Setting $z(n) = \Delta^{-1} v(n) + w(n)$, we see that $\Delta z(n) = v(n) \geq 0$, and hence $z(n)$ is nondecreasing. Consequently, $\Delta z(n) \leq g(n, z(n)) + \sum_{s=n_0}^{n-1} G(n, s, z(n))$, $z(n_0) = u_0$, which implies, by a comparison theorem (see Theorem 1.6.1 in Lakshmikantham and Trigiante [1]),

$$z(n) \leq \gamma(n), \quad n \geq n_0.$$

Since $u(n) \leq z(n)$, the proof of the lemma is complete. $\square$

Theorem 2.9.1 *Assume that there exist a $d \times d$ matrix $L(n,s)$ defined on $N_{n_0}^+ \times N_{n_0}^+$ and satisfying*

(2.9.2) $\qquad K(n,s) + L(n, s+1) - L(n,s) + L(n, s+1)A(s)$

$$+ \sum_{\sigma = s+1}^{n-1} L(n, \sigma+1) K(\sigma, s) = 0.$$

Then (2.9.1) is equivalent to the ordinary linear difference equation

(2.9.3) $\qquad \Delta y(n) = B(u)y(n) + L(n, n_0)x_0 + H(n), \quad y(n_0) = x_0$

where $B(n) = A(n) - L(n,n)$ and

(2.9.4) $\qquad H(n) = F(n) + \sum_{s=n_0}^{n-1} L(n, s+1) F(s).$

Proof We first prove that every solution of (2.9.1) satisfies (2.9.3). Let $x(n) = x(n, n_0, x_0)$ be the solution of (2.9.1). Setting $p(s) = L(n,s)x(s)$, we have

$$p(s+1) - p(s) = [L(n, s+1) - L(n, s)]x(s) + L(n, s+1)[x(s+1) - x(s)].$$

Substituting from (2.9.1), we get

(2.9.5) $\qquad p(s+1) - p(s) = [L(n, s+1) - L(n, s) + L(n, s+1)A(s)]x(s)$

$$+ L(n, s+1) \left[\sum_{\sigma = n_0}^{s-1} K(s, \sigma) x(\sigma) + F(s) \right].$$

Summing both sides of (2.9.5) from n_0 to $n-1$, we obtain

$$L(n,n)x(n) - L(n,n_0)x_0$$

$$= \sum_{s=n_0}^{n-1} \left[L(n,s+1) - L(n,s) + L(n,s+1)A(s) + \sum_{\sigma=s+1}^{n-1} L(n,\sigma+1)K(\sigma,s) \right] x(s)$$

$$+ \sum_{s=n_0}^{n-1} L(n,s+1)F(s).$$

Using Lemma 2.9.1, (2.9.2) and (2.9.4), and in view of (2.9.1), we obtain

$$\Delta x(n) = B(n)x(n) + L(n,n_0)x_0 + H(n),$$

which implies that $x(n)$ is a solution of (2.9.3).

To prove that every solution of (2.9.3) is also a solution of (2.9.1), let $y(n) = y(n,n_0,x_0)$ be any solution of (2.9.3) for $n \geq n_0$. Define

$$z(n) = \Delta y(n) - A(n)y(n) - F(n) - \sum_{s=s_0}^{n-1} K(n,s)y(s).$$

In view of (2.9.2), we then obtain

$$z(n) = \Delta y(n) - A(n)y(n) - F(n)$$

$$+ \sum_{s=s_0}^{n-1} \left[L(n,s+1) - L(n,s) + L(n,s+1)A(s) + \sum_{\sigma=s+1}^{n-1} L(n,\sigma+1)K(\sigma,s) \right] y(s).$$

By Lemma 2.9.1, it follows that

$$z(n) = \Delta y(n) - A(n)y(n) - F(n)$$

(2.9.6)
$$+ \sum_{s=n_0}^{n-1} \left[L(n,s+1) - L(s,n) + L(n,s+1), A(s) \right] y(s)$$

$$+ \sum_{s=s_0}^{n-1} L(n,s+1) \sum_{\sigma=n_0}^{s-1} K(s,\sigma)y(\sigma).$$

Setting $p(s) = L(n,s)y(s)$, we can obtain as before

$$L(n,n)y(n) - L(n,n_0)x_0 = \sum_{s=n_0}^{n-1} [L(n,s+1) - L(n,s)]y(s)$$

(2.9.7)
$$+ \sum_{s=n_0}^{n-1} L(s,s+1)\Delta y(s).$$

Using (2.9.3), (2.9.6) and (2.9.7), we see that

$$z(n) = -\sum_{s=n_0}^{n-1} L(n,s+1)\left[\Delta y(s) - A(s)y(s) - F(s) - \sum_{\sigma=n_0}^{s-1} K(s,\sigma)y(\sigma)\right]$$

which in view of the definition of $z(n)$, yields

$$z(n) = -\sum_{s=n_0}^{n-1}(n,s+1)z(s).$$

It is easy to see that $z(n_0) = 0$, and therefore it follows that $z(n) = 0$ for all $n \geq n_0$. Hence the proof is complete. □

Consider the linear nonhomogeneous difference equation

(2.9.8) $$\Delta x(n) = B(n)x(n) + P(n), \quad x(n_0) = x_0,$$

where $B(n)$ is a $d \times d$ matrix on $N_{n_0}^+$ and $P: N_{n_0}^+ \to R^d$. The solution $x(n) = x(n, n_0, x_0)$ of (2.9.8) is given by the variation of constants formula

(2.9.9) $$x(n) = \Phi(n,n_0)x_0 + \sum_{s=n_0}^{n-1} \Phi(n,s+1)P(s),$$

where $\Phi(n,n_0)$ is the fundamental matrix solution of the difference equation $\Delta x(n) = B(n)x(n)$ such that $\Phi(n_0,n_0)$ is an identity matrix.

We shall prove the following stability result for the linear difference equation of Volterra type (2.9.1).

Theorem 2.9.2 *Suppose that the assumptions of Theorem 2.9.1 are satisfied, and the following estimates for $n \geq n_0$, $0 < \alpha < 1$:*

(i) $|L(s,n)| \leq K_0 \alpha^{n-s};$
(ii) $|F(n)| \leq K_1 \alpha^n;$
(iii) $|\Phi(n,s)| \leq K_2 \alpha^{n-s};$

where K_0, K_1 and K_2 are positive constants. Then every solution of (2.9.1) tends to zero as $n \to \infty$.

Proof Let $y(n, n_0, x_0)$ be any solution of (2.9.1). Then by Theorem 2.9.1, it is also a solution of (2.9.3). Hence, in view of (2.9.9), we get

(2.9.10) $$|y(n,n_0,x_0)| \leq |\Phi(n,n_0)||x_0| + \sum_{s=n_0}^{n-1} |\Phi(n,s+1)||P(s)|$$

where $P(s) = H(s) + L(s, n_0)$.

Now by using the estimates (i), (ii) and (iii), we obtain successively

$$|P(s)| \leq |F(s)| + \sum_{\sigma = n_0}^{s-1} |L(s, \sigma+1)| \, |F(\sigma)| + |L(s, n_0)| \, |x_0|$$

$$\leq K_1 \alpha^s + \sum_{\sigma = n_0}^{s-1} K_0 \alpha^{s-\sigma-1} K_1 \alpha^\sigma + K_1 \alpha^{s-n_0} |x_0|.$$

Let $\widehat{K} = \max(K_0, K_1)$. Then we have

$$|P(s)| \leq \widehat{K} \alpha^s \left(1 + |x_0| \alpha^{-n_0} + \frac{1}{\alpha} \sum_{\sigma = n_0}^{s-1} \widehat{K} \right)$$

$$= \widehat{K} \alpha^s \left[1 + |x_0| \alpha^{-n_0} + \frac{\widehat{K}}{\alpha}(s - n_0) \right].$$

Hence (2.9.10) yields

$$|y(n, n_0, x_0)| \leq K_2 \alpha^{n - n_0} |x_0|$$

$$+ \sum_{s = n_0}^{n-1} K_2 \alpha^{n-s-1} \left\{ \widehat{K} \alpha^s \left[1 + |x_0| \alpha^{-n_0} + \frac{\widehat{K}}{\alpha}(s - n_0) \right] \right\}$$

which, after simplification, gives the estimate

$$|y(n, n_0, x_0)| \leq |x_0|(K_2 + K_3 n)\alpha^{n-n_0} + K_4(n + n^2)\alpha^{n-n_0},$$

where K_3 and K_4 are positive constants. It is then clear that the right-hand of the above inequality tends to zero as $n \to \infty$, and the proof is complete. □

We shall now consider the nonlinear difference equation of Volterra type given by

$$(2.9.11) \qquad \Delta x(n) = f(n, x(n)) + \sum_{s = n_0}^{n-1} g(n, s, x(s)), \quad x(n_0) = x_0$$

where $f: N_{n_0}^+ \times S(\rho) \to R^n$ and $g: N_{n_0}^+ \times N_{n_0}^+ \times S(\rho) \to R^n$, $f(n, x)$ and $g(n, s, x)$ are continuous in x, and $S(\rho) = \{x \in R^d : |x| < \rho\}$. Assume further that
 (i) $f(n, 0) \equiv 0$, $g(n, s, 0) \equiv 0$;
 (ii) f_x and g_x exist and are continuous in x.
Setting $f_x(n, 0) \equiv A(n)$, $g_x(n, s, 0) \equiv K(n, s)$ and using the mean value theorem, (2.9.11) can be written as

$$\text{(2.9.12)} \quad \Delta x(n) = A(n)x(n) + F(n,x(n))$$
$$+ \sum_{s=n_0}^{n-1} [K(n,s)x(s) + G(n,s,x(s))], \quad x(n_0) = x_0$$

where

$$F(n,x) = \int_0^1 [f_x(n,x\theta) - f_x(n,0)] d\theta x,$$

$$G(n,s,x) = \int_0^1 [g_x(n,s,x\theta) - g_x(n,s,0)] d\theta x.$$

Concerning (2.9.11), we shall give the following stability result.

Theorem 2.9.3 *Assume that all the conditions of Theorem 2.9.1 are satisfied. Suppose further that*

(H1) *the trivial solution of linear difference equation of Volterra type*

$$\text{(2.9.13)} \quad \Delta x(n) = A(n)x(n) + \sum_{s=n_0}^{n-1} K(n,s)x(s), \quad x(n_0) = x_0$$

is exponentially asymptotically stable,

(H2) *for* $(n,x) \in N_{n_0}^+ \times S(\rho)$, *we have* $|F(n,x)| \leq w_1(n,|x|)$ *and* $|G(n,s,x)| \leq w_2(n,s,|x|)$, *where* $w_1 : N_{n_0}^+ \times R_+ \to R_+$ *and* $w_2 : N_{n_0}^+ \times N_{n_0}^+ \times R_+ \to R_+$, *and* $w_1(n,u)$ *and* $w_2(n,s,u)$ *are continuous and nondecreasing in* u *for* n *and* (n,s) *respectively.*

Then the stability properties of the trivial solution of the difference equation

$$\text{(2.9.14)} \quad \Delta u = -u(n) + M\left[w_1(n,u(n)) + \sum_{s=n_0}^{n-1} w(n,s,u(s))\right], \quad u(n_0) = u_0 > 0$$

where M is a positive constant and

$$\text{(2.9.15)} \quad w(n,s,u(s)) = w_2(n,s,u(s)) + |L(n,s)| w_1(s,u(s))$$

$$+ \sum_{\sigma=n_0}^{s-1} |L(s,\sigma)| w_2(s,\sigma,u(\sigma))$$

in which $L(n,s)$ *is any solution of* (2.9.2), *imply the corresponding stability properties of the trivial solution of* (2.9.11).

LINEAR ANALYSIS

Proof By Theorem 2.9.1, (2.9.12) is equivalent to

$$(2.9.16) \quad \Delta x(n) = B(n)x(n) + L(n,n_0)x_0 + F(n,x(n))$$
$$+ \sum_{s=n_0}^{n-1}\left[L(n,s)F(s,x(s)) + G(n,s,x(s)) + \sum_{\sigma=n_0}^{s-1} L(s,\sigma)G(s,\sigma,x(\sigma))\right],$$

with $x(n_0) = x_0$, where $B(n) = A(n) - L(n,n)$ and $L(n,s)$ is a solution of (2.9.2). It is clear from Theorem 2.9.1 that (2.9.13) is equivalent to

$$(2.9.17) \quad \Delta x(n) = B(n)x(n) + L(n,n_0)x_0, \quad x(n_0) = x_0.$$

Hence, by hypothesis ($H1$), this implies that the trivial solution of (2.9.17) is also exponentially asymptotically stable. Consequently, there exists a Lyapunov-like function V such that

(a) $V: N_{n_0}^+ \times S(\rho) \to R_+$, $V(n,x)$ is Lipschitzian in x for a constant $M > 0$ and

$$|x| \leq V(n,x) \leq M|x|, \quad (n,x) \in N_{n_0}^+ \times S(\rho),$$

(b) $V(n+1, \hat{x}(n+1)) - V(n, \hat{x}(n)) \leq -V(n, \hat{x}(n))$, $(n,x) \in N_{n_0} \times S(\rho)$,
where $\hat{x}(n)$ is the solution of (2.9.17).

Let $x(n) = x(n,n_0,x_0)$ be any solution of (2.9.16). Then, using (a) and (b) and setting $p(n) = V(n,x(n))$, we get the difference inequality of Volterra type

$$\Delta p(n) \leq -p(n) + M\left[w_1(n,p(n)) + \sum_{s=n_0}^{n-1} w(n,s,p(s))\right],$$

where $w(n,s,u)$ is given by (2.9.15).

Setting $z(n) = p(n)\alpha^{-(n-n_0)+1}$, $0 < \alpha < 1$, we obtain

$$\Delta z(n) = M\alpha^{-(n-n_0)}\left[w_1(n, z(n)\alpha^{n-n_0-1}) + \sum_{s=n_0}^{n-1} w(n,s,z(s)\alpha^{n-n_0-1})\right]$$

which implies, by Lemma 2.9.12, that

$$(2.9.18) \quad z(n) = \gamma(n, n_0, z(n_0)), \quad n \geq n_0$$

where $\gamma(n, n_0, u_0)$ is the solution of

$$\Delta u(n) = M\alpha^{-(n-n_0)}\left[w_1(n, u(n)\alpha^{n-n_0-1}) + \sum_{s=n_0}^{n-1} w(n,s,u(n)\alpha^{n-n_0-1})\right]$$

with $u(n_0) = u_0$. Therefore, from (2.9.18), we have

$$V(n, x(n)) \leq \gamma(n, n_0, V(n_0, x_0))\alpha^{n-n_0-1}.$$

It is easy to verify that $R(n, n_0, u_0) = \gamma(n, n_0, u_0)\alpha^{n-n_0-1}$ is the solution of the difference equation (2.9.14). Hence, the desired stability properties of the trivial solution of (2.9.11) follow from the corresponding stability properties of the trivial solution of (2.9.14) and therefore the proof is complete. □

2.10 Impulsive Integro-differential Systems

Let $PC^+ = PC^+[R_+, R^n]$ be a class of piecewise continuous functions from R_+ into R^n with discontinuities of first kind only at $t = t_k$, $k = 1, 2, \ldots$ and left continuous at $t = t_k$.

Let us consider the linear impulsive integro differential system

(2.10.1)
$$\begin{cases} x'(t) = A(t)x(t) + \int_{t_k}^{t} K(t,s)x(s)\,ds + F(t), & t \neq t_k, \\ \Delta x(t_k) = B_k x(t_k), \\ x(t_0^+) = x_0, \end{cases}$$

where $0 \leq t_0 < t_1 < t_2 < \ldots < t_k < \ldots$ and $t_k \to \infty$ as $k \to \infty$, $A \in PC^+[R_+, R^{n^2}]$, $K \in PC^+[R_+^2, R^{n^2}]$, $F \in PC^+[R_+, R^n]$ and $B_k \geq 0$ is an $n \times n$ matrix for each k such that $(I + B_k)^{-1}$ exists, I being the identity matrix.

Theorem 2.10.1 *Assume that*

(H0) *the sequence $\{t_k\}$ satisfies $0 \leq t_0 < t_1 < t_2 < \ldots$, with $\lim_{k \to \infty} t_k = \infty$;*

(H1) $m \in PC^1[R_+, R]$ *and $m(t)$ is left continuous at t_k, $k = 1, 2, \ldots$;*

(H2) *for $k = 1, 2, \ldots$, $t \geq t_0$,*

(2.10.2) $$m'(t) \leq p(t)m(t) + q(t), \quad t \neq t_k$$

(2.10.3) $$m(t_k^+) \leq d_k m(t_k) + b_k$$

where $q, p \in C[R_+, R]$, $d_k \geq 0$ and b_k are constants.

Then

$$(2.10.4) \quad m(t) \leq m(t_0) \prod_{t_0 < t_k < t} d_k \exp\left(\int_{t_0}^{t} p(s)\,ds\right)$$

$$+ \sum_{t_0 < t_k < t} \left[\prod_{t_k < t_j < t} d_j \exp\left(\int_{t_k}^{t} p(s)\,ds\right)\right] b_k$$

$$+ \int_{t_0}^{t} \prod_{s < t_k < t} d_k \exp\left(\int_{s}^{t} p(\sigma)\,d\sigma\right) q(s)\,ds, \qquad t \geq t_0.$$

Proof Let $t \in [t_0, t_1]$. Then we get from (2.10.2)

$$\frac{d}{dt}\left[m(t)\exp\left(-\int_{t_0}^{t} p(\sigma)\,d\sigma\right)\right] \leq q(t)\exp\left(-\int_{t_0}^{t} p(\sigma)\,d\sigma\right),$$

which yields, after integrating from t_0 to t,

$$(2.10.5) \quad m(t) \leq m(t_0)\exp\left(\int_{t_0}^{t} p(\sigma)\,d\sigma\right) + \int_{t_0}^{t} q(s)\exp\left(\int_{s}^{t} p(\sigma)\,d\sigma\right)ds,$$

for $t_0 \leq t \leq t_1$. Hence (2.10.4) is true for $t \in [t_0, t_1]$. Now assume that (2.10.4) holds for $t \in [t_0, t_n]$ for some integer $n > 1$. Then, for $t \in (t_n, t_{n+1}]$, it follows from (2.10.2) and (2.10.5) that

$$(2.10.6) \quad m(t) \leq m(t_n^+)\exp\left(\int_{t_n}^{t} p(\sigma)\,d\sigma\right) + \int_{t_n}^{t} q(s)\exp\left(\int_{s}^{t} p(\sigma)\,d\sigma\right)ds.$$

Using (2.10.3), we obtain from (2.10.6)

$$(2.10.7) \quad m(t) \leq (d_n m(t_n) + b_n)\exp\left(\int_{t_n}^{t} p(\sigma)\,d\sigma\right) + \int_{t_n}^{t} q(s)\exp\left(\int_{s}^{t} p(\sigma)\,d\sigma\right)ds.$$

By the induction hypothesis, (2.10.7) can be reduced to

$$m(t) \leq d_n \exp\left(\int_{t_n}^{t} p(\sigma)\,d\sigma\right)\left\{m(t_0) \prod_{t_0 < t_k < t_n} d_k \exp\left(\int_{t_0}^{t_n} p(\sigma)\,d\sigma\right)\right.$$

$$+ \sum_{t_0 < t_k < t_n} \left[\prod_{t_k < t_j < t_n} d_j \exp\left(\int_{t_k}^{t_n} p(\sigma) d\sigma \right) \right] d_k$$

$$+ \int_{t_0}^{t_n} \left[\prod_{s < t_k < t_n} d_k \exp\left(\int_s^{t_n} p(\sigma) d\sigma \right) q(s) ds \right] \Bigg\}$$

$$+ b_n \exp\left(\int_{t_n}^{t} p(\sigma) d\sigma \right) + \int_{t_n}^{t} q(s) \exp\left(\int_s^t p(\sigma) d\sigma \right) ds,$$

which on simplification gives the estimate (2.10.4) for $t_0 \leq t \leq t_{n+1}$. This completes the proof. $\square$

To bring out the inherently rich behavior of the impulsive linear integro-differential system, we shall first discuss a method of finding an equivalent impulsive linear differential system, and then, exploiting this method, we investigate the stability properties of a nonlinear impulsive integro-differential system. With this in mind, let us prove the following important result.

Theorem 2.10.2 *Assume that there exists an $n \times n$ matrix function $L \in PC^+[R_+^2, R^{n^2}]$ such that $L_s(t,s)$ exists and is continuous for $t_{k-1} < s \leq t_k < t$ and satisfies*

$$(2.10.8) \quad \begin{cases} K(t,s) + L_s(t,s) + L(t,s)A(s) + \int_s^t L(t,\sigma)K(\sigma,s) d\sigma = 0, & s, t \neq t_k, \\ \\ L(t, t_k^+) = (I + B_k)^{-1} L(t, t_k). \end{cases}$$

Then system (2.10.1) is equivalent to the impulsive differential system

$$(2.10.9) \quad \begin{cases} y'(t) = C(t)y(t) + L(t, t_0)x_0 + H(t), & t \neq t_k, \\ \Delta y(t_k) = B_k y(t_k), \\ y(t_0^+) = x_0, \end{cases}$$

where $C(t) = A(t) - L(t,t)$ and $H(t) = F(t) + \int_{t_0}^{t} L(t,s)F(s) ds$.

LINEAR ANALYSIS

Proof Let $x(t)$ be any solution of (2.10.1) existing on $[t_0, \infty)$. Set $p(s) = L(t,s)x(s)$ for $t_{k-1} < s < t_k < t$, so that we have

$$p'(s) = L_s(t,s)x(s) + L(t,s)x'(s).$$

Substituting for $x'(s)$ from (2.10.1) and integrating from t_0 to t, we get

$$p(t) - p(t_0) - \sum_{t_0 < t_k < t} \Delta p(t_k)$$

$$= \int_{t_0}^{t} \left[L_s(t,s)x(s) + L(t,s)A(s)x(s) + L(t,s)\int_{t_0}^{s} K(s,\sigma)x(\sigma)\,d\sigma + L(t,s)F(s) \right] ds.$$

Then using Fubini's Theorem, (2.10.8) and (2.10.1), it follows that

$$L(t,t)x(t) - L(t,t_0)x_0 - \sum_{t_0 < t_k < t} \Delta p(t_k)$$

$$= \int_{t_0}^{t} \left[L_s(t,s) + L(t,s)A(s) + \int_{s}^{t} L(t,\sigma)K(\sigma,s)\,d\sigma \right] x(s)\,ds + \int_{t_0}^{t} L(t,s)F(s)\,ds$$

$$= -\int_{s}^{t} K(t,s)x(s)\,ds + \int_{t_0}^{t} L(t,s)F(s)\,ds$$

$$= -x'(t) + A(t)x(t) + F(t) + \int_{t_0}^{t} L(t,s)F(s)\,ds,$$

which implies

(2.10.10) $\quad x'(t) = C(t)x(t) + L(t,t_0)x_0 + H(t) + \sum_{t_0 < t_k < t} \Delta p(t_k), \quad t \neq t_k.$

Also, for $s = t_k$, we obtain, using (2.10.8) and (2.10.1),

(2.10.11) $\quad \Delta p(t_k) = L(t, t_k^+)x(t_k^+) - L(t, t_k)x(t_k)$

$\qquad\qquad\quad = [L(t, t_k^+)(I + B_k) - L(t, t_k)]x(t_k),$

and therefore (2.10.10) together with (2.10.11) shows that $x(t)$ satisfies (2.10.9).

Conversely, let $y(t)$ be any solution (2.10.9) existing on $[t_0, \infty)$. Define $z(t) = y'(t) - A(t)y(t) - \int_{t_0}^{t} K(t,s)y(s)\,ds - F(t)$ and note that

$$\Delta z(t_k) = \Delta y(t_k) - B_k y(t_k) = 0.$$

We show that $z(t) \equiv 0$, which proves that $y(t)$ satisfies (2.10.1). Now substituting for $y'(t)$ from (2.10.9), we get, using (2.10.8) together with Fubini's Theorem,

$$z(t) = [A(t) - L(t,t)]y(t) + L(t,t_0)x_0 + H(t) - A(t)y(t)$$

$$+ \int_{t_0}^{t} \left[L_s(t,s) + L(t,s)A(s) + \int_{s}^{t} L(t,\sigma)K(\sigma,s)\,d\sigma \right] y(s)\,ds - F(t), \quad t \neq t_k,$$

which yields

$$z(t) = -\left[L(t,t)y(t) - L(t,t_0)x_0 - \int_{t_0}^{t} L_s(t,s)y(s)\,ds \right]$$

$$+ \int_{t_0}^{t} L(t,s)\left[A(s)y(s) + \int_{t_0}^{t} L(s,\sigma)\,d\sigma + F(s) \right] ds, \quad t \neq t_k.$$

Setting $p(s) = L(t,s)y(s)$, we obtain, for $t_{k-1} < s < t_k < t$,

$$p'(s) = L_s(t,s)y(s) + L(t,s)y'(s),$$

which, by integrating from t_0 to t, yields

$$L(t,t)y(t) - L(t,t_0)x_0 - \sum_{t_0 < t_k < t} \Delta p(t_k) = \int_{t_0}^{t} [L_s(t,s)y(s) + L(t,s)y'(s)]\,ds.$$

Moreover, we have for $s = t_k$, using (2.10.8),

$$\Delta p(t_k) = [L(t,t_k^+)(I + B_k) - L(t,t_k)]y(t_k) = 0.$$

Hence it follows that

$$z(t) = \int_{t_0}^{t} L(t,s)\left[-y'(s) + A(s)y(s) + \int_{t_0}^{s} K(s,\sigma)y(\sigma)\,d\sigma + F(s) \right] ds$$

$$= -\int_{t_0}^{t} L(t,s)z(s)\,ds,$$

LINEAR ANALYSIS

and $\Delta z(t_k) = 0$. This implies $z(t) \equiv 0$ by uniqueness of the solutions of linear Volterra integral equations, and therefore the proof of the theorem is complete. $\square$

A couple of remarks are now in order.

(i) If $L(t,s)$ is the solution of the IVP

(2.10.12)
$$\begin{cases} L_s(t,s) + L(t,s)A(s) + \int_s^t L(t,\sigma)K(\sigma,s)\,d\sigma = 0, & s,t \neq t_k, \\ L(t,t_k^+) = (I + B_k)^{-1}L(t,t_k), & \text{for each } k, \end{cases}$$

such that $L(t,t) = I$ then it follows from Theorem 2.10.2 that the unique solution of (2.10.1) is given by

(2.10.13)
$$x(t) = L(t,t_0)x_0 + \int_{t_0}^t L(t,s)F(s)\,ds, \quad t \geq t_0,$$

where $L(t,s)$ is the corresponding resolvent kernel.

(ii) If we suppose that $K(t,s) \equiv 0$ in (2.10.1) so that we have only the linear impulsive differential system

(2.10.14)
$$\begin{cases} x'(t) = A(t)x + F(t), & t \neq t_k, \\ \Delta x(t_k) = B_k x(t_k), \\ x(t_0^+) = x_0, \end{cases}$$

then we get from Theorem 2.10.2 the corresponding variation of parameter formula in the form (2.10.13), where $L(t,s)$ is now the fundamental matrix solution such that $L(t_0,t_0) = I$.

We can now prove the following result concerning the linear impulsive integro-differential system:

(2.10.15)
$$\begin{cases} x'(t) = A(t)x(t) + \int_{t_0}^t K(t,s)x(s)\,ds, & t \neq t_k, \\ \Delta x(t_k) = B_k x(t_k), \\ x(t_0^+) = x_0. \end{cases}$$

Theorem 2.10.3 *Assume that*

(i) *there exists an $n \times n$ matrix function $L(t,s)$ satisfying (2.10.9);*

(ii) *$\mu[C(t)] \leq -\alpha$, $\alpha > 0$, $t \geq 0$, where $\mu(C)$ denotes the logarithmic norm of C;*

(iii) *$\int_{t_0}^{\infty} e^{\alpha s} |L(s,t_0)| \, ds < \infty$ and $\prod_{k=1}^{\infty} d_k$ converges where $|I + B_k| \leq d_k$.*

Then the trivial solution of (2.10.15) is exponentially asymptotically stable.

Proof Let $m(t) = |x(t)|$ where $x(t)$ is any solution of (2.10.15). By Theorem 2.10.2, $x(t)$ is also a solution of (2.10.9) with $H(t) \equiv 0$. Consequently, we obtain using (ii),

$$D^+ m(t) \leq -\alpha m(t) + |L(t,t_0)| \, |x_0|, \quad t \neq t_k$$

$$m(t_k^+) \leq d_k m(t_k),$$

$$m(t_0^+) = |x_0|.$$

An application of Theorem 2.10.1 yields, for $t \geq t_0$,

$$|x(t)| \leq |x_0| \left[\prod_{t_0 < t_k < t} e^{-\alpha(t-t_0)} d_k + \int_{t_0}^{t} \prod_{s < t_k < t} d_k e^{-\alpha(t-s)} |L(s,t_0)| \, ds \right],$$

which, by (iii), reduces to

$$|x(t)| e^{\alpha(t-t_0)} \leq N |x_0| \left[1 + \int_{t_0}^{t} e^{\alpha(s-t_0)} |L(s,t_0)| \, ds \right]$$

$$\leq N |x_0| [1 + M(t_0)], \quad t \geq t_0.$$

Hence the proof is complete.

Next we shall discuss the nonlinear impulsive integro-differential system

(2.10.16)
$$\begin{cases} x'(t) = f(t, x(t)) + \int_{t_0}^{t} g(t, s, x(s)) \, ds, & t \neq t_k, \\ \Delta x(t_k) = B_k x(t_k), \\ x(t_0^+) = x_0, \end{cases}$$

where $f \in C[R_+ \times R^n, R^n]$ and $g \in C[R_+^2 \times R^n, R^n]$ and B_k is $n \times n$ constant matrix for each k. Assume that f_x and g_x exist and are continuous, and $f(t,0) \equiv 0$ and $g(t,s,0) \equiv 0$. Then setting

$$f_x(t,0) = A(t)x, \qquad g_x(t,s,0) = K(t,s)$$

and using the mean value theorem, the system (2.10.16) can be written in the form

(2.10.17)
$$\begin{cases} x'(t) = A(t)x(t) + \int_{t_0}^t K(t,s)x(s)\,ds + F(t,x(t)) + \int_{t_0}^t G(t,s,x(s))\,ds, \quad t \neq t_k, \\ \Delta x(t_k) = B_k x(t_k), \\ x(t_0^+) = x_0, \end{cases}$$

where

$$F(t,x) = \int_0^1 [f_x(t,x\theta) - f_x(t,0)]\,d\theta\, x,$$

$$G(t,s,x) = \int_0^1 [g_x(t,s,x\theta) - g_x(t,s,0)]\,d\theta\, x.$$

We are now in a position to prove the following result concerning (2.10.16).

Theorem 2.10.4 *Assume that*
(i) *the conditions of Theorem 2.10.3 hold;*
(ii) $|F(t,x)| \leq \omega_1(t,|x|)$ *and* $|G(t,s,x)| \leq \omega_2(t,s,|x|)$, *where* $\omega_1 \in C[R_+^2, R_+]$ *and* $\omega_2 \in C[R_+^3, R^+]$ *and* $\omega_1(t,u)$ *and* $\omega_2(t,s,u)$ *are nondecreasing in* u.

Then the stability properties of the trivial solution of the differential equation

(2.10.18) $\quad u'(t) = -\alpha u(t) + M\left[\omega_1(t,u(t)) + \int_{t_0}^t \omega_2(t,s,u(t))\,ds\right], \quad u(t_0) = u_0 \geq 0,$

for some $\alpha > 0$ and $M > 0$, imply the corresponding stability properties of the trivial solution of (2.10.16).

Proof We first note that the assumption (i) implies, because of Theorem 2.10.3 that the trivial solution of (2.10.9) with $H(t) \equiv 0$ is exponentially asymptotically stable. By Theorem 2.10.2, this means the same kind of stability for the trivial solution of (2.10.15). As a result, if $x(t)$ is the solution of (2.10.15) and $R(t,s)$ is the corresponding resolvent kernel then $x(t) = R(t,t_0)x_0$, and therefore we obtain

(2.10.19) $\qquad |R(t,s)| \leq Me^{-\alpha(t-s)}, \quad t \geq s, \ \alpha > 0, \ M > 0.$

Furthermore, in view of (2.10.13), any solution $y(t)$ of (2.10.17) satisfies, for $t \geq t_0$,

$$y(t) = R(t,t_0)x_0 + \int_{t_0}^{t} R(t,s)\left[F(s,y(s)) + \int_{t_0}^{s} G(s,\sigma,y(\sigma))\,d\sigma\right]ds.$$

It therefore follows, using the condition (ii) and (2.10.19), that

$$|y(t)| \leq M|x_0|e^{-\alpha(t-t_0)} + M\int_{t_0}^{t} e^{-\alpha(t-s)}\left[\omega_1(s,y(s)) + \int_{t_0}^{s} \omega_2(s,\sigma,y(\sigma))\,d\sigma\right]ds,$$

$t \geq t_0$.

Setting $v(t) = |y(t)|e^{\alpha(t-t_0)}$, we get

(2.10.20) $\quad |v(t)| \leq M|x_0|$

$$+ M\int_{t_0}^{t} e^{\alpha(s-t_0)}\left[\omega_1(s,v(s)e^{-\alpha(s-t_0)}) + \int_{t_0}^{s} \omega_2(s,\sigma,v(\sigma)e^{-\alpha(\sigma-t_0)})\,d\sigma\right]ds, \quad t \geq t_0.$$

Denoting the right-hand side of (2.10.20) by $p(t)$, we see that $p'(t) \geq 0$ and hence it follows, using the monotone character of ω_1 and ω_2, that

$$p'(t) \leq Me^{\alpha(t-t_0)}\left[\omega_1(t,p(t)e^{-\alpha(t-t_0)}) + \int_{t_0}^{t} \omega_2(t,s,p(t)e^{-\alpha(s-t_0)})\,ds\right].$$

We then get, by the comparison theorem,

(2.10.21) $\qquad p(t) \leq r(t,t_0,M|x_0|), \quad t \geq t_0,$

where $r(t, t_0, u_0)$ is the maximal solution of the differential equation

$$u'(t) = Me^{\alpha(t-t_0)}\left[w_1(t, u(t)e^{-\alpha(t-t_0)}) + \int_{t_0}^{t} w_2(t, s, u(t)e^{-\alpha(s-t_0)})\,ds\right],$$

$$u(t_0) = u_0 \geq 0.$$

Hence it follows that

(2.10.22) $\qquad |y(t)| \leq R(t, t_0, M|x_0|), \quad t \geq t_0,$

where $R(t, t_0, u_0) = r(t, t_0, u_0)e^{-\alpha(t-t_0)}$ is the maximal solution of (2.10.18).

The conclusion of the theorem is therefore complete in view of (2.10.22). □

2.11 Periodic Solutions

It is well known that for ordinary differential equations and functional differential equations with fixed finite delay, periodic solutions occur when solutions are uniformly bounded and uniformly ultimately bounded. It is also known that periodic solutions occur in a natural way for Volterra integro-differential equations

(2.11.1) $\qquad x'(t) = Ax(t) + \int_{-\infty}^{t} K(t-s)x(s)\,ds + F(t)$

in which A is an $n \times n$ constant matrix, K is an $n \times n$ matrix function continuous on $(-\infty, \infty)$, $F \in [R, R^n]$ and $F(t+T) = F(t)$ for all $t \in R$ and some $T > 0$. In fact, for (2.11.1), it is easy to verify that if $x(t)$ is any solution then so is $x(t+T)$, which is one of the crucial requirements for the existence of periodic solutions. However, such a requirement may not hold for the corresponding integro-differential equation

(2.11.2) $\qquad y'(t) = Ay(t) + \int_{0}^{t} K(t-s)y(s)\,ds + F(t).$

Nonetheless, we find that $y(t) = \cos t + \sin t$ is a solution of

$$y'(t) = y(t) + \int_{0}^{t} e^{-(t-s)} y(s)\,ds - 3\sin t,$$

and consequently the study of periodic solutions of (2.11.2) is interesting. In fact, we shall show that asymptotically periodic solutions exist for (2.11.2).

Definition 2.11.1 An asymptotically T-periodic function f is a bounded continuous function for which there exists a continuous T-periodic function q and an integer n such that $f(t+nT) - q(t) \to 0$ uniformly on $[0,T]$ as $n \to \infty$. This is equivalent to the condition that $f(t) - q(t) \to 0$ as $t \to \infty$.

Recall that, by the variation of parameters formula, the solution $y(t)$ of (2.11.2) is given by

$$(2.11.3) \qquad y(t) = Z(t)y(0) + \int_0^t Z(t-s)F(s)\,ds,$$

where $Z(t)$ is the differentiable resolvent corresponding to the kernel $K(t)$ and is an $n \times n$ matrix function satisfying the IVP

$$(2.11.4) \qquad Z'(t) = AZ(t) + \int_0^t K(t-s)Z(s)\,ds, \qquad Z(0) = I.$$

We shall see that the system (2.11.2) is closely related to its perturbation (or limiting) system (2.11.1), and hence its investigation depends on the properties of solutions of (2.11.1) with bounded continuous initial functions $\phi: (-\infty, 0] \to R^n$.

We use the following basic results in our subsequent discussions.

Lemma 2.11.1 *If $K(t)$ and $Z(t) \in L^1[0,\infty)$ then $Z(t) \to 0$ as $t \to \infty$.*

Proof Since A is a constant matrix and $Z(t) \in L^1[0,\infty)$, it is clear that $AZ(t) \in L^1[0,\infty)$. Also, the convolution of two functions in $L^1[0,\infty)$ is in $L^1[0,\infty)$, and hence $\int_0^t K(t-s)Z(s)\,ds$ is in $L^1[0,\infty)$. Therefore from (2.11.4), $Z'(t) \in L^1[0,\infty)$. This implies that $Z(t)$ has a limit as $t \to \infty$. But $Z(t) \in L^1[0,\infty)$. Hence $Z(t) \to 0$ as $t \to \infty$ and this completes the proof. $\square$

Lemma 2.11.2 *Assume that*
  (i)   *$K \in L^1[0,\infty)$ and $F(t)$ is continuous for $t \in R$ with $F(t+T) = F(t)$ for all $t \in R$ and some constant $T > 0$;*
  (ii) *$y(t)$ is a solution of (2.11.2) that exists and bounded on R^+.*

Then there exists a sequence of integers $\{\eta_j\}$, with $\eta_j \to \infty$ as $j \to \infty$, such that $y(t + \eta_j T) \to x(t)$ where $x(t)$ is a solution of (2.11.1) on R and the convergence is uniform on any compact subset of R.

Proof Let M be a bound for $|y(t)|$ on $0 \le t < \infty$. Then from (2.11.2) we have

$$|y'(t)| \le |A||y(t)| + \int_0^t |K(t-s)||y(s)|\,ds$$

$$\le M|A| + M \int_0^t |K(s)|\,ds.$$

Since $K \in L^1[0, \infty)$, it is clear that $|y'(t)|$ is bounded on R_+. Therefore, by the mean value theorem, $y(t)$ is uniformly continuous on R_+. Let $\{\eta_j\}$ be a sequence of integers such that $\eta_j \to \infty$ as $j \to \infty$. Then by Ascoli's theorem, the sequence $\{y(t + \eta_j T)\}$ contains a subsequence such that $y(t + \eta_{j_m} T) \to x(t)$ as $m \to \infty$ with convergence uniform for t on compact subsets of $t \in R$. Further, from the assumption (ii), it is clear that $F(t + \eta_{j_m} T) = F(t)$ for all $t \in R$.

Let $y_m(t) = y(t + \eta_{j_m} T)$ and $F_m(t) = F(t + \eta_{j_m} T)$. Then from (2.11.2) for any m, any $t \ge -\tau_{j_m}$, we have

$$y'_m(t) = A y_m(t) + \int_{-\tau_{j_m}}^t K(t-s) y_m(s)\,ds + F_m(t).$$

This implies that

(2.11.5) $$y_m(t) = y_m(0) + \int_0^t A y_m(s)\,ds + \int_0^t F_m(s)\,ds$$

$$+ \int_0^t \left[\int_{-\tau_{j_m}}^s K(s - \sigma) y_m(\sigma)\,d\sigma \right] ds.$$

Let $\max\{|y_m(s) - x(s)| : -m \le s \le m\} < 1/m$. Then for any real number s, if $s + \tau_{j_m} > 0$, it follows that

$$\left| \int_{-\infty}^{s} K(s-\sigma)x(\sigma)\,d\sigma - \int_{-\tau_{j_m}}^{s} K(s-\sigma)y_m(\sigma)\,d\sigma \right|$$

$$\leq M \int_{-\infty}^{-m} |K(s-\sigma)|\,d\sigma + M \int_{-\tau_{j_m}}^{-m} |K(s-\sigma)|\,d\sigma$$

$$+ \int_{-mm}^{s} |K(s-\sigma)| \, |x(\sigma) - y_m(\sigma)|\,d\sigma$$

$$\leq 2M \int_{s+m}^{\infty} |K(\sigma)|\,d\sigma$$

$$+ \left[\int_{0}^{s+m} |K(\sigma)|\,d\sigma \right] \max\{|x(\sigma) - y_m(\sigma)| : -m \leq \sigma \leq m\}$$

$\to 0$ as $m \to \infty$.

Therefore we take the limit in (2.11.5) as $m \to \infty$ and obtain

$$x(t) = x(0) + \int_0^t Ax(s)\,ds + \int_0^t F(s)\,ds + \int_0^t \left[\int_{-\infty}^s K(s-\sigma)x(\sigma)\,d\sigma \right] ds.$$

On differentiating this equation with respect to t, we get

$$x'(t) = Ax(t) + \int_{-\infty}^t K(t-s)x(s)\,ds + F(t)$$

and hence $x(t)$ is a solution of (2.11.1). This completes the proof. □

Theorem 2.11.1 *Let the assumptions of Lemmas 2.11.1 and 2.11.2 hold. Then the equation (2.11.1) has a T-periodic solution $\int_{-\infty}^{t} Z(t-s)F(s)\,ds$, and all solutions of (2.11.1) defined for $t \geq 0$ with bounded continuous initial functions on $(-\infty, 0]$ tend to this T-periodic solution as $t \to \infty$.*

Proof Since K and $Z \in L^1[0,\infty)$, it follows by Lemma 2.11.1 that $\lim_{t \to \infty} Z(t) = 0$. Therefore, by (2.11.3), all solutions of (2.11.2) are bounded. On the other hand, from (2.11.3) we have

$$y(t+\eta_j T) = Z(t+\eta_j T)y(0) + \int_0^{t+\eta_j T} Z(t+\eta_j T - s)F(s)\,ds$$

$$= Z(t+\eta_j T)y(0) + \int_{-\eta_j T}^{t} Z(t-u)F(u+\eta_j T)\,du.$$

Since $F(u+\eta_j T) = F(u)$, we obtain

$$y(t+\eta_j T) = Z(t+\eta_j T)y(0) - \int_{-\eta_j T}^{t} Z(t-u)F(u)\,du$$

and hence $y(t+\eta_j T) \to \int_{-\infty}^{t} Z(t-u)F(u)\,du$ as $j\to\infty$.

By Lemma 2.11.2, $\int_{-\infty}^{t} Z(t-s)F(s)\,ds$ is a solution of (2.11.1) and this is obviously a periodic function of period T. Suppose that $x(t)$ is a solution of (2.11.1) defined for $t \geq 0$ with bounded initial function ϕ on $(-\infty, 0]$. Then we have

$$x'(t) = Ax(t) + \int_0^t K(t-s)x(s)\,ds + \int_{-\infty}^0 K(t-s)\phi(s)\,ds + F(t).$$

From the variation of parameters formula, we obtain

(2.11.6) $\quad x(t) = Z(t)x(0) + \int_0^t Z(t-s)\left[F(s) + \int_{-\infty}^0 K(s-\sigma)\phi(\sigma)\,d\sigma\right]ds.$

In view of the boundedness of ϕ and $K \in L^1[0,\infty)$, it follows that

$$\left|\int_{-\infty}^0 K(s-\sigma)\phi(\sigma)\,d\sigma\right| \leq \sup |\phi(\sigma)| \int_s^\infty |K(u)|\,du,$$

which tends to zero as $s\to\infty$. Thus the last term on the right-hand side of (2.11.6) is the convolution of an L^1-function and a function tending to zero, and therefore the convolution tends to zero as $t\to\infty$. Further, $Z(t)\to 0$ as $t\to\infty$ by Lemma 2.11.1. Hence it follows from (2.11.6) that $x(t)\to \int_{-\infty}^{t} Z(t-s)F(s)\,ds$ as $t\to\infty$, and this completes the proof. □

Theorem 2.11.2 *Suppose all the assumptions of Theorem 2.11.1 are satisfied. If the equation (2.11.2) has a periodic solution $\eta(t)$ then $\eta(t)$ is the unique periodic solution of (2.11.1).*

Proof By Theorem 2.11.1, the equation (2.11.1) has a periodic solution $\psi(t)$. Set $w(t) = \psi(t) - \eta(t)$. Then from (2.11.1) and (2.11.2) we have

$$(2.11.7) \qquad w'(t) = Aw(t) + \int_0^t K(t-s)w(s)\,ds + \int_{-\infty}^0 K(t-s)\psi(s)\,ds.$$

By the variation of parameters, any solution $w(t)$ of (2.11.7) can be expressed as

$$(2.11.8) \qquad w(t) = Z(t)w(0) + \int_0^t Z(t-s)\left[\int_{-\infty}^0 K(s-\sigma)\psi(\sigma)\,d\sigma\right]ds.$$

In view of the boundedness of $\psi(t)$ and $K \in L^1[0,\infty)$, it follows that

$$\left|\int_{-\infty}^0 K(s-\sigma)\psi(\sigma)\,d\sigma\right| \le \sup |\psi(\sigma)| \int_s^\infty |K(u)|\,du,$$

which tends to zero as $s\to\infty$. Thus the last term on the right-hand side of (2.11.8) is the convolution of an L^1-function with a function tending to zero, and hence the convolution tends to zero. Moreover, application of Lemma 2.11.1 yields that $Z(t)\to 0$ as $t\to\infty$. This shows that $w(t)\to 0$ as $t\to\infty$. Since $\psi(t)$ is $\eta(t)$-periodic, $\psi(t) \equiv \eta(t)$. □

The following corollaries are direct consequences of Theorem 2.11.1.

Corollary 2.11.1 *In addition to the assumptions of Theorem 2.11.1, suppose that $\psi(t)$ is a periodic solution of (2.11.1) and $\eta(t)$ is an asymptotically periodic solution of (2.11.2). Then $\psi(t) \equiv \eta(t)$.*

Corollary 2.11.2 *Let the conditions of Theorem 2.11.1 hold and let $\eta(t)$ be a periodic solution of (2.11.1) and (2.11.2). Then*

$$(2.11.9) \qquad \int_{-\infty}^0 K(t-s)\eta(s)\,ds = 0.$$

Remark 2.11.1 In view of Corollary 2.11.2, our study reveals that the periodicity of the right-hand side of (2.11.2) for a periodic y can be achieved through a certain type of orthogonality property (2.11.9).

Corollary 2.11.3 *Let the assumptions of Theorem 2.11.1 hold. If the equation (2.11.2) has a periodic solution then*

$$(2.11.10) \qquad h(t) = \int_t^\infty Z(s)F(t-s)\,ds$$

is a solution of

$$(2.11.11) \qquad h'(t) = Ah(t) + \int_0^t K(t-s)h(s)\,ds.$$

We shall now consider the nonlinear integro-differential equation

$$(2.11.12) \qquad y'(t) = Ay(t) + \int_0^t K(t-s)g(s,y(s))\,ds + F(t)$$

and the perturbed (or limiting) equation

$$(2.11.13) \qquad x'(t) = Ax(t) + \int_{-\infty}^t K(t-s)g(s,x(s))\,ds + F(t),$$

where $g \in C[R, R^n]$, A is an $n \times n$ constant matrix, K is an $n \times n$ matrix function continuous for $t \in R$, and $F \in C[R, R^n]$, $F(t+T) = F(t)$ and $g(t+T,x) = g(t,x)$ for all $t \in R$ and some $T > 0$.

Theorem 2.11.3 *Let $0 < \alpha < \beta$. Assume that*
- (i) *the equation (2.11.13) has a unique T-periodic solution $x(t)$ satisfying $\alpha \leq x(t) \leq \beta$;*
- (ii) *for any bounded set $B \subset R^n$, the function $g(s,x)$ is bounded and uniformly continuous on $R_+ \times B$;*
- (iii) *$K \in L^1[0,\infty)$ and the equation (2.11.12) has a bounded solution $y(t)$ satisfying $\alpha \leq y(t) \leq \beta$ for all sufficiently large t.*

Then for an integer n,

$$(2.11.14) \qquad y(t+nT) - x(t) \to 0 \quad \text{uniformly on } [0,T] \text{ as } n \to \infty.$$

That is, $y(t)$ is asymptotically T-periodic.

Proof We claim that (2.11.14) holds. Suppose that this is not true. Then there exists an integer sequence $\{\eta_j\}$, with $\eta_j \to \infty$ as $j \to \infty$, and a sequence $\{t_j\} \in [0,T]$ such that $|y(t_j + \eta_j T) - x(t_j)| > d$ for some $d > 0$. Since for large t, $\alpha \leq y(t) \leq \beta$, we may assume that $\alpha \leq y(t_j + \eta_j T) \leq \beta$ for all j. By Lemma 2.11.2, there is a subsequence η_{j_k}, labeled η_k and functions y^* and g^* such that

$$y(t + \eta_k T) \to y^*(t) \quad \text{and} \quad g(t + \eta_k T, y) \to g^*(t, y)$$

as $k \to \infty$, with uniform convergence on compact subsets of $t \in R$. Now following a similar argument as in Lemma 2.11.2, we obtain

$$y^*(t) = Ay^*(t) + \int_{-\infty}^{t} K(t-s) g^*(s, y^*(s)) \, ds + F(t).$$

Since $\alpha \leq y(t_j + \eta_j T) \leq \beta$, the same is true for $y^*(t)$. Hence, since $y^*(t)$ and $x(t)$ satisfy (2.11.13), we have $x(t) = y^*(t)$. This implies that $|y(t + \eta_j T) - x(t)| < d$ for sufficiently large η_j. This is a contradiction, so in fact $y(t + \eta_j T) \to x(t)$ uniformly on $[0, T]$. Thus, by Definition 2.11.1, $y(t)$ is an asymptotically T-periodic solution of (2.11.12).

Remark 2.11.3 If $K(t) = a(t)$ for $t \leq L$ and $K(t) = 0$ for $t > L$ then the limiting equation (2.11.13) is equivalent to

(2.11.15) $$x'(t) = Ax(t) + \int_{t-L}^{t} a(t-s) g(s, x(s)) \, ds + F(t).$$

Suppose that the conditions (ii) and (iii) (with $K(t) = a(t)$) of Theorem 2.11.3 are satisfied, so that (2.11.15) has a unique T-periodic solution $x(t)$ satisfying $\alpha \leq x(t) \leq \beta$. Then the bounded solution $y(t)$ of (2.11.12) is asymptotically T-periodic.

2.12 Notes and Comments

Section 2.1 deals with elementary properties of linear integro-differential equations. Theorem 2.2.1 is due to Miller [15]. Theorems 2.2.2 and 2.2.23 are taken from Hara, Yoneyama and Itoh [1]. See also Rama Mohana Rao and Sanjay [1]. The material covered in Section 2.3 is found in Hara, Yoneyama and Itoh [2]. Section 2.4 contains the work of Rama Mohana Rao and Srinivas [3]. See also Burton [3] and Friedman [1]. The results of Section 2.5 are taken from Lakshmikantham and Rama Mohana Rao [1].

Section 2.6 is due to Elaydi and Rama Mohana Rao [1]. For semilinear results on ordinary differential equations, see Dannan and Elaydi [1]. Theorems 2.7.1 and 2.7.2 are found in Cushing [1]. Theorem 2.7.3 is taken from Nohel [3] and Theorem 2.7.4 is due to Morchazo [1, 2]. See also Nohel [2]. The results on ultimate behavior of solutions discussed in Section 2.8 are taken from Corduneanu [4]. See also Cassago and Corduneanu [1] and Moser [1]. The result on difference equations covered in Section 2.9 may be found in Zouyousefain and Leela [1]. An excellent monograph on difference equations is that by Lakshmikantham and Trigiante [1].

The theory of impulsive integro-differential equations contained in Section 2.10 is due to Rama Mohana Rao, Sathananthan and Sivasundaram [1]. Lemmas 2.11.1 and 2.11.2 are taken from Miller [3]. The rest of the material of Section 2.11 is adapted from the works of Burton [1], Becker, Burton and Krisztin [1] and Islam [1]. For further results on periodic and almost periodic solutions in this direction, see Corduneanu [2], Langenhop [1], Hale [3], Seifert [1], Gustafson and Schmitt [1], Marcus and Mizel [1], Grimmer [2] and Wu [2].

3 LYAPUNOV STABILITY

3.0 Introduction

It is well known that Lyapunov's second method is one of the basic tools for investigating the qualitative behavior of solutions of nonlinear differential equations. In extending this method to either differential equations with delay or to Volterra integro-differential equations, one has the choice of employing Lyapunov functionals or Lyapunov functions. Using a Lyapunov functional demands prior knowledge of the solutions of the equations under consideration. On the other hand, utilizing a Lyapunov function depends crucially on choosing appropriate minimal sets of a suitable space of continuous functions, along which the derivative of the Lyapunov function admits a convenient estimate. If we examine, however, the Lyapunov functionals constructed in applications, we find that one always employs a combination of a Lyapunov function and a functional in such a way that the corresponding derivative can be estimated suitably without demanding the minimal class of functions or prior knowledge of solutions. This observation leads one to develop the method of Lyapunov functions on product spaces.

In this chapter we shall discuss, both of these techniques to investigate the stability and boundedness properties of solutions of integro-differential equations.

3.1 Method of Lyapunov Functionals

In this section, we shall consider the stability properties of linear integro-differential equations by constructing Lyapunov functionals. We shall also utilize in this set-up the equivalent linear differential as well as linear integro-differential equations to discuss stability properties.

Let us begin with a simple scalar equation

$$(3.1.1) \qquad u' = \alpha u + \int_0^t a(t-s)u(s)\,ds,$$

where α is a constant and $a(t)$ is continuous for $0 \leq t < \infty$, with a strong sign condition. The following result for (3.1.1) is not only interesting in itself but also useful in subsequent discussions.

Theorem 3.1.1 *Suppose that $\alpha < 0$, $a(t) > 0$ and $\alpha + \int_0^\infty a(t)\,dt \neq 0$. Then the following statements are equivalent:*

(a) *All the solutions of (3.1.1) tend to zero;*

(b) $\alpha + \int_0^\infty a(t)\,dt < 0;$

(c) *Each solution of (3.1.1) is in $L^1(R_+)$;*

(d) *The zero solution of (3.1.1) is uniformly asymptotically stable;*

(e) *The zero solution of (3.1.1) is asymptotically stable.*

Proof It is clear from Theorem 2.2.2 that (c) implies (d). By definition, (d) implies (e). Certainly (e) implies (a). We shall show that (a) implies (b), and (b) implies (c). Suppose that (a) holds, but $\alpha + \int_0^\infty a(t)\,dt > 0$. Choose $t_0 > 0$ so large so that $\alpha + \int_0^{t_0} a(t)\,dt \geq 0$. Let $u(t, t_0, \phi)$ be any solution of (3.1.1) with the initial function $\phi(t) = 2$ on $[0, t_0]$. We claim that $u(t) = u(t, t_0, \phi) > 1$ on $[t_0, \infty)$. If not, there exists a first $t_1 > t_0$ with $u(t_1) = 1$ and hence $u'(t_1) \leq 0$. But it follows from (3.1.1) that

$$u'(t_1) = \alpha u(t_1) + \int_0^{t_1} a(t_1 - s)u(s)\,ds$$

$$= \alpha + \int_0^{t_1} a(s)u(t_1 - s)\,ds$$

$$\geq \alpha + \int_0^{t_1} a(s)\,ds$$

$$> \alpha + \int_0^{t_0} a(s)\,ds \geq 0,$$

which is a contradiction. Hence (a) implies (b). To prove that (b) implies (c), we choose a Lyapunov functional

$$V(t, u(\,\cdot\,)) = |u| + \int_0^t \int_t^\infty a(\tau - s)\,d\tau\, |u(s)|\,ds.$$

If $u(t)$ is a solution of (3.1.1) then, for $u(t) \not\equiv 0$, it follows that

$$V'_{(3.1.1)}(t, u(\,\cdot\,)) \leq \alpha |u| + \int_0^t a(t-s)|u(s)|\,ds$$

$$+ \int_t^\infty a(\tau - t)\,d\tau\, |u| - \int_0^t a(t-s)|u(s)|\,ds$$

$$= \left[\alpha + \int_t^\infty a(\tau - t)\,d\tau\right]|u|$$

$$= \left[\alpha + \int_0^\infty a(\tau)\,d\tau\right]|u| = -\beta|u|$$

for some constant $\beta > 0$. Thus we have

$$0 \leq V(t, u(\,\cdot\,)) \leq V(t_0, \phi(\,\cdot\,)) - \beta \int_{t_0}^t |u(s)|\,ds,$$

which implies that $\int_{t_0}^\infty |u(s)|\,ds < \infty$, and hence $u \in L^1(R_+)$. This completes the proof. □

By slightly modifying the Lyapunov functional used in the proof of Theorem 3.1.1, one can relax the sign condition of $a(t)$, as the following result shows.

Theorem 3.1.2 *Suppose that*

(C_1): $\alpha + \int_0^\infty |a(t)|\,dt < 0.$

Then the zero solution of (3.1.1) is uniformly asymptotically stable.

Proof Choose a Lyapunov functional

$$V(t, u(\cdot)) = |u| + \int_0^t \int_t^\infty |a(\tau - s)|\, d\tau\, |u(s)|\, ds.$$

If $u(t) = u(t, t_0, \phi)$ is a solution of (3.1.1) then, for $u(t) \not\equiv 0$, we obtain

$$\begin{aligned}V'_{(3.1.1)}(t, u(\cdot)) &\leq \alpha |u| + \int_0^t |a(t-s)|\, |u(s)|\, ds \\ &\quad + \int_t^\infty |a(\tau - s)|\, d\tau\, |u| - \int_0^t |a(t-s)|\, |u(s)|\, ds \\ &\leq \left[\alpha + \int_0^\infty |a(\tau)|\, d\tau\right] |u| = -\beta |u|,\end{aligned}$$

for some positive constant β. Hence $u \in L^1(R_+)$. By Theorem 3.1.1, this shows that the zero solution of (3.1.1) is uniformly asymptotically stable, and thus the proof of the theorem is complete. $\square$

The condition (C_1) in Theorem 3.1.2 cannot be replaced by a weaker condition such as

$$(C_2): \alpha + |\int_0^\infty a(t)\, dt| < 0.$$

In fact, the following example shows that the zero solution of (3.1.1) is unstable under the condition (C_2) even if $a(t) \in L^1(R_+)$ and bounded.

Example 3.1.1: Let $\alpha = -1$ and

(3.1.2) $$a(t) = \begin{cases} m \sin t & \text{if } 0 \leq t \leq 2\pi, \\ 0 & \text{if } t \geq 2\pi \end{cases}$$

in (3.1.1), with constant $m > 0$. Then (3.1.1) takes the form

$$u' = \begin{cases} -u + m \int_0^t \sin(t-s) u(s)\, ds & \text{if } 0 \leq t \leq 2\pi, \\ -u + m \int_{t-2\pi}^t \sin(t-s) u(s)\, ds & \text{if } t \geq 2\pi. \end{cases}$$

Let $t_0 = 2\pi$ and $\phi(t) = pe^{t-2\pi}$ on $[0, t_0]$, where $p > 0$ is a constant. Since $\int_{t-2\pi}^{t} e^s \sin(t-s)\,ds = \frac{1}{2}(1 - e^{-2\pi})e^t$, it is clear that $u(t) = pe^{t-t_0}$ with $m = 4e^{2\pi}/(e^{2\pi} - 1)$ and $u(t) = \phi(t)$ on $[0, t_0]$ is a solution of (3.1.1). Notice that $a(t)$ is bounded, $a \in L^1(R_+)$, the and condition (C_2) holds.

We shall now consider a more general scalar equation, namely

(3.1.3) $$u' = \alpha(t)u + \int_0^t a(t,s)u(s)\,ds, \quad t \in R^+,$$

and given necessary and sufficient conditions for the stability of the zero solution of (3.1.3).

Theorem 3.1.3 *Assume that*

(i) $\alpha: R_+ \to R$ *is continuous and a is continuous for $0 \leq s \leq t < \infty$;*

(ii) *the integral $\int_t^\infty |a(\tau,t)|\,d\tau$ is defined and finite for all $t \geq 0$, and suppose that there is a positive real number β such that*

(3.1.4) $$\int_0^t |a(t,s)|\,ds + \int_t^\infty |a(\tau,t)|\,d\tau - 2|\alpha(t)| \leq -\beta.$$

Then the zero solution of (3.1.3) is stable if and only if $\alpha(t) < 0$.

Proof Suppose that $\alpha(t) < 0$, and consider a Lyapunov functional

$$V(t, u(\cdot)) = u^2 + \int_0^t \int_t^\infty |a(\tau,s)|\,d\tau\, u^2(s)\,ds.$$

Then the time derivative of $V(t, u(\cdot))$ along the solutions of (3.1.3) is given by

$$V'_{(3.1.3)}(t, u(\cdot)) \leq 2\alpha(t)u^2 + 2\int_0^t |a(t,s)|\,|u(s)|\,|u|\,ds$$

$$+ \int_t^\infty |a(\tau,t)|\,d\tau\, u^2 - \int_0^t |a(t,s)|\,u^2(s)\,ds.$$

Since $2|u(s)|\,|u| \leq u^2(s) + u^2$, it follows that

$$V'_{(3.1.3)}(t, u(\cdot)) \leq 2\alpha(t)u^2 + \int_0^t |a(t,s)|\,[u^2(s) + u^2]\,ds$$

$$+ \int_t^\infty |a(\tau,t)|\, d\tau\, u^2 - \int_0^t |a(t,s)|\, u^2(s)\, ds$$

$$= \left[2\alpha(t) + \int_0^t |a(t,s)|\, ds + \int_t^\infty |a(\tau,t)|\, d\tau\right] u^2.$$

This, in view of (3.1.4), yields

$$V'_{(3.1.3)}(t,u(\cdot)) \leq -\beta u^2.$$

Since V is positive definite and $V'_{(3.1.3)}(t,u(\cdot)) \leq 0$, it follows that the zero solution of (3.1.3) is stable. Suppose $\alpha(t) \geq 0$ and consider the functional

(3.1.5) $$W(t,u(\cdot)) = u^2 - \int_0^t \int_t^\infty |a(\tau,s)|\, d\tau\, u^2(s)\, ds.$$

Then we have

$$W'_{(3.1.3)}(t,u(\cdot)) \geq 2\alpha(t)u^2 - 2\int_0^t |a(t,s)|\,|u(s)|\,|u|\, ds$$

$$- \int_t^\infty |a(\tau,t)|\, d\tau\, u^2 + \int_0^t |a(t,s)|\, u^2(s)\, ds$$

$$\geq 2\alpha(t)u^2 - \int_0^t |a(t,s)|\left[u^2(s) + u^2\right] ds$$

$$- \int_t^\infty |a(\tau,t)|\, d\tau\, u^2 + \int_0^t |a(t,s)|\, u^2(s)\, ds$$

$$= \left[2\alpha(t) - \int_0^t |a(t,s)|\, ds - \int_\tau^\infty |a(\tau,t)|\, d\tau\right] u^2.$$

Hence (3.1.4) gives

(3.1.6) $$W'_{(3.1.3)}(t,u(\cdot)) \geq \beta u^2.$$

Now, given any $t_0 \geq 0$ and any $\delta > 0$, we can find a continuous function $\phi:[0,t_0] \to R$ with $|\phi(t)| < \delta$ and $W(t_0,\phi(\cdot)) > 0$ such that if $u(t) = u(t,t_0,\phi)$ is a solution of (3.1.3) then we obtain from (3.1.5) and (3.1.6) that

(3.1.7) $\quad u^2(t) \geq W(t, u(\cdot)) \geq W(t_0, \phi(\cdot)) + \beta \int_{t_0}^{t} u^2(s)\,ds$

$$\geq W(t_0, \phi(\cdot)) + \beta \int_{t_0}^{t} W(t_0, \phi(\cdot))\,ds$$

$$= W(t_0, \phi(\cdot)) + \beta W(t_0, \phi(\cdot))(t - t_0).$$

Hence $|u(t)| \to \infty$ as $t \to \infty$, and the proof is complete. □

Corollary 3.1.1 *If* (3.1.4) *holds and* $\alpha(t) < 0$ *and bounded then the zero solution of* (3.1.3) *is asymptotically stable.*

Proof From (3.1.6), we have

$$V'_{(3.1.3)}(t, u(\cdot)) \leq -\beta u^2.$$

This implies that $u^2 \in L^1(R_+)$ and $u^2(t)$ is bounded on R_+. Therefore it follows from (3.1.3) and (3.1.4) that $u'(t)$ is bounded on R_+. Hence $u(t) \to 0$ as $t \to \infty$. □

Corollary 3.1.2 *If* (3.1.4) *holds and* $\alpha(t) > 0$ *then the zero solution of* (3.1.3) *is completely unstable. Furthermore, for any* $t_0 \geq 0$ *and any* $\delta > 0$, *there is a continuous function* $\phi : [0, t_0] \to R$ *and a solution* $u(t, t_0, \phi)$ *of* (3.1.3) *with* $|\phi(t)| < \delta$ *and*

$$|u(t, t_0, \phi)| \geq [c_1 + c_2(t - t_0)]^{1/2},$$

where c_1 *and* c_2 *are positive constants depending on* t_0 *and* ϕ.

Proof This corollary is an immediate consequence of (3.1.7). □

We shall now obtain sufficient conditions for the asymptotic stability of (3.1.3) in which $\alpha(t)$ is not necessarily negative and $a(t, s)$ need not be integrable on $R_+ \times R_+$.

From Theorem 1.7.3 ($n = 1$), it is clear that (3.1.3) is equivalent to

(3.1.8) $\quad v'(t) = \mu(t)v(t) + \int_{0}^{t} b(t,s)v(s)\,ds + c(t), \quad t \in R_+$

where $\mu(t) = \alpha(t) - q(t, t)$,

$$b(t,s) = a(t,s) + \frac{\partial q}{\partial s}(t,s) + q(t,s)\alpha(s) + \int_s^t q(t,\tau)a(\tau,s)\,d\tau,$$

and $c(t) = q(t,0)v(0)$, $q(t,s)$ being a continuously differentiable function for $0 \leq s \leq t < \infty$.

Assume that there exists a continuously differentiable function $b(t,s)$ for $0 \leq s \leq t < \infty$ such that

(C_3) $|q(t,0)| \to 0$ as $t \to \infty$ and $\int_0^\infty |q(t,0)|\,dt < \infty$; and

(C_4) $\int_t^\infty |q(\tau,t)|\,d\tau$ is defined and finite for all $t \geq 0$.

Then we have the following results, which improve and include Theorem 3.1.3 and Corollary 3.1.1.

Theorem 3.1.4 *Suppose that (C_3) and (C_4) holds and there is a constant $\beta > 0$ such that*

(C_5) $\int_0^t |b(t,s)|\,ds + \int_t^\infty |b(\tau,t)|\,d\tau + |q(t,0)| - 2|\mu(t)| \leq -\beta.$

If $\mu(t) < 0$ and bounded then the zero solution of (3.1.3) is asymptotically stable.

Proof Since (3.1.3) is equivalent to (3.1.8), the proof is exactly similar to that of Corollary 3.1.1 with Lyapunov functional

$$V(t, u(\,\cdot\,)) = u^2 + \int_0^t \int_t^\infty |b(\tau,s)|\,d\tau\, u^2(s)\,ds.$$

□

Theorem 3.1.5 *Suppose that (C_3) and (C_4) with $q(t,s) = q(t-s)$ and $b(t,s) = b(t-s)$ hold. If $\mu + \int_0^\infty |b(t)|\,dt < 0$ then the zero solution of (3.1.1) is uniformly asymptotically stable.*

Example 3.1.2 Consider the scalar equation

(3.1.9) $$u' = -5u - \int_0^t 12e^{2(t-s)}u(s)\,ds.$$

Here $\alpha = -5$ and $a(t) = -12e^{2t}$ in (3.1.1). Clearly, $a(t)$ is not in $L^1(R_+)$. Choose $q(t-s) = -3e^{-(t-s)} = q(t,s)$. Then

$$\mu(t) = \alpha(t) - q(t,t) = -5 + 3 = -2 < 0$$

and

$$b(t,s) = a(t,s) + \frac{\partial q(t,s)}{\partial s} + q(t,s)\alpha(s) + \int_s^t q(t,\tau)a(\tau,s)\,d\tau$$

$$= -12e^{2(t-s)} + 12e^{-(t-s)} + 36\int_s^t e^{-t+2s+3\tau}\,d\tau = 0.$$

Thus all the conditions of Theorem 3.1.5 are satisfied, and hence the zero solution of (3.1.9) is uniformly asymptotically stable.

Example 3.1.3 Consider the scalar equation

(3.1.10) $$u' = \frac{u}{6+t} - \int_0^t \frac{6e^{-5(t-s)}(2+s)}{(2+t)^2}(3+s)u(s)\,ds.$$

Here

$$\alpha(t) = \frac{1}{6+t}, \quad a(t,s) = \frac{-6e^{-5(t-s)}(2+s)(3+s)}{(2+t)^2}.$$

Select

$$q(t,s) = \frac{2e^{-2(t-s)}(2+s)^2}{(2+t)^2};$$

then $\mu(t) = \alpha(t) - q(t,t) = 1/(6+t) - 2 < 0$ for all $t \geq 0$. Furthermore,

$$b(t,s) = a(t,s) + \frac{\partial q(t,s)}{\partial s} + q(t,s)\alpha(s) + \int_s^t q(t,\tau)a(\tau,s)\,d\tau$$

$$= \frac{-2e^{-5(t-s)}(2+s)(3+s)}{(2+t)^2} + \frac{2e^{-2(t-s)}(2+s)^2}{(2+t)^2(6+t)}.$$

Therefore

$$\int_0^t |b(t,s)|\,ds + \int_t^\infty |b(\tau,t)|\,d\tau + |q(t,0)| - 2|\mu(t)|$$

$$\leq \frac{118}{150} + \frac{23}{30} + 2 - 2\left|2 - \frac{1}{6+t}\right| \leq -\frac{17}{150},$$

and hence $\beta = \frac{17}{150}$ in the condition (C_5). Thus all the conditions of Theorem 3.1.4 are satisfied, and hence the zero solution of (3.1.10) is asymptotically stable. However, Corollary 3.1.1 cannot be applied to equation (3.1.10) since

$\alpha(t) = 1/(6+t)$ is positive for all $t \geq 0$.

We shall next extend Theorem 3.1.4 to the system

$$(3.1.11) \qquad x'(t) = A(t)x(t) + \int_0^t K(t,s)x(s)\,ds, \quad t \geq 0,$$

where $A(t)$ is an $n \times n$ continuous matrix for $t \geq 0$ and $K(t,s)$ is an $n \times n$ continuous matrix for $0 \leq s \leq t < \infty$. By Theorem 1.7.3, (3.1.11) is equivalent to the system

$$(3.1.12) \qquad y'(t) = B(t)y(t) + \int_0^t L(t,s)y(s)\,ds + F(t), \quad t \in R_+$$

where $B(t) = A(t) - \phi(t,t)$,

$$L(t,s) = K(t,s) + \frac{\partial \phi(t,s)}{\partial s} + \phi(t,s)A(s) + \int_s^t \phi(t,u)K(u,s)\,du,$$

and $F(t) = \phi(t,0)y_0$, $\phi(t,s)$ being an $n \times n$ matrix continuously differentiable for $0 \leq s \leq t < \infty$.

Let $H(t)$ be an $n \times n$ real symmetric matrix bounded and continuously differentiable for $0 \leq t < \infty$. We shall also assume that there exists a constant $\delta > 0$ such that

$$(3.1.13) \qquad y^T\bigl[H'(t) + H(t)B(t) + B^T(t)H(t)\bigr]y \leq -\delta\,|y|^2$$

for $y \in R^n$. Let $\int_t^\infty |\phi(u,t)|\,du$ be defined for all $t \geq 0$.

Theorem 3.1.6 *Suppose that there exists an $n \times n$ matrix $\phi(t,s)$ continuously differentiable for $0 \leq s \leq t < \infty$ satisfying the following conditions:*

(i) $|\phi(t,0)| \to 0$ as $t \to \infty$ and $\int_0^\infty |\phi(t,0)|\,dt < \infty$;

(ii) $\int_t^\infty |\phi(u,t)|\,du$ and $\int_t^\infty |L(u,t)|\,du$ are defined for all $t \geq 0$, and

(iii) there is constant $\alpha_0 > 0$ such that

$$(3.1.14) \qquad H_0\left[\int_0^t |L(t,s)|\,ds + \int_t^\infty |L(u,t)|\,du + |\phi(t,0)|\right] \leq \alpha_0,$$

where $H_0 = \sup_{t \geq 0} |H(t)|$.

If $\delta > \alpha_0$, (3.1.13) holds and $|B(t)|$ is bounded then the zero solution of (3.1.11) is asymptotically stable.

Proof Consider the functional

$$V(t, y(\cdot)) = y^T H(t) y + 2 H_0 \int_0^t \int_t^\infty |L(u,s)| \, du \, |y(s)|^2 \, ds$$

so that along the solution $y(t)$ of (3.1.12) (equivalent to (3.1.11)), we have

$$\begin{aligned} V'_{(3.1.12)}(t, y(\cdot)) \quad &\leq y^T \big[H'(t) + B^T(t) H(t) + H(t) B(t) \big] y \\ &\quad + 2 H_0 |y| \, |\phi(t,0)| \, |y_0| \\ &\quad + 2 H_0 \int_0^t |L(t,s)| \, |y(s)| \, |y| \, ds \\ &\quad + H_0 \int_t^\infty |L(u,t)| \, du \, |y|^2 + 2 H_0 \\ &\quad - H_0 \int_0^t |L(t,s)| \, |y(s)|^2 \, ds. \end{aligned}$$

From (3.1.13) and the fact that $2 |y| \, |y_0| \leq |y|^2 + |y_0|^2$, it follows that

$$\begin{aligned} V'_{(3.1.12)}(t, y(\cdot)) \quad &\leq -\delta |y|^2 + H_0 |\phi(t,0)| \big(|y|^2 + |y_0|^2 \big) \\ &\quad + H_0 \int_0^t |L(t,s)| \big(|y(s)|^2 + |y|^2 \big) ds \\ &\quad + H_0 \int_t^\infty |L(u,t)| \, du \, |y|^2 - H_0 \int_0^t |L(t,s)| \, |y(s)|^2 \, ds. \end{aligned}$$

Thus (3.1.14) gives

$$V'_{(3.1.12)}(t, y(\cdot)) \leq (-\delta + \alpha_0) |y|^2 + H_0 |\phi(t,0)| \, |y_0|^2.$$

Since $\delta > \alpha_0$, it follows from the assumption (i) that $|y(t)|$ is in $L^2[0, \infty)$. Further it can be easily seen from the assumptions of the theorem and equation (3.1.12) that $|y'(t)|$ is bounded. Thus the application of Barbălat's lemma yields that $|y(t)| \to 0$ as $t \to \infty$, and the proof is complete. □

The following special cases of Theorem 3.1.6 are important.

(a) Suppose $\phi(t,s) \equiv 0$. Then $B(t) = A(t)$, $L(t,s) = K(t,s)$ and $F(t) \equiv 0$. Further, if $A(t) = A$, a constant matrix, then we look for a real symmetric $n \times n$ constant matrix H such that

(3.1.15) $$A^T H + HA = -I,$$

where I is an identity matrix. In this case, $\delta = 1$ in (3.1.13) and the condition (3.1.14) reduces to

(3.1.16) $$|H|\left[\int_0^t |K(t,s)|\,ds + \int_t^\infty |K(u,t)|\,du\right] \leq \alpha_0.$$

If $\alpha_0 < 1$ in (3.1.16) then the application of Theorem 3.1.6 yields that the zero solution of

(3.1.17) $$x' = Ax + \int_0^t K(t,s)x(s)\,ds$$

is asymptotically stable. For the convolution kernel $K(t,s) = K(t-s)$, the condition (3.1.16) takes the form

(3.1.18) $$|H|\left[\int_0^\infty |K(\tau)|\,d\tau\right] \leq \alpha_0.$$

If $\alpha_0 < 1$ in (3.1.18) then the zero solution is uniformly asymptotically stable.

(b) Suppose that there exists an $n \times n$ continuous matrix function $\phi(t,s)$ on $R_+ \times R_+$ such that $\partial \phi/\partial s$ exists, continuous and satisfies

$$K(t,s) + \frac{\partial \phi(t,s)}{\partial s} + \phi(t,s)A(s) + \int_s^t \phi(t,u)K(u,s)\,du = 0.$$

Then the integro-differential system (3.1.11) is equivalent to an ordinary differential system

$$y'(t) = B(t)y(t) + \phi(t,0)y_0, \quad y(0) = y_0.$$

Select $V(t,y) = y^T H(t) y$, where $H(t)$ satisfies the condition (3.1.13). Then the assumptions of Theorem 3.1.6 with $L(t,s) \equiv 0$ are sufficient for asymptotic stability of zero solution of (3.1.11).

(c) Let $\phi(t,s) \equiv 0$ and $A(t) = A$, a constant matrix. Suppose that (3.1.15) holds for some real symmetric constant matrix H and there exists a positive constant $\alpha_0 < 1$ such that (3.1.16) is satisfied. Then the zero solution of (3.1.17) is stable if and only if H is positive definite. It is well know that the matrix equation (3.1.15) has a positive definite matrix solution if and only if A is a stable matrix.

(d) Let $\phi(t,s) \equiv 0$ and $A(t) = A$, a constant matrix. Suppose that H is a positive definite symmetric matrix satisfying $A^T H + HA = I$. If (3.1.16) holds with $\alpha_0 < 1$, then the zero solution of (3.1.17) is completely unstable.

3.2 Equations with Unbounded Delay

The theory of equations with unbounded delay has emerged in recent years as an independent branch of modern research because of its connections to many fields such as continuum mechanics, population dynamics, ecology, systems theory and nuclear reactor dynamics.

In this section, we shall consider the related integro-differential system with infinite memory of the form

$$(3.2.1) \qquad x' = A(t)x + \int_{-\infty}^{t} K(t,s)x(s)\,ds + f(t,x),$$

and show that the arguments similar to the previous sections can be utilized to discuss the stability properties of (3.2.1). Here A is an $n \times n$ matrix continuous for $-\infty < t < \infty$, K is an $n \times n$ matrix continuous for $-\infty < s \leq t < \infty$, $f \in C[R \times S(\rho), R^n]$, $S(\rho) = \{x \in R^n : |x| < \rho\}$ and $f(t,0) \equiv 0$. Let $x(t) = x(t,0,g)$ be a solution of the integro-differential system (3.2.1) with continuous initial function $g: (-\infty, 0] \to R^n$. Since $x(t) = x(t,0,g)$ is a solution of (3.2.1) with initial function $g(t)$ on $(-\infty, 0]$, (3.2.1) can be written as

$$(3.2.2) \qquad x'(t) = A(t)x(t) + \int_0^t K(t,s)x(s)\,ds + f(t,x(t)) + F(t),$$

where $F(t) = \int_{-\infty}^{0} K(t,s)g(s)\,ds$.

It is convenient to list below a modification of Theorem 1.7.3 to suit our present situation.

Theorem 3.2.1 Let $\phi(t,s)$ be an $n \times n$ matrix continuously differentiable for $-\infty < s \leq t < \infty$. Then (3.2.2) is equivalent to

(3.2.3)
$$\begin{cases} y'(t) = B(t)y(t) + \int_0^t L(t,s)y(s)\,ds + G(t,y(t)), \\ y(0) = g(0) = x_0, \end{cases}$$

where $B(t) = A(t) - \phi(t,t)$,
$$L(t,s) = K(t,s) + \frac{\partial \phi(t,s)}{\partial s} + \phi(t,s)A(s) + \int_s^t \phi(t,u)K(u,s)\,du,$$

and
$$G(t,y(t)) = f(t,y(t)) + F(t) + \phi(t,0)x_0 + \int_0^t \phi(t,s)f(s,y(s))\,ds$$
$$+ \int_0^t \phi(t,s)F(s)\,ds.$$

Recall that if $\phi(t,s)$ is the differentiable resolvent corresponding to the kernel $K(t,s)$ (that is, $\phi(t,s)$ satisfies the adjoint equation
$$\frac{\partial \phi(t,s)}{\partial s} = -\phi(t,s)A(s) - \int_s^t \phi(t,u)K(u,s)\,du, \quad \phi(t,t) = I)$$

then (3.2.3) gives the usual linear variation of parameters formula
$$y(t) = \phi(t,0)x_0 + \int_0^t \phi(t,s)f(s,y(s))\,ds + \int_0^t \phi(t,s)F(s)\,ds.$$

We now list a number of assumptions before we proceed further.

(H0) There exists a positive continuous function $\lambda(t)$ for $0 \leq t < \infty$ such that $|f(t,x)| \leq \lambda(t)|x|$ for $(t,x) \in R^+ \times S(\rho)$ with $\lambda(t) \to 0$ as $t \to \infty$;

(H1) There exists an $n \times n$ symmetric matrix, bounded and continuously differentiable for $-\infty < s \leq t < \infty$ and a constant $\alpha > 0$ such that

$$y^T\left[H'(t) + H(t)B(t) + B^T(t)H(t)\right]y \leq -\alpha |y|^2 \quad \text{for } y \in R^n;$$

(H2) There exists an $n \times n$ matrix $\phi(t,s)$ that is differentiable for $-\infty < s \leq t < \infty$ and satisfies the conditions $|\phi(t,0)| \to 0$ as $t \to \infty$ and $\int_0^\infty |\phi(t,0)| \, dt < \infty$;

(H3) $\int_t^\infty |\phi(u,t)| \, du$ and $\int_t^\infty |L(u,t)| \, du$ are defined for all $t \in R$;

(H4) There exists a continuous function $\eta(t)$ with the property $\eta(t) \to 0$ as $t \to \infty$ and $\eta(t) \in L^1(R_+)$, where

$$\eta(t) = \int_{-\infty}^0 |K(t,s)| \, ds + \int_0^t |\phi(t,s)| \int_{-\infty}^0 |K(s,\tau)| \, d\tau \, ds,$$

and there is a constant $\delta_0 > 0$ such that

$$H_0\left[\int_0^t |L(t,s)| \, ds + \int_t^\infty |L(u,t)| \, du + |\phi(t,0)| + \eta(t)\right]$$

$$+ \lambda_0 H_0\left[\int_0^t |\phi(t,s)| \, ds + \int_t^\infty |\phi(u,t)| \, du\right] \leq \delta_0$$

where $\lambda_0 = \sup_{t \geq 0} [\lambda(t)]$ and $H_0 = \sup_{t \geq 0} |H(t)|$.

Theorem 3.2.2 *Assume that* $(H0)-(H4)$ *hold. If* $\alpha > \delta_0 + 2\lambda_0 H_0$, $|B(t)|$ *is bounded and* $|g(t)| < \delta$ *on* $(-\infty, 0]$ *then the zero solution of* (3.2.1) *is asymptotically stable.*

Proof In view of Theorem 3.2.1, it is enough to show that the zero solution of (3.2.3) is asymptotically stable. To this end, let us consider the Lyapunov functional

$$V(t, y(\cdot)) = y^T H(t)y + H_0 \int_0^t \int_t^\infty |L(u,s)| \, du \, |y(s)|^2 \, ds$$

$$+ \lambda_0 H_0 \int_0^t \int_t^\infty |\phi(u,s)| \, du \, |y(s)|^2 \, ds.$$

Then the time derivative of V along the solution of (3.2.3) is given by

$$V'_{(3.2.3)}(t, y(\cdot)) = y^T\left[H'(t) + B^T(t)H(t) + H(t)B(t)\right]y$$

$$+ 2H_0|y||G(t,y)| + 2H_0\int_0^t |L(t,s)||y(s)||y|\,ds$$

$$+ H_0\int_t^\infty |L(u,t)|\,du\,|y|^2 - H_0\int_0^t |L(t,s)||y(s)|^2\,ds$$

$$+ \lambda_0 H_0\int_t^\infty |\phi(u,t)|\,du\,|y|^2$$

$$- \lambda_0 H_0\int_0^t |\phi(t,s)||y(s)|^2\,ds.$$

From the definition of $G(t, y)$, hypotheses $(H0)$, $(H1)$ and $(H4)$ and the fact that $2|y(s)||y| \leq |y(s)|^2 + |y|^2$, we obtain

$$V'_{(3.2.3)}(t, y(\cdot))$$

$$\leq (-\alpha + 2\lambda_0 H_0)|y|^2$$

$$+ H_0\left[|\phi(t,0)| + \int_t^\infty |L(y,t)|\,du + \int_0^t |L(t,s)|\,ds\right]|y|^2$$

$$+ \lambda_0 H_0\left[\int_0^t |\phi(t,s)|\,ds + \int_t^\infty |\phi(u,t)|\,du\right]|y|^2 + H_0|\phi(t,0)||x_0|^2$$

$$+ H_0\left[\int_{-\infty}^0 |K(t,s)|\,ds + \int_0^t |\phi(t,s)|\int_{-\infty}^0 |K(s,\tau)|\,d\tau\,ds\right]|y|^2$$

$$+ H_0\left[\int_{-\infty}^0 |K(t,s)||g(s)|^2\,ds + \int_0^t |\phi(t,s))|\int_{-\infty}^0 |K(s,\tau)||g(\tau)|^2\,d\tau\,ds\right].$$

This implies that

$$V'_{(3.2.3)}(t,y,(\cdot)) \leq (-\alpha + 2\lambda_0 H_0 + \delta_0)|y|^2$$
$$+ H_0\delta^2\eta(t) + H_0|\phi(t,0)| \, |x_0|^2.$$

Since $\alpha > \delta + 2\lambda_0 H_0$, it follows from $(H2)$ and $(H4)$ that $|y(t)|$ is in $L^2[0,\infty)$. Also, it can be shown from (3.2.3) that $|y'(t)|$ is bounded. Thus the application of Barbalat's lemma yields $|y(t)| \to 0$ as $t \to \infty$. Hence the proof is complete. □

Example 3.2.1 Consider the scalar equation

$$(3.2.4) \qquad u' = -3u + 3\int_{-\infty}^{t} e^{-5(t-s)}u(s)\,ds + \tfrac{1}{9}e^{-2t}x^3.$$

Select $\phi(t,s) = 3e^{-2(t-s)}$, $H(t) = \tfrac{1}{12}$ and $\rho = 1$. Then $B(t) = -6$, $L(t,s) = 0$, $\eta(t) = \tfrac{3}{5}e^{-t}$, $\lambda_0 = \tfrac{1}{9}$, $H_0 = \tfrac{1}{12}$ and $\delta_0 = \tfrac{41}{120}$. Thus $\delta_0 + 2\lambda_0 H_0 \simeq 0.36 < 1$ $(= \alpha)$. Hence the zero solution of (3.2.4) is asymptotically stable.

The following special cases of Theorem 3.2.2 are interesting. We shall reduce the study of linear integro-differential system to that of an ordinary differential system and investigate the stability properties.

(a) Consider the linear system obtained from (3.2.1) by setting $f(t,x) \equiv 0$, namely,

$$(3.2.5) \qquad x' = A(t)x + \int_{-\infty}^{t} K(t,s)x(s)\,ds,$$

which is, because of Theorem 3.2.1, equivalent to

$$(3.2.6) \qquad y' = B(t)y + \int_{0}^{t} L(t,s)y(s)\,ds + D(t), \qquad y(0) = x_0,$$

where $D(t) = F(t) + \phi(t,0)x_0 + \int_0^t \phi(t,s)F(s)\,ds$.

Assume that there exists an $n \times n$ matrix function $\phi(t,s)$ continuously differentiable for $-\infty < s \leq t < \infty$ and satisfying the equation

$$K(t,s) + \frac{\partial \phi(t,s)}{\partial s} + \phi(t,s)A(s) + \int_{s}^{t} \phi(t,u)K(u,s)\,du = 0.$$

Then (3.2.6) reduces to an ordinary differential equation

$$(3.2.7) \qquad y' = B(t)y + D(t), \qquad y(0) = g(0) = x_0.$$

By the variation of parameters formula, any solution $y(t)$ with $y(0) = x_0$ of (3.2.7) can be written as

$$(3.2.8) \qquad y(t) = \psi(t,0)x_0 + \int_0^t \psi(t,s)D(s)\,ds,$$

where $\psi(t,s)$ is a fundamental matrix solution of $y' = B(t)y$. Then we have the following result.

Theorem 3.2.3 *Assume that*

(i) *there exists positive constants M, K, p and q such that*

$$|\psi(t,s)| \leq Me^{-p(t-s)}, \quad |\phi(t,s)| \leq Ke^{-q(t-s)} \quad \text{for } 0 \leq s \leq t < \infty;$$

(ii) *there exists a $\delta > 0$ sufficiently small such that $|F(t)| \to 0$ as $t \to \infty$ whenever $|g(t)| < \delta$ on $(-\infty, 0]$.*

Then the zero solution of (3.2.5) is asymptotically stable.

Example 3.2.2 Consider the scalar equation

$$(3.2.9) \qquad u' = \tfrac{1}{2}u - 9 \int_{-\infty}^t e^{-7(t-s)} u(s)\,ds.$$

Choose $\phi(t,s) = 6e^{-(t-s)}$. Then $B(t) = -\tfrac{11}{2}$, $\psi(t,s) = e^{-11(t-s)/2}$, $L(t,s) \equiv 0$ and $F(t) \to 0$ as $t \to \infty$ whenever $|g(t)| \leq \delta$ on $(-\infty, 0]$, δ being sufficiently small. Thus in view of Theorem 3.2.3, the zero solution of (3.2.9) is asymptotically stable.

(b) Consider the scalar equation ($n=1$ and $f(t,x) \equiv 0$ in (3.2.1))

$$(3.2.10) \qquad x' = A(t)x + \int_{-\infty}^t K(t,s)x(s)\,ds.$$

Equation (3.2.10) is equivalent to the scalar equation

$$(3.2.11) \qquad y' = B(t)y + \int_0^t L(t,s)y(s)\,ds + D(t)$$

where $D(t) = F(t) + \phi(t,0)x_0 + \int_0^t \phi(t,s)F(s)\,ds$.

We make the following hypotheses to start our result.

(C_1) There exists a continuous function $\beta(t)$ with the properties

$\beta(t) \to 0$ as $t \to \infty$ and $\beta(t) \in L^1(R_+)$, where $\beta(t)$ is given by

$$\beta(t) = \int_{-\infty}^{0} |K(t,s)|\, ds + \int_{0}^{t} |\phi(t,s)| \int_{-\infty}^{0} |K(s,\tau)|\, d\tau\, ds.$$

(C_2) There exists a constant $m > 0$ such that

$$\int_{0}^{t} |L(t,s)|\, ds + \int_{t}^{\infty} |L(u,t)|\, du + |\phi(t,0)| + \beta(t) - 2|B(t)| < -m.$$

Theorem 3.2.4 *Assume that (C_1), (C_2) and $(H2)$, $(H3)$ with $n=1$ hold. Suppose that $B(t) < 0$ and is bounded, and $|g(t)| < \delta$ on $(-\infty, 0]$. Then the zero solution of (3.2.10) is asymptotically stable.*

Proof Consider the functional

$$V(t, y(\cdot)) = y^2 + \int_{0}^{t}\int_{t}^{\infty} |L(u,s)|\, du\, y^2(s)\, ds.$$

Then the time derivative of V along the solutions of (3.2.11) is given by

$$V'_{(3.2.11)}(t, y(\cdot)) = 2yy' + \int_{t}^{\infty} |L(u,t)|\, du\, y^2(t) - \int_{0}^{t} |L(t,s)|\, y^2(s)\, ds$$

$$\leq 2|y|\left[|B(t)||y| + \int_{0}^{t} |L(t,s)||y(s)|\, ds + |D(t)|\right]$$

$$+ \int_{t}^{\infty} |L(u,t)|\, du\, y^2(t) - \int_{0}^{t} |L(t,s)|\, y^2(s)\, ds.$$

Using the definition of $D(t)$ and the fact that $2|y(s)||y| \leq y^2(s) + y^2$, we obtain

$$V'_{(3.2.11)}(t, y(\cdot)) \leq \left[2|B(t)| + |\phi(t,0)| + \int_{0}^{t} |L(t,s)|\, ds + \int_{t}^{\infty} |L(u,t)|\, du\right.$$

$$\left. + \int_{-\infty}^{0} |K(t,s)|\, ds + \int_{0}^{t} |\phi(t,s)| \int_{-\infty}^{0} |K(s,\tau)|\, d\tau\, ds\right]|y|^2$$

$$+ \delta^2 \left[\int_{-\infty}^{0} |K(t,s)|\, ds + \int_{0}^{t} |\phi(t,s)| \int_{-\infty}^{0} |K(s,\tau)|\, d\tau\, ds\right]$$

$$+ |\phi(t,0)| |x_0|^2.$$

This implies, in view of (C_1) and (C_2), that

$$V'_{(3.2.11)}(t,y(\cdot)) \leq -m|y|^2 + \delta^2\beta(t) + |\phi(t,0)| |x_0|^2.$$

Now following the proof of Theorem 3.2.2, it is easy to see that the zero solution of (3.2.10) is asymptotically stable, and thus the proof is complete. □

(c) We shall next consider the scalar equation (3.2.10) and select a proper Lyapunov functional to obtain the asymptotic stability property directly without using Theorem 3.2.1.

Theorem 3.2.5 *Suppose that there exists a constant $\delta_0 > 0$ such that*

$$\int_{-\infty}^{t} |K(t,s)|\, ds + \int_{t}^{\infty} |K(u,t)|\, du - 2|A(t)| \leq -\delta_0.$$

If $A(t) < 0$ and is bounded then the zero solution of (3.2.10) is asymptotically stable.

Proof Consider the functional

$$V(t,x(\cdot)) = x^2 + \int_{-\infty}^{t}\int_{t}^{\infty} |K(u,s)|\, du\, x^2(s)\, ds$$

so that along the solutions of (3.2.10), we get

$$V'_{(3.2.10)}(t,x(\cdot)) = 2xx' + \int_{t}^{\infty} |K(u,s)|\, du\, x^2(t) - \int_{-\infty}^{t} |K(t,s)|\, x^2(s)\, ds$$

$$\leq \left[2A(t) + \int_{-\infty}^{t} |K(t,s)|\, ds + \int_{t}^{\infty} |K(u,t)|\, du\right]|x|^2.$$

This implies, in view of the assumptions of the theorem, that

$$V'_{(3.2.10)}(t,x(\cdot)) \leq -\delta_0 |x|^2.$$

This together with Barbalat's lemma implies the desired result. Thus the proof is complete. □

3.3 Perturbed Systems

In this section, we shall consider some results on perturbed systems. We shall first discuss the required converse theorems, and, utilizing them, we shall study boundedness properties of perturbed systems.

For any $\phi \in C[R_+, R^n] = C(R_+)$, define $|\phi|_t = \max\{|\phi(s)| : 0 \leq s \leq t\}$. Consider the Volterra system of equations

$$(3.3.1) \qquad x'(t) = F(t, x(\,\cdot\,)),$$

where $F: R_+ \times C(R_+) \to R^n$ is continuous and $x(\,\cdot\,)$ represents the function x on $[0, t]$, with the values of t always being determined by the first coordinate of F in (3.3.1). The solution of (3.3.1) with initial values (t_0, ϕ) will be denoted by $x(t, t_0, \phi)$, where $t_0 \geq 0$ and $\phi: [0, t_0] \to R^n$ is a continuous function. We assume that $F(t, 0) \equiv 0$, so that $x(t) \equiv 0$ is a solution of (3.3.1).

Equation (3.3.1) contains the following Volterra integro-differential systems

$$(3.3.2) \qquad x'(t) = Ax(t) + \int_0^t C(t-s)x(s)\,ds,$$

$$(3.3.3) \qquad x'(t) = A(t)x(t) + \int_0^t B(t,s)x(s)\,ds,$$

and the perturbed systems

$$(3.3.4) \qquad y'(t) = Ay(t) + \int_0^t C(t-s)y(s)\,ds + g(t, y(\,\cdot\,)),$$

$$(3.3.5) \qquad y'(t) = A(t)y(t) + \int_0^t B(t,s)y(s)\,ds + g(t, y(\,\cdot\,)),$$

$$(3.3.6) \qquad y'(t) = A(t)y(t) + \int_0^t B(t,s)y(s)\,ds + h(t),$$

where $A(t)$ and $C(t)$ are $n \times n$ continuous matrices for $0 \leq t < \infty$, A is an $n \times n$ constant matrix, $B(t,s)$ is an $n \times n$ continuous matrix for $0 \leq s \leq t < \infty$, $g(t, \phi)$ is a continuous functional on $R_+ \times C(R_+)$ and $h(t)$ is an n-vector that is continuous on R_+.

We need the following notions.

Definition 3.3.1 Let g be a given functional with domain $R_+ \times C(R_+)$ and range in R^n. This functional $g(t,\phi)$ will be called nonanticipative if and only if for each $t \geq 0$, one has $g(t,\phi) = g(t,\psi)$ whenever ϕ and ψ are continuous functions such that $\phi(s) = \psi(s)$ on $0 \leq s \leq t$.

Definition 3.3.2 Let $g(t,\phi)$ be a continuous functional on $R_+ \times C(R_+)$ into R^n. Then

(i) $g(t,\phi)$ is locally Lipschitz continuous in ϕ if and only if given any pair of positive constants a and b there exists a constant $K > 0$ such that

$$|g(t,\phi) - g(t,\psi)| \leq K |\phi - \psi|_t$$

whenever $0 \leq t \leq a$ and both $|\phi|_t$ and $|\psi|_t \leq b$;

(ii) $g(t,\phi)$ is locally Lipschitz continuous in ϕ uniformly in t if it is locally Lipschitz continuous in ϕ and the Lipschitz constant K can be chosen independent of a.

We note that if $g(t,\phi)$ is continuous in (t,ϕ) and locally Lipschitz continuous in ϕ then it is automatically nonanticipative.

Let $V(t,\phi): R_+ \times C(R_+) \to R$ be a continuous functional satisfying the property of locally Lipschitz continuity in ϕ.

Definition 3.3.3 The derivative of $V(t,\phi)$ with respect to (3.3.1) is defined by

$$V'_{(3.3.1)}(t,\phi) = \lim_{h \to 0^+} \sup \left[\frac{V(t+h,\phi^*) - V(t,\phi)}{h} \right]$$

where

$$\phi^*(s) = \begin{cases} \phi(s) & \text{on } 0 \leq s \leq t, \\ \phi(s) + F(t,\phi)(s-t) & \text{on } t \leq s \leq t+h. \end{cases}$$

It is well known that

$$V'_{(3.3.1)}(t_0,\phi) = \lim_{h \to 0^+} \sup \left[\frac{V(t_0+h, x(\cdot,t_0,\phi)) - V(t_0,\phi)}{h} \right],$$

where $x(\cdot, t_0, \phi)$ is the unique solution of (3.3.1) with initial values $(t_0, \phi) \in R_+ \times C(R_+)$.

Let $\mathcal{K}$ be the class of functions $b(r)$, defined and continuous on R_+, $b(0) = 0$ and monotone increasing in r.

Definition 3.3.4 A functional $V: R_+ \times C(R_+) \to R_+$ is said to be
(i) positive definite if $V(t, 0) \equiv 0$ and there exists a function $b \in \mathcal{K}$ such that $b(|\phi|) \leq V(t, \phi)$, $(t, \phi) \in R_+ \times C(R_+)$;
(ii) decrescent if there exists a function $a \in \mathcal{K}$ such that $V(t, \phi) \leq a(|\phi|)$, $(t, \phi) \in R_+ \times C(R_+)$.

Definition 3.3.5 The zero solution of (3.3.1) is eventually uniformly stable $(EvUS)$ if, for any $\epsilon > 0$, there exists $\delta = \delta(\epsilon) > 0$ and $\alpha(\epsilon) > 0$ such that $t_0 \geq \alpha(\epsilon)$ and $|\phi|_{t_0} \leq \delta$ imply $|x(t, t_0, \phi)| < \epsilon$ for all $t \geq t_0$.

Definition 3.3.6 The zero solution of (3.3.1) is uniformly attractive (UA) if there is a $\delta_0 > 0$ and, for any given $\eta > 0$, there exists a $T = T(\eta) > 0$ such that $t_0 \geq 0$ and $|\phi|_{t_0} \leq \delta_0$ imply $|x(t, t_0, \phi)| < \eta$ for all $t \geq t_0 + T$.

Definition 3.3.7 The zero solution of (3.3.1) is eventually uniformly asymptotically stable $(EvUAS)$ if it is $EvUS$ and UA.

Definition 3.3.8 The zero solution of (3.3.1) is uniformly asymptotically stable (UAS) if it is US and UA.

The following basic results are useful in our study of perturbation problems.

Theorem 3.3.1 Suppose that $C(t) \in L^1(R_+)$. The zero solution of (3.3.2) is exponentially asymptotically stable $(ExAS)$ if and only if there exists a continuous functional $V(t, \phi)$ defined for all $t \geq 0$ and $\phi \in C([0, t], R^n)$, $\lambda > 0$ and $K > 0$ such that
(i) $|\phi(t)| \leq V(t, \phi) \leq K|\phi|_t$;
(ii) $|V(t, \phi) - V(t, \psi)| \leq K|\phi - \psi|_t$ for $\phi, \psi \in C([0, t], R^n)$;
(iii) $V'_{(3.3.2)}(t, \phi) \leq -\lambda V(t, \phi)$.

Proof The sufficiency part can be proved by standard arguments for the functional $V(t, \phi)$, so we shall indicate only the necessity part. Since the zero solution of (3.3.2) is $ExAS$, it follows by definition that there exist positive constants λ and K such that

$$|x(t,t_0,\phi)| \le Ke^{-\lambda(t-t_0)}|\phi|_{t_0}, \quad \text{for all } t_0 \ge 0 \text{ and } t \ge t_0.$$

Let $V(t,\phi) = \sup_{t_0 \ge 0} |x(t+t_0,t,\phi)|e^{\lambda t_0}$. Then it is easy to verify (see Yoshizawa [1]) that the functional $V(t,\phi)$ satisfies the conditions (i), (ii) and (iii), and hence the proof of the theorem is complete. □

Let $t \ge 0$, $\phi \in C[[0,t,R^n]$ and $p(t,\phi) = \sup_{t-r \le s \le t} |\phi(s)|$, where $r \ge 0$ and $\phi(s) = \phi(0)$ for $s \le 0$.

Theorem 3.3.2 *Let $A(t) \in L^1(R_+)$ and $B(t,s) \in L^1(R_+ \times R_+)$. The zero solution of (3.3.3) is exponentially asymptotically stable (ExAS) if and only if there exists a continuous functional $W(t,\phi)$ defined for $t \ge 0$ and $\phi \in C([0,t],R^n)$, $\lambda > 0$ and $K > 0$ such that*
  (i) $p(t,\phi) \le W(t,\phi) \le K|\phi|_t$;
  (ii) $|W(t,\phi) - W(t,\psi)+ \le K|\phi - \psi|_t$ *for* $\phi, \psi \in C([0,t],R^n)$;
  (iii) $W'_{(3.3.3)}(t,\phi) \le -\lambda W(t,\phi)$.

Proof The sufficiency is proved using the standard arguments for the functional $W(t,\phi)$. Necessity is proved by defining

$$W(t,\phi) = \sup_{t_0 \ge 0} p(t+t_0, x(\cdot,t,\phi))e^{\lambda t_0},$$

and proceeding as in the proof of Theorem 3.3.1. □

Theorem 3.3.3 *Let $C \in L^1(R_+)$. Suppose that the zero solution of (3.3.2) is uniformly asymptotically stable (UAS). Then there exists a Lyapunov functional $V(t,\phi)$ with the following properties:*
  (i) $V(t,\phi)$ *is locally Lipschitz continuous in ϕ uniformly in t*;
  (ii) $V(t,0) \equiv 0$ *for all $t \ge 0$*;
  (iii) *there exists a function $b \in \mathcal{K}$ such that $b(|\phi|) \le V(t,\phi)$, $(t,\phi) \in R_+ \times C(R_+)$*;
  (iv) *the derivative of $V(t,\phi)$ with respect to (3.3.2) satisfies $V'_{(3.3.2)}(t,\phi) \le -a(|\phi|)$, $a \in \mathcal{K}$.*

Proof The proof is exactly similar to that of the converse theorem of Massera. Let K and T_m be positive real numbers such that if $|\phi|_{t_0} \le 1$ then $|x(t,t_0,\phi)| \le K$ for all $t \ge t_0$ and $|x(t+T_m+t_0,t_0,\phi)| < 1/m$ for all $t \ge 0$. Let $g(t)$ be a continuous, nonincreasing, positive function such that $g(t) = K$

on $0 \le t \le T_1$ and $g(T_m) = 1/(m-1)$ for $m = 2, 3, 4, \ldots$. Then $|x(t+t_0, t_0, \phi)| \le g(t) \to 0$ as $t \to \infty$ whenever $t_0 \ge 0$ and $|\phi|_{t_0} \le 1$. For this $g(t)$, there exists a function $G(y) \in C^1(R_+)$ such that $G(y) > 0$, $G'(y) > 0$ for all $y > 0$, $G(0) = G'(0) = 0$, $G'(y)$ is increasing in y and for any constant $c_1 > 0$, the integrals $\int_0^\infty G(c_1 g(s))\,ds$ and $\int_0^\infty G'(c_1 g(s))\,ds$ are finite.

Let us now define the functional

$$(3.3.7) \qquad V(t, \phi) = \int_0^\infty G(|x(s+t, t, \phi)|)\,ds$$

where $x(t, t_0, \phi)$ is the unique solution of (3.3.2) with initial values (t_0, ϕ). Since $x(t, t_0, \phi)$ is continuous on $R_+ \times R^n \times C(R_+)$ and the integral in (3.3.7) converges uniformly for $t \ge 0$ and $|\phi|_t \le A$ for any fixed constant $A > 0$ then it is clear that $V: R_+ \times C(R_+) \to R_+$ is continuous and nonanticipative. To see that $V(t, \phi)$ is locally Lipschitz continuous in ϕ uniformly in $t \ge 0$, let ϕ_1 and ϕ_2 be such that $|\phi_1|_t, |\phi_2|_t \le B_1$ for some fixed constant $B_1 > 0$. Since $G'(y)$ is increasing, for any vectors y_1 and y_2 we have by the mean value theorem $G(|y_1|) - G(|y_2|) \le G'(q|y_1| + |1-q||y_2|)(|y_1| - |y_2|)$ for some q in $0 < q < 1$. Therefore

$$(3.3.8) \qquad |G(|y_1|) - G(|y_2|)| \le G'(|y_1| + |y_2|)|y_1 - y_2|.$$

By the stability of the zero solution of (3.3.2), there exists a constant $K > 0$ such that if $|\phi|_t \le 1$, then $|x(s+t, t, \phi)| \le K$ for all $t, s \ge 0$. Thus from (3.3.7) and (3.3.8), we obtain

$$|V(t, \phi_1) - V(t, \phi_2)|$$

$$\le \int_0^\infty |G(|x(s+t, t, \phi_1)|) + |x(s+t, t, \phi_2)|)| |x(s+t, t, \phi_1 - \phi_2)|\,ds$$

$$\le \int_0^\infty G'(2B_1 g(s)) K |\phi_1 - \phi_2|_t\,ds$$

$$= \left[K \int_0^\infty G'(2B_1 g(s))\,ds \right] |\phi_1 - \phi_2|_t.$$

This proves the Lipschitz continuity of $V(t,\phi)$ in ϕ uniformly in t. To prove that $V(t,\phi)$ is positive definite, let $t \geq 0$ be fixed and let B_1 be a given constant. By stability, if $|\phi|_t \leq B_1$ then $x(s+t,t,\phi)$ is uniformly bounded. Hence $|x'(s+t,t,\phi)| \leq \alpha(B_1)$ is uniformly bounded for $t, s \geq 0$ and $|\phi|_t < B_1$. This implies that

$$|x(s+t,t,\phi) - \phi(t)| \leq s\alpha(B_1) \leq \tfrac{1}{2}|\phi|$$

if $0 \leq s \leq |\phi|/2\alpha(B_1)$. This means that

$$\tfrac{1}{2}|\phi(t)| \leq |x(t+s,t,\phi)| \leq \tfrac{3}{2}|\phi(t)|.$$

Therefore it follows from (3.3.7) and the increasing property of G that

$$V(t,\phi) \geq \int_0^{w(|\phi(t)|)} G\bigl(\tfrac{1}{2}|\phi(t)|\bigr) dt = b(|\phi(t)|), \qquad b \in \mathcal{K}$$

if $|\phi|_t \leq B_1$ and $w(y) = y/2\alpha(B_1)$.

Finally, from uniqueness of the solutions we obtain

$$V(t, x(t, t_0, \phi)) = \int_0^\infty G(|x(s+t, t, x(t, t_0, \phi))|) ds$$

$$= \int_0^\infty G(|x(s+t, t_0, \phi)|) ds$$

$$= \int_t^\infty G(|x(s, t_0, \phi)|) ds,$$

and hence the derivative of V with respect to (3.3.2) is

$$V'_{3.3.2}(t, \phi) = -G(|\phi(t)|) = -a(|\phi(t)|), \qquad a \in \mathcal{K}.$$

This completes the proof. □

The existence of V with the properties $(i) - (iv)$ guarantees that (3.3.2) can be perturbed with a certain class of perturbation functionals to recover a certain type of stability property for the perturbed system, as the following result shows.

Theorem 3.3.4 Suppose that $C \in L^1(R_+)$ and the zero solution of (3.3.2) is uniformly asymptotically stable. Then the perturbed system (3.3.4) has the following type of stability. Given $\epsilon > 0$, there exist $\eta_1 > 0$ and $\eta_2 > 0$ such that for any initial values (t_0, ϕ), if $|\phi|_{t_0} \leq \eta_1$ and if $g(t, \phi)$ is any continuous, nonanticipative functional with $|g(t, \phi)| \leq \eta_2$ on the set $\{t \geq 0, |\phi|_t \leq \epsilon\}$ then any solution $y(t, t_0, \phi)$ of (3.3.4) has the estimate $|y(t, t_0, \phi)| < \epsilon$ for all $t \geq t_0$.

Proof Given $\epsilon > 0$, let $a, b \in \mathcal{K}$ be the functions given in Theorem 3.3.3. Let L be the Lipschitz constant for $V(t, \phi)$, where $|\phi|_t \leq \epsilon$. Define $m = b(\epsilon)$. Since $V(t, \phi) \leq L|\phi|_t$ whenever $|\phi|_t \leq \epsilon$, it follows that $V(t, \phi) \leq m$ if $|\phi|_t \leq \eta_1$ and $\eta_1 = \min(\epsilon/2, m/2L)$. Choose $\alpha = a(\eta_1)$ and set $\eta_2 = \alpha/2L$. Let ϕ and $g(t, \phi)$ be majorized by these η_1 and η_2 respectively. Let $\psi(t) = y(t + s, s, \phi)$. Since $\psi(t)$ is continuous and $|\psi(0)| = |\phi(0)| \leq \eta_1 < \epsilon$, it is clear that $|\psi(t)| < \epsilon$ for t sufficiently small. If $\psi(t)$ gets into the region $R(\eta_1, \epsilon) = \{\eta_1 \leq |\psi(t)| \leq \epsilon\}$ then in this region $R(\eta_1, \epsilon)$ the derivative of V along the solutions of (3.3.4) satisfies

$$V'_{(3.3.4)}(t + s, \psi) \leq V'_{(3.3.2)}(t + s, \psi) + L|g(t + s, \psi)|$$

$$\leq -a(|\psi(t)|) + L\eta_2 \leq -\alpha + \frac{L\alpha}{2L} = -\tfrac{1}{2}\alpha < 0.$$

This shows that $V(t + s, \psi)$ is decreasing in the region $R(\eta_1, \epsilon)$. In particular,

$$V(t + s, \psi) \leq \max\{V(t, \phi): t \geq 0, |\phi|_t \leq \eta_1\} < m,$$

so that $|\psi(t)| < \epsilon$. Since $|\psi(t)|$ can never reach the circle $|\psi(t)| = \epsilon$, the proof is complete. □

We need the following well-known result in our subsequent discussion (see Coppel [1]).

Lemma 3.3.1 For any $n \times m$ matrix function $A(t)$ on R_+,

(i) $\lim_{t \to \infty} \int_t^{t+1} |A(s)| ds = 0$, if and only if $\lim_{t \to \infty} e^{-t} \int_0^t e^{\lambda s} |A(s)| ds = 0$ for any $\lambda > 0$;

(ii) $\sup_{t \geq 0} \int_t^{t+1} |A(s)| ds < \infty$ if and only if $\sup_{t \geq 0} e^{-\lambda t} \int_0^t e^{\lambda s} |A(s)| ds < \infty$ for any $\lambda > 0$.

Theorem 3.3.5 *Suppose that the solutions of* (3.3.3) *are uniformly bounded* (UB) *and uniformly ultimately bounded* (UUB) *and*

(3.3.9) $$|g(t,\phi(\cdot))| \leq \delta(t) \sup_{t-r \leq s \leq t} |\phi(s)|$$

for all $t \geq 0$ *and* $\phi \in C(R_+)$ *where* $r \geq 0$ *and*

(3.3.10) $$\lim_{t \to \infty} \sup \int_t^{t+1} \delta(s)\, ds$$

is sufficiently small. Then the solutions of (3.3.5) *are* UB *and* UUB.

Proof Since the solutions of (3.3.3) are UB and UUB, it follows that the zero solution of (3.3.3) is exponentially asymptotically stable $(ExAS)$. Thus there exists a Lyapunov functional $W(t,\phi)$ satisfying the conditions $(i)-(iii)$ of Theorem 3.3.2. Let $0 < \epsilon < \lambda K^{-1}$. Then, in view of (3.3.10), there exists a $T > 0$ such that

(3.3.11) $$\int_t^{t+1} \delta(s)\, ds < \epsilon \quad \text{for all } t \geq T.$$

From Theorem 3.3.2, we obtain

$$\begin{aligned} W'_{(3.3.5)}(t,y(\cdot)) &\leq W'_{(3.3.3)}(t,y(\cdot)) + K|g(t,y(\cdot))| \\ &\leq -\lambda W(t,y(\cdot)) + K\delta(t)p(t,y(\cdot)) \\ &\leq -\lambda W(t,y(\cdot)) + K\delta(t)W(t,y(\cdot)) \\ &= [K\delta(t) - \lambda]W(t,y(\cdot)) \quad \text{for all } t \geq t_0. \end{aligned}$$

Hence we have

$$W(t,y(\cdot)) \leq W(t_0,\phi) \exp\left[\int_{t_0}^t \delta(s)\, ds - \lambda(t-t_0)\right].$$

Thus, for all $t \geq t_0$, it follows that

(3.3.12) $$|y(t,t_0,\phi)| \leq p(t,y(\cdot)) \leq K|\phi|_{t_0} \exp\left[\int_{t_0}^t \delta(s)\, ds - \lambda(t-t_0)\right].$$

For $t_0 \geq T$, from (3.3.11) and (3.3.12) we obtain

$$|y(t,t_0,\phi)| \leq K|\phi|_{t_0} e^{K\epsilon} e^{-(\lambda - K\epsilon)(t-t_0)} \quad \text{for all } t \geq t_0.$$

For $t_0 < T$, again from (3.3.11) and (3.3.12) we have

$$|y(t,t_0,\phi)| \leq K\exp\left\{K\left[\epsilon + \int_0^T \delta(s)\,ds\right]\right\}|\phi|_{t_0} e^{-(\lambda - K\epsilon)(t-t_0)}$$

for all $t \geq t_0$. Thus the solutions of (3.3.5) are UB and UUB. This completes the proof. □

Remark 3.3.3 It is easy to verify that the following examples for $g(t,\phi)$ satisfy the conditions of Theorem 3.3.5:

(i) $g(t,x) \in C(R_+ \times R^n, R^n)$ satisfying $|g(t,x)| \leq \delta(t)|x|$;
(ii) $g(t,\phi(\cdot)) = H(t)\phi(t-r)$;
(iii) $g(t,\phi(\cdot)) = \int_{t-r}^{t} H(s)\phi(s)\,ds$.

In (ii) and (iii), $r > 0$ and $H(t)$ is an $n \times n$ matrix function such that $\lim\sup_{t\to\infty} \int_t^{t+1} |H(s)|\,ds$ is sufficiently small.

Theorem 3.3.6 *Suppose that the solutions of (3.3.3) are UB and UUB and $|g(t,\phi(\cdot))| = \delta(t) \sup_{t-r \leq s \leq t} |\phi(s)|^\beta$ for all $t \geq 0$ and $\phi \in C(R_+)$ where $0 \leq \beta < 1$, $r \geq 0$ and*

(3.3.13) $$\sup_{t \geq 0} \int_t^{t+1} \delta(s)\,ds < \infty.$$

Then all the solutions of (3.3.5) are UB and UUB.

Proof Since the solutions of (3.3.3) are UB and UUB, as in Theorem 3.3.5, there exists a Lyapunov function $W(t,\phi)$ satisfying the conditions (i) − (iii) of Theorem 3.3.2. Thus we have

$$\begin{aligned}W'_{(3.3.5)}(t,y(\cdot)) &\leq -\lambda W(t,y(\cdot)) + K\delta(t)[p(t,y(\cdot))]^\beta \\ &\leq -\lambda W(t,y(\cdot)) + K\delta(t)[W(t,y(\cdot))]^\beta\end{aligned}$$

for all $t \geq t_0$.

Let $V(t,\phi(\cdot)) = [W(t,\phi(\cdot)) + e^{-\lambda t}]^{1-\beta}$. Then $V(t,\phi(\cdot))$ satisfies a local Lipschitz condition with respect to ϕ as $W(t,\phi(\cdot)) + e^{-\lambda t} > 0$. Thus we obtain

$$V'_{(3.3.5)}(t,y(\cdot)) = (1-\beta)\left[-\lambda e^{-\lambda t} + W'_{(3.3.5)}(t,y(\cdot))\right]\left[W(t,y(\cdot)) + \epsilon^{-\lambda t}\right]^{-\beta}$$

$$\leq -\lambda(1-\beta)V(t,y(\cdot)) + K(1-\beta)\delta(t).$$

Let $\sigma = \lambda(1-\beta) > 0$. Then we have

(3.3.14) $\quad V(t,y(\cdot)) \leq V(t_0,\phi)e^{-\sigma(t-t_0)} + K(1-\beta)e^{-\sigma t}\int_{t_0}^{t} e^{\sigma s}\delta(s)\,ds.$

Therefore, from Theorem 3.3.2, Lemma 3.3.1, (3.3.13) and (3.3.14), there exists an $L > 0$ such that

$$|y(t,t_0,\phi)|^{1-\beta} \leq \left[K|\phi|_{t_0} + 1\right]^{1-\beta} e^{-\sigma(t-t_0)} + K(1-\beta)L$$

for all $t \geq t_0$, and hence the solutions of (3.3.5) are UB and UUB. This completes the proof. □

Theorem 3.3.7 *Suppose that*

(3.3.15) $\quad \sup_{t \geq 0}\int_{t}^{t+1}|A(s)|\,ds < \infty, \quad \sup_{t \geq 0}\int_{0}^{t}|B(t,s)|\,ds < \infty$

and the solutions of (3.3.3) are UB and UUB. Then the solutions of (3.3.6) are UB and UUB if only if

(3.3.16) $\quad \sup_{t \geq 0}\left|e^{-t}\int_{0}^{t}e^{s}h(s)\,ds\right| < \infty.$

Proof We first show sufficiency. Let $y(t) = y(t,t_0,\phi)$ be the solutions of (3.3.6) with initial values (t_0,ϕ). Let $H(t) = e^{-t}\int_0^t e^s h(s)\,ds$ and $x(t) = y(t) - H(t)$. Then (3.3.16) yields that $|H(t)|$ is bounded for all $t \geq 0$. From (3.3.6), we have

(3.3.17) $\quad x'(t) = A(t)x(t) + \int_{0}^{t}B(t,s)x(s)\,ds + \int_{0}^{t}B(t,s)H(s)\,ds$

$\qquad\qquad + [A(t) + I]H(t).$

Let $\widehat{h}(t) = (A(t) + I)H(t) + \int_0^t B(t,s)H(s)\,ds$. Since $|H(t)|$ is bounded for all $t \geq 0$, it follows from (3.3.15) and Lemma 3.3.1 that

$$\sup_{t \geq 0} e^{-t} \int_0^t e^s |\widehat{h}(s)| \, ds < \infty.$$

Thus $\widehat{h}(t)$ satisfies the conditions of Theorem 3.3.6 for $\beta = 0$. Hence the solutions of (3.3.17) are UB and UUB. Since $y(t) = x(t) + H(t)$, the solutions of (3.3.6) are UB and UBB.

We shall now prove the necessity. From (3.3.15) and Lemma 3.3.1, there exists positive numbers L_1 and L_2 such that

(3.3.18) $$e^{-t} \int_0^t |A(s) + I| \, ds \leq L_1$$

and

(3.3.19) $$e^{-t} \int_0^t e^s \left[\int_0^s |B(s,\tau)| \, d\tau \right] ds \leq L_2 \quad \text{for all } t \geq 0.$$

Let $y(t)$ be a solution of (3.3.5) with $y(0) = 0$. Since the solutions of (3.3.5) are UB and UUB, there exists an $M > 0$ such that $|y(t)| \leq M$ for all $t \geq 0$. Since $h(t) = y'(t) - A(t)y(t) - \int_0^t B(t,s)y(s)\,ds$, it follows, by using integration by parts, that

$$\int_0^t e^s h(s)\,ds = e^t y(t) - \int_0^t [A(s) + I] e^s y(s)\,ds$$

$$- \int_0^t e^s \left[\int_0^s B(s,\tau) y(\tau)\,d\tau \right] ds.$$

From (3.3.18), (3.3.19) and the fact that $|y(t)| \leq M$, $t \geq 0$, we obtain $|e^{-t} \int_0^t e^s h(s)\,ds| \leq (1 + L_1 + L_2)M$ for all $t \geq 0$. This implies (3.3.16), and hence the proof is complete. □

Theorem 3.3.8 *Suppose that the zero solution of* (3.3.3) *is UAS and*

$$|g(t, \phi(\,\cdot\,))| \leq \delta(t) \sup_{t-r \leq s \leq t} |\phi(s)| \quad \text{for } t \geq 0$$

and $\phi \in S(\rho)$ for some $\rho > 0$, where $r \geq 0$ and $\lim_{t \to \infty} \sup \int_t^{t+1} \delta(s)\,ds$ is sufficiently small. Then the zero solution of (3.3.5) *is UAS.*

Proof Since the zero solution of (3.3.3) is UAS, it is $ExAS$. Therefore by the same argument as in the proof of Theorem 3.3.5, there exists an $M > 0$ such that $|y(t, t_0, \phi)| \le M |\phi|_{t_0} e^{-(\lambda - K\epsilon)(t-t_0)}$ for $t \ge t_0$ and $|\phi|_{t_0} < \rho/M$. Hence the zero solution of (3.3.5) is UAS. This completes the proof. □

Theorem 3.3.9 *Suppose that the zero solution of (3.3.3) is UAS and there exist $M > 0$ and $\beta > 1$ such that*

(3.3.20) $\qquad |g(t, \phi(\,\cdot\,))| \le M \sup_{t-r \le s \le t} |\phi(s)|^\beta \qquad \text{for all } t \ge 0$

and $\phi \in S(\rho)$ for some $\rho > 0$. Then the zero solution of (3.3.5) is UAS.

Proof Since the zero solution (3.3.3) is UAS, it is $ExAS$. Hence there exists a Lyapunov functional $W(t, \phi)$ satisfying the conditions $(i) - (iii)$ of Theorem 3.3.2. Let $\epsilon_0 = \lambda/K$, $\delta(\epsilon_0) = \min[(\epsilon_0/M)^{1/\beta - 1}, \rho]$. Then if $|\phi(t)| \le \delta(\epsilon)$, we have $|g(t, \phi(\,\cdot\,))| \le \epsilon_0 |\phi(t)|$. Now consider the solution $y(t)$ of (3.3.5) such that $|\phi|_{t_0} \le \delta(\epsilon_0)/K$. We assume that there exists a $t_1 \ge t_0$ such that $|y(t_1, t_0, \phi)| = \delta(\epsilon_0)$ and $|y(t, t_0, \phi)| < \delta(\epsilon_0)$ on $[t_0, t_1)$. Then, as in the proof of Theorem 3.3.5, we have

$$W'_{(3.3.5)}(t, y(\,\cdot\,)) \le W'_{(3.3.2)}(t, y(\,\cdot\,)) + K |g(t, y(\,\cdot\,))|$$

$$\le -(\lambda - K\epsilon_0) W(t, y(\,\cdot\,))$$

for all $t \in [t_0, t_1]$. Hence

$$|y(t_1, t_0, \phi)| \le K |\phi|_{t_0} e^{-(\lambda - K\epsilon_0)(t_1 - t_0)} < \delta(\epsilon_0).$$

This is a contradiction. Thus we have

$$|y(t, t_0, \phi)| \le K |\phi|_{t_0} e^{-(\lambda - K\epsilon_0)(t - t_0)} \qquad \text{for all } t \ge t_0.$$

Therefore the zero solution of (3.3.5) is UAS. This completes the proof. □

Remark 3.3.4 The growth condition (3.3.20) on $g(t, \phi)$ in Theorem 3.3.9 cannot be replaced by the following condition

(H): $|g(t, \phi(\,\cdot\,))| \le \delta(t) \sup_{t-r \le s \le t} |\phi(s)|^\beta$ for $t \ge 0$ and $\phi \in S(\rho)$ for some $\rho > 0$, where $r \ge 0$, $0 < \beta < 1$ and

$$\lim_{t \to \infty} \int_t^{t+1} \delta(s)\, ds = 0$$

as the following example shows. However, the eventual uniform asymptotic stability can be preserved.

Example 3.3.1 Consider the scalar equation

(3.3.21) $$y'(t) = -y(t) + \delta(t)y^{1/2}(t),$$

with

$$\delta(t) = \begin{cases} c & \text{if } 0 \leq t < 1, \\ -ct + 2c & \text{if } 1 \leq t < 2, \\ 0 & \text{if } t \geq 2, \end{cases}$$

where c is any positive real number. Here $n = 1$, $A(t) \equiv -1$, $B \equiv 0$ and $g(t, y) = \delta(t)y^{1/2}$ in (3.3.5). It is clear that g satisfies the hypothesis (H). Moreover, the solution $y(t, t_0, 0)$ of (3.3.21) is given by

$$y(t, t_0, 0) = e^{-t} \left[c(e^{-t/2} - e^{-t_0/2}) \right]^2$$

for $0 \leq t_0 \leq t \leq 1$. Hence the zero solution of (3.3.21) is not unique. Thus the zero solution of (3.3.21) is not US.

Theorem 3.3.10 *Suppose that the zero solution of* (3.3.3) *is UAS and* $g(t, \phi(\cdot))$ *satisfies the hypothesis* (H). *Then the zero solution of* (3.3.5) *is EvUAS.*

Proof As in the proof of Theorem 3.3.6, we obtain

$$|y(t, t_0, \phi)|^{1-\beta} \leq \left[K|\phi|_{t_0} + e^{-\lambda t_0} \right]^{1-\beta} e^{-\sigma(t-t_0)}$$

$$+ K(1-\beta)e^{-\sigma t} \int_{t_0}^{t} e^{\lambda s} \delta(s) \, ds.$$

From Lemma 3.3.1 for any $\epsilon > 0$, there exists $\alpha(\epsilon) > 0$ such that $e^{-\sigma t} \int_{t_0}^{t} e^{\sigma s} \delta(s) \, ds < \epsilon$ and $e^{-\lambda t_0} < \epsilon$ for all $t \geq t_0 \geq \alpha(\epsilon)$. Hence the zero solution of (3.3.5) is $EvUS$. It is clear that the zero solution of (3.3.5) is UA. This completes the proof. □

3.4 Method of Lyapunov Functions

In this section, we wish to employ the method of Lyapunov functions to discuss the stability properties of a general class of integro-differential equations. This important technique, which permits us to reduce the study of integro-differential systems to the study of a scalar differential equation, depends crucially on selecting appropriate minimal subsets of a suitable space of continuous functions, along which the derivative of the Lyapunov function allows a convenient estimate.

Consider the integro-differential system

(3.4.1) $\qquad x'(t) = F(t, x(t), (Tx)(t)), \quad t \geq t_0, \, x(t_0) = x_0,$

where $(Tx)(t) = \int_{t_0}^{t} g(t, s, x(s)) \, ds$, $F \in C[R_+ \times S(\rho) \times R^n, R^n]$ and $g \in C[R_+ \times R_+ \times S(\rho), R^n]$. Let us assume that $F(t, 0, 0) \equiv 0$ and $g(t, s, 0) \equiv 0$, so that the system (3.4.1) admits the zero solution. As indicated earlier, when we use Lyapunov functions, we have to seek appropriate minimal class of functions along which we can estimate the derivative of the Lyapunov functions so that the behavior of solutions of integro-differential equations can be reduced to that of scalar ordinary differential equations. To this end, let $V \in C[R_+ \times S(\rho), R_+]$, and for $x \in [R_+, R^n]$,

$$D_-V(t, x(t)) = \lim_{h \to 0^-} \inf \frac{1}{h}[V(t+h, x(t) + hF(t, x(t), (Tx)(t))) - V(t, x(t))].$$

We shall now define certain subsets of $C[R, R^n]$ along which we try to estimate $D_-V(t, x(t))$ subject to the demands we impose on the solutions of (3.4.1)

$E_\alpha\{x \in C[R_+, R^n]: V(, x(s))\alpha(s) \leq V(t, x(t))\alpha(t), \qquad t_0 \leq s \leq t\},$

$E_1\{x \in C[R_+, R^n]: V(s, x(s)) \leq V(t, x(t)), \qquad t_0 \leq s \leq t\},$

$E_0 = \{x \in C[R_+, R^n]: V(s, x(s)) \leq f(V(t, x(t)), \qquad t_1 \leq s \leq t, t_1 \geq t_0\},$

where

(i) $\alpha(t) > 0$ is a continuous function on R_+;

(ii) $f(r)$ is continuous on R_+, nondecreasing in r and $f(r) > r$ for $r > 0$.

We are now in a position to prove some basic comparison theorems that will be useful in our subsequent discussion.

Theorem 3.4.1 *Let $V \in C[R_+ \times S(\rho), R_+]$ and let $V(t,x)$ be locally Lipschitzian in x. Assume that for $t \geq t_0$ and $x \in E_1$,*

(3.4.2) $$D_-V(t,x(t)) \leq w(t, V(t,x(t))),$$

where $w \in C[R_+ \times R_+, R_+]$. Let $r(t) = r(t, t_0, u_0)$ be the maximal solution of the scalar ordinary differential equation

(3.4.3) $$u' = w(t,u), \quad u(t_0) = u_0 \geq 0,$$

existing on $t_0 \leq t < \infty$. Let $x(t) = x(t, t_0, x_0)$ be any solution of (3.4.1) such that $x(t) \in S(\rho)$ for $t \in [t_0, t_1]$ satisfying

(3.4.4) $$V(t_0, x_0) \leq u_0.$$

Then

(3.4.5) $$V(t, x(t)) \leq r(t) \quad \text{for all } t \in [t_0, t_1].$$

Proof Let $x(t) = x(t, t_0, x_0)$ be any solution of (3.4.1) such that $x(t) \in S(\rho)$ for $t \in [t_0, t_1]$. Define $m(t) = V(t, x(t))$, $t \in [t_0, t_1]$. For sufficiently small $\epsilon > 0$, consider the differential equation

$$u' = w(t,u) + \epsilon, \quad u(t_0) = u_0 + \epsilon,$$

whose solutions $u(t, \epsilon) = u(t, t_0, u_0, \epsilon)$ exist as far as $r(t)$ exists, to the right of t_0.

It follows from continuity that $\lim_{\epsilon \to 0} u(t, \epsilon) = r(t)$. Thus the desired inequality (3.4.5) is immediate if we can establish that

(3.4.6) $$m(t) < u(t, \epsilon), \quad t \in [t_0, t_1].$$

Suppose (3.4.6) is not true. Then there exists a $t_2 \in (t_0, t_1)$ such that
    (i) $m(t) \leq u(t, \epsilon), \quad t_0 \leq t \leq t_2$,
    (ii) $m(t_2) = u(t_2, \epsilon)$.
From (i) and (ii), we get

(3.4.7) $$D_-m(t_2) \geq u'(t_2, \epsilon) = w(t_2, u(t_2, \epsilon)) + \epsilon.$$

It is clear from the assumption of w that the solutions $u(t,\epsilon)$ are monotonically increasing functions of t. Consequently, it follows from (i) and (ii) and the definition of $m(t)$ that

$$V(s,x(s)) \leq V(t_2, x(t_2)), \qquad t_0 \leq s \leq t_2,$$

and hence $x(t) \in E_1$ for $t_0 \leq t \leq t_2$.

In view of the Lipschitz character of $V(t,x)$ and the assumption (3.4.2), it therefore follows for $t = t_2$ that

$$D_- m(t_2) \leq w(t_2, m(t_2)),$$

which is a contradiction to (3.4.7). Hence we have

$$m(t) < u(t,\epsilon), \qquad t \in [t_0, t_1].$$

This completes the proof. □

Corollary 3.4.1 *Let $V \in C[R_+ \times S(\rho), R^n]$ and let $V(t,x)$ be locally Lipschitzian in x. Assume that*

$$D_- V(t, x(t)) \leq 0 \quad \text{for } t > t_0 \text{ and } x \in E_0.$$

Let $x(t) = x(t, t_0, x_0)$ be any solution of (3.4.1) such that $x(t) \in S(\rho)$ for $t \in [t_0, t_1]$. Then $V(t, x(t)) \leq V(t_0, x_0)$, $t \in [t_0, t_1]$.

Proof Proceeding as in Theorem 3.4.1 with $w \equiv 0$, we obtain the inequality $V(s, x(s)) \leq V(t_2, x(t_2))$, $t_2 \in (t_0, t_1)$. Since $V(t_2, x(t_2)) = V(t_0, x_0) + \epsilon + \epsilon[t_2 - t_0] > 0$, the assumptions on $f(r)$ imply that

$$V(s, x(s)) < f(V(t, x(t))).$$

This shows that $x(t) \in E_0$ for $t_0 \leq t \leq t_2$. The rest of the proof is similar to that of Theorem 3.4.1. □

The next result is somewhat general than Theorem 3.4.1, and it is useful to ascertain the asymptotic stability of the zero solution of (3.4.1) from a simple stability of (3.4.3).

Theorem 3.4.2 *Assume that the hypotheses of Theorem 3.4.1 hold except that the inequality (3.4.2) is replaced by*

(3.4.8) $\qquad \alpha(t) D_- V(t, x(t)) + V(t, x(t)) D_- \alpha(t) \leq w(t, V(t, x(t))\alpha(t))$

for $t > t_0$ and $x \in E_\alpha$, where $\alpha(t) > 0$, is continuous on R_+ and

$$D_-\alpha(t) = \lim_{h \to 0}\inf h^{-1}[\alpha(t+h) - \alpha(t)].$$

Then $\alpha(t_0)V(t_0, x_0) \leq u_0$ implies the estimate $\alpha(t)V(t, x(t)) \leq r(t)$, $t \geq t_0$.

Proof We set $L(t, x(t)) = \alpha(t)V(t, x(t))$. Let $t > t_0$ and $x \in E_\alpha$. For sufficiently small $h < 0$, we have

$$L(t+h, x(t) + hF(t, x(t), (Tx)(t)) - L(t, x(t))$$
$$= V(t+h, x(t) + hF(t, x(t), (Tx)(t)))[\alpha(t+h) - \alpha(t)]$$
$$+ \alpha(t)[V(t+h, x(t) + hF(t, x(t), (Tx)(t)) - V(t, x(t))],$$

and therefore, in view of the assumption (3.4.8), it follows that $D_-L(t, x(t)) \leq w(t, L(t, x(t)))$ for $t \in [t_0, t_1]$, $x \in E_1$, where E_1, in this case, is to be defined with $L(t, x)$ replacing $V(t, x)$ in the definition of the set E_1. It is clear that $L(t, x)$ is locally Lipschitzian in x, and thus all the assumptions of Theorem 3.4.1 are satisfied, with $L(t, x)$ in place of $V(t, x)$. Hence the conclusion of the theorem follows from the proof of Theorem 3.4.1. □

We shall now give sufficient conditions for the stability and asymptotic stability of the zero solution of (3.4.1).

Theorem 3.4.3 *Assume that there exist functions $V(t, x)$ and $w(t, x)$ satisfying the following conditions:*

(i) $w \in C[R_+ \times R_+, R_+]$ and $w(t, 0) \equiv 0$;

(ii) $V \in C[R_+ \times S(\rho), R_+]$, and $V(t, 0) \equiv 0$, $V(t, x)$ is positive definite and locally Lipschitzian in x;

(iii) *for $t > t_0$ and $x \in E_1$*

$$D_-V(t, x(t)) \leq w(t, V(t, x(t))).$$

Then stability of the zero solution of (3.4.3) implies stability of the zero solution of (3.4.1).

Proof Let $0 < \epsilon < \rho$ and $t_0 \in R_+$ be given. Suppose that the zero solution of (3.4.3) is stable. Then, given $b(\epsilon) > 0$, $t_0 \in R_+$, there exists a $\delta = \delta(t_0, \epsilon) > 0$ such that whenever $u_0 \leq \delta$, we have

(3.4.9) $$u(t) < b(\epsilon), \quad t \geq t_0,$$

where $u(t) = u(t, t_0, u_0)$ is any solution of (3.4.3). From the positive definiteness of $V(t, x)$, we have

(3.4.10) $$b(|x|) \leq V(t, x), \quad (t, x) \in R_+ \times S(\rho), \, b \in \mathcal{K}.$$

Choose $u_0 = V(t_0, x_0)$. Since $V(t, x)$ is continuous and $V(t, 0) \equiv 0$, it is possible to find a positive function $\delta_1 = \delta_1(t_0, \epsilon) > 0$ such that $|x_0| \leq \delta_1$ and $V(t_0, x_0) \leq \delta$ hold simultaneously. We claim that if $|x_0| \leq \delta_1$ then $|x(t)| < \epsilon$ for all $t \geq t_0$. Suppose this is not true. Then there exists a solution $x(t) = x(t, t_0, x_0)$ of (3.4.1) satisfying the properties $|x(t_2)| = \epsilon$ and $|x(t)| \leq \epsilon$ for $t_0 \leq t \leq t_2$, $t_2 \in (t_0, t_1)$. This implies from (3.4.10) that

(3.4.11) $$V(t_2, x(t_2)) \geq b(\epsilon).$$

Furthermore, $x(t) \in S(\rho)$ for $t \in [t_0, t_2]$. Hence the choice of $u_0 = V(t_0, x_0)$ and the condition (iii) give, as a consequence of Theorem 3.4.1, the estimate

(3.4.12) $$V(t, x(t)) \leq r(t), \quad t \in [t_0, t_2],$$

where $r(t) = r(t, t_0, u_0)$ is the maximal solution of (3.4.3). Now the relations (3.4.9), (3.4.11) and (3.4.12) lead to the contradiction $b(\epsilon) \leq V(t_2, x(t_2)) \leq r(t_2) < b(\epsilon)$. The proof is therefore complete. □

The next result is concerned with asymptotic stability of the zero solution of (3.4.1), and depends basically on Theorem 3.4.2.

Theorem 3.4.4 *Assume that there exists functions $V(t, x)$, $w(t, u)$ and $\alpha(t)$ satisfying the following properties:*
   (i) $\alpha(t) > 0$ is continuous for $t \in R_+$ and $\alpha(t) \to \infty$ as $t \to \infty$;
   (ii) $w \in C[R_+ \times R_+, R_+]$ and $w(t, 0) \equiv 0$;
   (iii) $V \in C[R_+ \times S(\rho), R_+]$, $V(t, 0) \equiv 0$ and $V(t, x)$ is positive definite and locally Lipschitzian in x;
   (iv) $\alpha(t) D_- V(t, x(t)) + V(t, x(t)) D_- \alpha(t) \leq w(t, V(t, x(t)) \alpha(t))$ for $t > t_0$ and $x \in E_\alpha$.

Then if the zero solution of (3.4.3) is stable, the zero solution of (3.4.1) is asymptotically stable.

LYAPUNOV STABILITY

Proof Let $0 < \epsilon < \rho$ and $t_0 \in J$ be given. Let $\alpha_0 = \min_{t \in R_+} \alpha(t)$. By the assumption (i), we have $\alpha_0 > 0$. Since $V(t,x)$ is positive definite, there exists a function $b \in \mathcal{K}$ such that

(3.4.13) $$b(|x|) \leq V(t,x) \text{ for } (t,x) \in R_+ \times S(\rho).$$

Define $\epsilon_1 = \alpha_0 b(\epsilon)$. From the definition of stability of the zero solution of (3.4.3), it follows that for a given $\epsilon_1 > 0$, $t_0 \in R_+$, there is a $\delta = \delta(t_0, \epsilon_1) > 0$ such that $u_0 \leq \delta$ implies $u(t) < \epsilon_1$ for all $t \geq t_0$, where $u(t) = u(t, t_0, u_0)$ is any solution of (3.4.3). Choose $u_0 = \alpha(t_0)V(t_0, x_0)$. Then proceeding as in the proof of Theorem 3.4.3 with ϵ_1 instead of $b(\epsilon)$, it is easy to prove that the zero solution of (3.4.1) is stable. Now, let $x(t) = x(t, t_0, x_0)$ be any solution of (3.4.1) such that $|x_0| \leq \delta_0$, where $\delta_0 = \delta(t_0, \frac{1}{2}\rho)$. Then it follows from stability that $|x(t)| < \frac{1}{2}\rho$ for $t \geq t_0$. Since $\alpha(t) \to \infty$, as $t \to \infty$, there exists a number $T = T(t_0, \epsilon) > 0$ such that

(3.4.14) $$b(\epsilon)\alpha(t) > \epsilon \text{ for all } t \geq t_0 + T.$$

Now, Theorem 3.4.2 and the relation (3.4.13) yield the inequality

(3.4.15) $$\alpha(t)b(|x(t)|) \leq \alpha(t)V(t, x(t)) \leq r(t), \quad t \geq t_0,$$

where $x(t)$ is any solution of (3.4.1) such that $|x_0| \leq \delta_0$. If there exists a sequence $\{t_k\}$, $t_k \geq t_0 + T$ and $t_k \to \infty$ as $k \to \infty$ such that $|x(t_k)| \geq \epsilon$, for some solution $x(t)$ satisfying $|x_0| \leq \delta_0$, then we obtain from the inequality (3.4.15) $\alpha(t_k)b(\epsilon) < \epsilon_1$ which is a contradiction to (3.4.14). Thus it is clear that the zero solution of (3.4.1) is asymptotically stable and the proof is complete. □

We shall now give sufficient conditions for the uniform asymptotic stability of the zero solution of (3.4.1).

Theorem 3.4.5 *Assume that there exists a function $V(t,x)$ satisfying the following properties:*
  (i) $V \in C[R_+ \times S(\rho), R_+]$ and $V(t,x)$ is positive definite, decrescent and locally Lipschitzian in x;
  (ii) $D_- V(t, x(t)) \leq -\phi(|x(t)|)$ for $t > t_0$ and $x \in E_0$, where $\phi \in \mathcal{K}$.

Then the zero solution of (3.4.1) is uniformly asymptotically stable.

Proof Since $V(t,x)$ is positive definite and decrescent, there exists functions $a, b \in \mathcal{K}$ such that

(3.4.16) $\qquad b(|x|) \leq V(t,x) \leq a(|x|), \qquad (t,x) \in R_+ \times S(\rho).$

Let $0 < \epsilon < \rho$ and $t_0 \in R_+$ be given. Choose $\delta = \delta(\epsilon) > 0$ such that

(3.4.17) $\qquad\qquad\qquad a(\delta) < b(\epsilon).$

Now we claim that if $|x_0| \leq \delta$ then $|x(t)| < \epsilon$ for all $t \geq t_0$, where $x(t) = x(t, t_0, x_0)$ is any solution of (3.4.1). Suppose this is not true. Then there exists a solution $x(t)$ of (3.4.1) with $|x_0| \leq \delta$ and $t_2 > t_0$ such that $|x(t_2)| = \epsilon$ and $|x(t)| \leq \epsilon$ for $t \in [t_0, t_2]$. Thus, in view of (3.4.16), we obtain

(3.4.18) $\qquad\qquad\qquad V(t_2, x(t_2)) \geq b(\epsilon).$

Furthermore, it is clear that $x(t) \in S(\rho)$ for $t \in [t_0, t_2]$. Hence the choice $u_0 = V(t_0, x_0)$ and the condition $D_- V(t, x(t)) \leq 0$ for $t > t_0$, $x \in E_0$ give, because of Corollary 3.4.1, the estimate

(3.4.19) $\qquad\qquad V(t, x(t)) \leq V(t_0, x_0), \quad t \in [t_0, t_2].$

Now the relations (3.4.16) − (3.4.19) lead to the contradiction

$$b(\epsilon) \leq V(t_2, x(t_2)) \leq V(t_0, x_0) \leq a(|x_0|) \leq a(\delta) < b(\epsilon).$$

This proves the uniform stability. Now let $x(t) = x(t, t_0, x_0)$ be any solution of (3.4.1) such that $|x_0| \leq \delta_0$, where $\delta_0 = \delta(\frac{1}{2}\rho)$, δ being the same as before. It then follows from uniform stability that $|x(t)| < \frac{1}{2}\rho$ for $t \geq t_0$, and hence $x(t) \in S(\rho)$ for all $t \geq t_0$. Let $0 < \eta < \delta_0$ be given. Clearly, we have $b(\epsilon) \leq a(\delta_0)$. In view of the assumptions on $f(r)$, which occurs in the definition of E_0, it is possible to find a $\beta = \beta(\eta) > 0$ such that

(3.4.20) $\qquad\qquad f(r) > r + \beta \quad \text{if } b(\eta) \leq r \leq a(\delta_0).$

Furthermore, there exists a positive integer $N = N(\eta)$ satisfying the inequality

(3.4.21) $\qquad\qquad\qquad b(\eta) + N\beta > a(\delta_0).$

If, for some $t \geq t_0$, we have $V(t, x(t)) \geq b(\eta)$, it follows from (3.4.16) that there exists a $\delta_2 = \delta_2(\eta) > 0$ such that $|x(t)| \geq \delta_2$. This in turn implies that

(3.4.22) $$\phi(|x(t)|) \geq \phi(\delta_2) = \delta_3.$$

Obviously, δ_3 depends on η. With the positive integer chosen previously, let us construct $N+1$ numbers $t_k = t_k(t_0, \eta)$ such that $t_0(t_0, \eta) = t_0$, $t_{k+1}(t_0, \eta) = t_k(t_0, \eta) + \beta/\delta_0$. It then turns out that $t_k(t_0, \eta) = t_0 + k\beta/\delta_3$, and consequently, letting $T(\eta) = N(\beta/\delta_3)$, we have

$$t_k(t_0, \eta) = t_0 + T(\eta).$$

Now to prove uniform asymptotic stability, we still have to prove that $|x(t)| < \eta$ for all $t \geq t_0 + T(\eta)$. It is therefore sufficient to show that

(3.4.23) $\quad V(t, x(t)) < b(\eta) + (N-k)\beta, \quad t \geq t_k, \quad$ for $k = 0, 1, 2, \ldots, N$.

For $k = 0$, this result follows from the first part of the proof and the choice of N. We now wish to prove the desired inequality (3.4.23) by the method of induction. Suppose that, for some k, we have

(3.4.24) $$V(s, x(s)) < b(\eta) + (N-k)\beta, \quad s \geq t_k$$

and, if possible,

$$V(t, x(t)) \geq b(\eta) + (N-k-1)\beta, \quad t \in [t_k, t_{k+1}].$$

It then follows that $b(\eta) \leq V(t, x(t)) \leq a(\delta_0)$, $t \in [t_k, t_{k+1}]$, and therefore we derive from (3.4.20) that

$$f(V(t, x(t)) > V(t, x(t)) + \beta$$
$$\geq b(\eta) + (N-k)\beta > V(s, x(s))$$

for $t_k \leq s \leq t$, $t \in [t_k, t_{k+1}]$. This implies that $x(t) \in E_0$ for $t_k \leq s \leq t$, $t \in [t_k, t_{k+1}]$, and hence we obtain from the assumption (ii) and (3.4.24) that

$$V(t_{k+1}, x(t_{k+1})) \leq V(t_k, x(t_k)) - \int_{t_k}^{t_{k+1}} \phi(|x(s)|)\, ds$$
$$< b(\eta) + (N-k)\beta - \delta_3(t_{k+1} - t_k)$$
$$< b(\eta) + (N-k-1)\beta < V(t_k, x(t_k)).$$

This absurdity shows that there exists a $t^* \in [t_k, t_{k+1}]$ such that

(3.4.25) $$V(t^*, x(t^*)) < b(\eta) + (N-k+1)\beta.$$

Next we shall show that (3.4.25) implies

$$V(t, x(t)) < b(\eta) + (N - k + 1)\beta, \quad t \geq t^*.$$

If this is not true, there exists a $t_1 > t^*$ such that

$$V(t_1, x(t_1)) = b(\eta) + (N - k - 1)\beta,$$

and, for small $h > 0$,

$$V(t_1 + h, x(t_1 + h)) < b(\eta) + (N - k - 1)\beta.$$

Then we have

(3.4.26) $$D_- V(t_1, x(t_1)) \geq 0.$$

Also, arguing as before, we show that $x(t) \in E_0$ for $t^* \leq s \leq t_1$, and so $D_- V(t_1, x(t_1)) \leq -\delta_3 < 0$. This contradicts (3.4.26), and hence

$$V(t, x(t)) < b(\eta) + (N - k - 1)\beta, \quad t \geq t_{k+1}.$$

The proof of the theorem is now complete. □

Example 3.4.1 Consider the scalar integro-differential equation

(3.4.27) $$x'(t) = -ax(t) + \int_{t_0}^{t} k(t,s)x(s)\,ds, \quad a > 0,$$

where $k \in C[R_+ \times R_+, R_+]$. Take $L(t,x) = \alpha(t)V(t,x) = e^{\delta t}x^2$, $\delta > 0$. Then the set E_α is given by

$$E_\alpha = \{x \in C[R_+, R] : x^2(s)e^{\delta s} \leq x^2(t)e^{\delta t}, t \geq t_0\}.$$

Hence

$$D_- L(t, x(t)) = \delta e^{\delta t}x^2(t) + 2x(t)\left[-ax(t) + \int_{t_0}^{t} k(t,s)x(s)\,ds\right]e^{\delta t}$$

$$\leq L(t, x(t))\left[\delta - 2a + 2\int_{t_0}^{t} k(t,s)e^{(\delta/2)(t-s)}\,ds\right].$$

We wish to apply Theorem 3.4.4 with $w \equiv 0$, which implies that

(3.4.28) $$\int_{t_0}^{t} k(t,s)e^{(\delta/2)(t-s)}\,ds \leq \frac{2a - \delta}{2}.$$

This shows, from Theorem 3.4.4, that the zero solution of (3.4.27) is exponentially asymptotically stable. Since δ is arbitrary, on letting $\delta \to 0$, the condition (3.4.28) reduces to

$$\int_{t_0}^{t} k(t,s)\,ds \leq a,$$

which is a sufficient condition for uniform stability of the zero solution of (3.4.28) by Theorem 3.4.3.

Finally, we state the following general result, which offers various stability criteria in a single set-up.

Theorem 3.4.6 *Assume that there exists a function $V(t,x)$ satisfying the following properties:*

(i) $V \in C[R_+ \times S(\rho), R_+]$, $V(t,x)$ *is locally Lipschitzian in x, and* $b(|x|) \leq V(t,x) \leq a(|x|)$, $a,b \in \mathcal{K}$, $(t,x) \in R_+ \times S(\rho)$;

(ii) w_0 *and* $w \in C[R_+ \times R_+, R]$, $w_0(t,u) \leq w(t,u)$, $\eta(t,t^0,v_0)$ *is the left maximal solution of*

$$v' = w_0(t,v), \quad v(t^0) = v_0 \geq 0,$$

existing on $t_0 \leq t \leq t^0$, and $r(t,t_0,u_0)$ is the right maximal solution of (3.4.3) existing on $[t_0, \infty)$;

(iii) $D_- V(t,x(t)) \leq w(t, V(t,x(t)))$ *on Ω, where $\Omega = \{x \in C[R_+, R^n]: V(s,x(s)) \leq \eta(s,t,V(t,x(t))), t_0 \leq s \leq t\}$.*

Then the stability properties of the zero solution of (3.4.3) imply the corresponding stability properties of the zero solution of (3.4.1).

In order to prove Theorem 3.4.6, it is enough to obtain the estimate

(3.4.29) $\qquad V(t, x(t,t_0,x_0)) \leq r(t,t_0,V(t_0,x_0)), \quad t > t_0,$

where $x(t) = x(t,t_0,x_0)$ is any solution of (3.4.1) and $r(t,t_0,u_0)$ is the maximal solution of (3.4.3). Then using the condition (i) and the stability properties of the trivial solution of (3.4.3), one can prove the corresponding stability properties of the trivial solution (3.4.1) by the standard arguments. Hence we shall show (3.4.29). For this purpose, we need the following lemma.

Lemma 3.4.1 Let $w_0, w \in C[R_+ \times R_+, R]$ satisfy

(3.4.30) $$w_0(t, u) \leq w(t, u), \quad (t, u) \in R_+ \times R_+.$$

Then the right maximal solution $r(t, t_0, u_0)$ of (3.4.3) and the left maximal solution $\eta(t, T, v_0)$ of

(3.4.31) $$u' = w_0(t, u), \quad u(T) = v_0 \geq 0,$$

satisfy the relation

(3.4.32) $$r(t, t_0, u_0) \leq \eta(t, T, v_0), \quad t \in [t_0, T]$$

whenever

(3.4.33) $$r(T, T_0, u_0) \leq v_0.$$

Proof It is known that

(3.4.34) $$\lim_{\epsilon \to 0} u(t, \epsilon) = r(t, t_0, u_0),$$

and

(3.4.35) $$\lim_{\epsilon \to 0} v(t, \epsilon) = \eta(t, T, v_0),$$

where $u(t, \epsilon)$ is a solution of

(3.4.36) $$u' = w(t, u) + \epsilon, \quad u(t_0) = u_0 + \epsilon,$$

existing to the right of t_0 and $v(t, \epsilon)$ is a solution of

(3.4.37) $$u' = w_0(t, u) - \epsilon, \quad u(T) = v_0,$$

existing to the left of T, where $\epsilon > 0$ is sufficiently small. Note that the assertion (3.4.32) follows from (3.4.34) and (3.4.35) if we establish the inequality

(3.4.38) $$u(t, \epsilon) < v(t, \epsilon), \quad t_0 \leq t < T.$$

By using (3.4.30) and (3.4.33), it can be seen that for a sufficiently small $\delta > 0$, we have $u(t, \epsilon) < v(t, \epsilon)$, $T - \delta \leq t < T$, and in particular $u(T - \delta, \epsilon) < v(T - \delta, \epsilon)$.

We claim

(3.4.39) $$u(t, \epsilon) < v(t, \epsilon), \, t_0 \leq t \leq T - \delta.$$

If this is not true, there exists a $t^* \in [t_0, T-\delta)$ such that

(3.4.40) $$u(t,\epsilon) < v(t,\epsilon), \qquad t^* < t \leq T-\delta$$

and

(3.4.41) $$u(t^*,\epsilon) = v(t^*,\epsilon).$$

Now the relations (3.4.40), (3.4.41), (3.4.36), (3.4.37) and (3.4.30) lead to the contradiction

$$w(t^*, u(t^*,\epsilon)) + \epsilon = u'(t^*,\epsilon) \leq v'(t^*,\epsilon) = w_0(t^*, v(t^*,\epsilon)) - \epsilon.$$

Hence (3.4.39) holds for any sufficiently small $\delta > 0$. Consequently, inequality (3.4.38) is established and the proof is complete. $\square$

To prove (3.4.29), we set $m(t) = V(t, x(t,t_0,x_0))$, $t \geq t_0$ so that $m(t_0) \leq u_0$. Because $r(t,t_0,u_0) = \lim_{\epsilon \to 0^+} u(t,\epsilon)$ where $u(t,\epsilon)$ is any solution of

$$u' = w(t,u) + \epsilon, \qquad u(t_0) = u_0 + \epsilon,$$

for sufficiently small $\epsilon > 0$, it is enough to prove that $m(t) < u(t,\epsilon)$ for $t \geq t_0$. If this is not true, there exists a $t_1 > t_0$ such that

$$m(t_1) = \epsilon, \qquad m(t) < u(t,\epsilon), \qquad t_0 \leq t < t_1.$$

This implies that

(3.4.42) $$D_- m(t_1) \geq u'(t_1,\epsilon) = w(t_1, m(t_1)) + \epsilon.$$

Consider now the left maximal solution $\eta(s, t_1, m(t_1))$ of (3.4.31) with $v(t_1) = m(t_1)$ on the interval $t_0 \leq s \leq t_1$. By Lemma 3.4.1, we have $r(s,t_1,u_0) \leq \eta(s,t_1,m(t_1))$, $s \in [t_0,t_1]$. Since $r(t_1,t_0,u_0) = \lim_{\epsilon \to 0^+} u(t_1,\epsilon) = m(t_1) = \eta(t_1,t_1,m(t_1))$ and $m(s) \leq u(s,\epsilon)$, for $t_0 \leq s \leq t_1$, it follows that

$$m(s) \leq r(s,t_0,u_0) \leq \eta(s,t_1,m(t_1)), \qquad t_0 \leq s \leq t_1.$$

This inequality implies that (iii) holds for $x(s,t_0,x_0)$ on $t_0 \leq s \leq t_1$, and, as a result, standard computation yields

$$D_- m(t_1) \leq w(t_1, m(t_1)),$$

which contradicts the relation (3.4.42). Thus $m(t) \leq r(t,t_0,u_0)$, $t \geq t_0$ and the proof is complete. $\square$

We shall now show how Theorem 3.4.6 unifies various stability results discussed early. For this purpose, the following special cases are important.

(a) Suppose $w_0(t,u) \equiv 0$. Then $\eta(s,t^0,v_0) = v_0$, and hence Ω reduces to E_1.

(b) Suppose $w_0(t,u) = -[\alpha'(t)/\alpha(t)]u$ where $\alpha(t) > 0$ is continuously differentiable on R_+ and $\alpha(t) \to \infty$ as $t \to \infty$. Let $w(t,u) = w_0(t,u) + [1/\alpha(t)]w_1(t,\alpha(t)u)$, with $w_1 \in C[R_+ \times R_+, R_+]$, then $\eta(s,t^0,v_0) = v_0[\alpha(t^0)/\alpha(s)]$. Thus $\Omega = E_\alpha$.

(c) Let $w_0 = w = -c(u)$, $c \in \mathcal{K}$. Then it is easy to show that $\eta(s,t^0,v_0) = \phi^{-1}[\phi(v_0) - (s - t^0)]$, $t_0 \leq s \leq t^0$ where $\phi(u) = \phi(u_0) + \int_{u_0}^{u}[ds/c(s)]$ and ϕ^{-1} is the inverse function of ϕ. Since $\eta(s,t^0,v_0)$ is increasing in s to the left of t^0, on choosing a fixed $s_0 \leq t^0$ and defining $f(u) = \eta(s_0, t^0, v_0)$, it is clear that $f(u) > u$ for $u > 0$. Thus $f(u)$ is continuous and increasing in u. Hence $\Omega = E_0$.

(d) Let $w_0 = w = -\delta u$, $\delta > 0$. Then $\eta(s,t^0,v_0) = v_0 e^{\delta(t-s)}$, and hence $\Omega = \{x \in C(R_+, R): V(s, x(s)) \leq V(t, x(t))e^{\delta(t-s)}, t_0 \leq s \leq t\}$, and this is the subset we used in Example 3.4.1.

Remark 3.4.1 The foregoing discussion clearly demonstrates that the hypothesis (ii) in Theorem 3.4.6 is instrumental in unifying the selection of minimal classes that are required to estimate the derivative of the Lyapunov function and extract various stability results.

Remark 3.4.2 In the entire discussion of 3.4, we considered the system (3.4.1) with initial condition that, for simplicity, is a point $x(t_0) = x_0 \in R^n$, and thus all the solutions of (3.4.1) start at $t = t_0$. However, it is always possible to consider the system

(3.4.43) $$x'(t) = F(t, x(t), (Tx)(t))$$

where $(Tx)(t) = \int_0^t g(t,s,x(s))\,ds$ with initial value (t_0, ϕ), $t_0 \geq 0$, $\phi:[0,t_0] \to R^n$ and ϕ is a continuous function, and obtain various stability results similar to that of Theorems 3.4.3 − 3.4.6 with obvious modifications.

3.5 Lyapunov Functions on Product Spaces

If we examine the Lyapunov functionals constructed for all the examples that have been discussed, we find that we have inadvertently employed a combination of a Lyapunov function and a functional in such a way that the corresponding derivative can be estimated suitably without demanding a minimal class of functions or prior knowledge of solutions. This observation leads us to consider the method of Lyapunov functions on product spaces to discuss stability and boundedness properties of solutions of integro-differential systems.

Let $x_t(\cdot) \in C\{[0,t], R^n\}$. If $x \in C\{R_+, R^n\} = C(R_+)$ then for each $t \in R_+$, $x_t(\cdot)$ is the restriction of $x(s)$ given by $x_t(s) = x(s)$, $0 \leq s \leq t$, and the norm is defined by

$$|x_t(\cdot)| = \sup_{0 \leq s \leq t} |x(s)|.$$

It is clear that $|x(t)| \leq |x_t(\cdot)|$.

Consider the system of equations of the form

(3.5.1) $$x'(t) = F(t, x(t), x_t(\cdot)), \quad t \in R_+$$

where $F \in C[R_+ \times R^n \times C(R_+), R^n]$ and $F(t, 0, 0) \equiv 0$. Let $x(t) = x(t, t_0, x_{t_0}(\cdot))$ be a solution of (3.5.1) with initial values $(t_0, x_{t_0}(\cdot))$ existing for all $t \geq t_0 \geq 0$. A special case of (3.5.1) is the Volterra integro-differential equations

(3.5.2) $$x'(t) = f(t, x(t)) + \int_0^t g(t, s, x(s))\, ds, \quad t \in R_+$$

where $f \in C[R_+ \times R^n, R^n]$, $g \in C[R_+ \times R_+ \times C(R_+), R^n]$, $f(t, 0) \equiv 0$ and $g(t, s, 0) \equiv 0$.

We wish to employ Lyapunov functions on the product space $R^n \times C(R_+)$ and develop the stability theory for the system (3.5.1).

If $V \in C[R_+ \times R^n \times C(R_+), R^n]$ then we define

(3.5.3) $D^+V(t, x, x_t(\cdot))$

$$= \lim_{h \to 0^+} \sup \frac{1}{h}[V(t+h, x + hF(t, x, x_t(\cdot)), x_{t+h}(\cdot)) - V(t, x, x_t(\cdot))].$$

Also, if we assume that $V(t,x,x_t(\cdot))$ is locally Lipschitzian in x and $x(t) = x(t,t_0,x_{t_0}(\cdot))$ is a solution of (3.5.1) then (3.5.3) is equivalent to

$$(3.5.4) \quad D^+V_{(3.5.1)}(t,x(t),x_t(\cdot))$$

$$= \lim_{h \to 0^+} \sup \frac{1}{h}[V(t+h,x(t+h),x_{t+h}(\cdot)) - V(t,x(t),x_t(\cdot))].$$

We shall now give sufficient conditions guaranteeing uniform asymptotic stability of zero solution of (3.5.1).

Theorem 3.5.1 *Assume that there exists a function $V(t,x,x_t(\cdot)) \in C[R_+ \times S(\rho) \times C(R_+), R^+]$ such that*
   (i) $a(|x(t)|) \leq V(t,x,x_t(\cdot)) \leq b(|x_t(\cdot)|)$ *where* $a,b \in \mathcal{K}$;
   (ii) $D^+V_{(3.5.1)}(t,x(t),x_t(\cdot)) \leq 0$.
Then the zero solution of (3.5.1) is uniformly stable.

Proof Let $0 < \epsilon < \rho$ be given. Choose $\delta = \delta(\epsilon) > 0$ such that

$$(3.5.5) \quad b(\delta) < a(\epsilon).$$

Let $x(t,t_0,x_{t_0}(\cdot))$ be a solution of (3.5.1) with initial values $(t_0,x_{t_0}(\cdot))$ existing for all $t \geq t_0 \geq 0$. We claim that the zero solution of (3.5.1) is uniformly stable. If this is false then there exists a $t_1 > t_0$ such that $|x_{t_0}(\cdot)| \leq \delta$ and

$$(3.5.6) \quad |x(t_1,t_0,x_{t_0}(\cdot))| = \epsilon.$$

It follows from (ii) that $V(t,x(t),x_t(\cdot)) \leq V(t_0,x(t_0),x_{t_0}(\cdot))$. Hence the condition (i) and the relations (3.5.5) and (3.5.6) lead to

$$a(\epsilon) \leq a(|x(t_1)|) \leq V(t_1,x(t_1),x_{t_1}(\cdot)) \leq V(t_0,x(t_0),x_{t_0}(\cdot))$$
$$\leq b(|x_{t_0}(\cdot)|) \leq b(\delta) < a(\epsilon),$$

which is a contradiction, and this completes the proof. □

Theorem 3.5.2 *Suppose that there exists a function $V(t,x,x_t(\cdot)) \in C[R_+ \times S(\rho) \times C(R_+), R_+]$ such that*
   (i) $a(|x(t)|) \leq V(t,x,x_t(\cdot)) \leq b(|x_t(\cdot)|)$, *where* $a,b \in \mathcal{K}$;
   (ii) $D^+V_{(3.5.1)}(t,x(t),x_t(\cdot)) \leq -c(|x_t(\cdot)|)$, *where* $c \in \mathcal{K}$.
Then the zero solution of (3.5.1) is uniformly asymptotically stable.

Proof By Theorem 3.5.1, the zero solution of (3.5.1) is uniformly stable. Thus, taking $\epsilon = \rho$, $t_0 \in R_+$, there exists a $\delta_0 = \delta(\rho) > 0$ such that $|x_{t_0}(\cdot)| \leq \delta_0$ implies $|x(t, t_0, x_{t_0}(\cdot))| < \rho$ for all $t \geq t_0$. From the assumption (ii), we have

$$(3.5.7) \qquad V(t, x(t), x_t(\cdot)) \leq V(t_0, x(t_0), x_{t_0}(\cdot)) - \int_{t_0}^{t} c(|x_s(\cdot)|)\,ds.$$

Now let $0 < \eta < \rho$ and $T(\eta) = b(\delta_0)/c(\delta(\eta))$, where $\delta(\eta)$ corresponds to η in uniform stability. We claim that every solution $x(t, t_0, x_{t_0}(\cdot))$ of (3.5.1) with $|x_{t_0}(\cdot)| \leq \delta_0$ satisfy the relation

$$(3.5.8) \qquad |x_t(\cdot)| < \delta(\eta) \quad \text{for some } t_1 \in [t_0, t_0 + T],$$

and hence it follows from uniform stability of the zero solution of (3.5.1) that

$$|x(t)| < \eta \quad \text{for all } t \geq t_1.$$

Suppose (3.5.8) is not true. Then we have

$$(3.5.9) \qquad |x_t(\cdot)| \geq \delta(\eta) \quad \text{for all } t \in [t_0, t_0 + T]$$

and $t_2 > t_1$, $t_1 \in [t_0, t_0 + T]$ such that

$$|x(t_2)| = \eta.$$

Hence for $t_1 = t_0 + T$, $t_2 > t_1$ and $t_2 - t_0 > T$, the assumption (i) and the relations (3.5.7) and (3.5.9) lead to

$$0 < a(\eta) = a(|x(t_2)|) \leq V(t_2, x(t_2), x_{t_2}(\cdot))$$

$$\leq V(t_0, x(t_0), x_{t_0}(\cdot)) - \int_{t_0}^{t_2} c(|x_s(\cdot)|)\,ds$$

$$\leq V(t_0, x(t_0), x_{t_0}(\cdot)) - \int_{t_0}^{t_0+T} c(|x_s(\cdot)|)\,ds,$$

$$\leq b(\delta_0) - c(\delta(\eta))T = 0$$

which is a contradiction to (3.5.9). Thus at some t_1 such that $t_0 \leq t_1 \leq t_0 + T$, we must have from (3.5.8) that

$$|x_{t_1}(\cdot)| < \delta(\eta).$$

Therefore, the uniform stability of the zero solution of (3.5.1) implies

$$|x(t, t_0, x_{t_0}(\cdot))| < \eta \qquad \text{for all } t \geq t_1$$

and in particular for $t \geq t_0 + T$. Hence we have

$$|x(t, t_0, x_{t_0}(\cdot))| < \eta \qquad \text{for all } t \geq t_0 + T$$

whenever $|x_{t_0}(\cdot)| \leq \delta_0$. This shows that the zero solution of (3.5.1) is uniformly attracting, and hence the proof is complete.

Define

$$\Gamma = Ch \in C[R_+ \times R^n, R_+]: \inf_{x \in R^n} h(t, x) = 0 \quad \text{for each } t \in R_+].$$

For $h, h^0 \in \Gamma$ and $x_t(\cdot) \in C(R_+)$, let

(3.5.10) $$h_0(t, x_t(\cdot)) = \sup_{0 \leq s \leq t} h^0(s, x(s)),$$

$$\tilde{h}(t, x_t(\cdot)) = \sup_{0 \geq s \geq t} h(s, x(s)).$$

It is clear that

$$h^0(t, x(t)) \leq h_0(t, x_t(\cdot)) \quad \text{and} \quad h(t, x(t)) \leq \tilde{h}(t, x_t(\cdot)).$$

Definition 3.5.1 Let $h, h^0 \in \Gamma$ and let $\tilde{h}$ and h_0 be defined by (3.5.10). Then h_0 is said to be finer than $\tilde{h}$ if there exists a $\lambda > 0$ and $\phi \in \mathcal{K}$ such that

$$h_0(t, x_t(\cdot)) < \lambda \quad \text{implies} \quad \tilde{h}(t, x_t(\cdot)) \leq \phi(h_0(t, x_t(\cdot))).$$

Definition 3.5.2 Let $V \in C[R_+ \times R^n \times C(R_+), R_+]$. Then V is said to be

(i) h-positive definite if there exist a $\rho > 0$ and $a \in \mathcal{K}$ such that $h(t, x) < \rho$ implies $a(h(t, x)) \leq V(t, x, x_t(\cdot))$;

(ii) h_0-decrescent if there exist $\rho_0 > 0$ and $b \in \mathcal{K}$ such that $h_0(t, x) < \rho_0$ implies $V(t, x, x_t(\cdot)) \leq b(h_0(t, x_t(\cdot)))$.

Let $x(t) = x(t, t_0, x_{t_0}(\cdot))$ be any solution of (3.5.1) with initial values $(t_0, x_{t_0}(\cdot))$ existing for all $t \geq t_0 \geq 0$.

Definition 3.5.3 The system (3.5.1) is said to be (h_0, h)-uniformly stable if, given $\epsilon > 0$ and $t_0 \in R_+$, there exists a $\delta = \delta(\epsilon) > 0$ such that $t_0 \geq 0$ and $h_0(t_0, x_{t_0}(\cdot)) < \delta$ imply $h(t, x(t)) < \epsilon$ for all $t \geq t_0$.

Based on Definition 3.5.3, and the usual stability notions, it is easy to formulate the other kinds of stability concepts in terms of two measures (h_0, h). We give here a few choices of (h_0, h) to demonstrate the generality of Definition 3.5.3:

(a) the well-known uniform stability of the zero solution of (3.5.1) if $h^0(t,x) = h(t,x) = |x|$ and consequently $h_0(t, x_t(\cdot)) = |x_t(\cdot)|$ where $|x|$ is the usual Euclidean norm and $|x_t(\cdot)| = \sup_{0 \le s \le t} |x(s)|$;

(b) stability of the prescribed solution $y(t) = y(t, t_0, y_{t_0}(\cdot))$ of (3.5.1) if $h(t,x) = |x - y(t)|$ and $h_0(t, x_t(\cdot)) = |x_t(\cdot) - y_t(\cdot)|$;

(c) partial stability of the zero solution of (3.5.1) if $h(t,x) = |x|_k$, $1 < k < n$, and $h_0(t, x_t(\cdot)) = |x_t(\cdot)|$;

(d) stability of the conditionally invariant set B with respect to A where $A \subset B \subset R^n$ if $h(t,x) = d(x, B)$ and $h_0(t, x_t(\cdot)) = \sup_{0 \le s \le t} d(x(s), A)$, d being the distance function;

(e) eventual stability of (3.5.1) if $h(t,x) = |x|$ and $h_0(t, x_t(\cdot)) = |x_t(\cdot)| + \sigma(t)$, $\sigma \in L$ where $L = \{\sigma \in C[R_+, R_+]: \sigma(t)$ is decreasing with $\lim_{t \to \infty} \sigma(t) = 0\}$.

For any $h \in \Gamma$, define $S(h, \rho) = \{(t,x): h(t,x) < \rho\}$, and for any $h^* \in C[R_+ \times C(R_+), R_+]$, let $S(h^*, \rho) = \{(t, x_t(\cdot)): h^*(t, x_t(\cdot)) < \rho\}$.

Theorem 3.5.3 *Assume that*

(i) *$h, h^0 \in \Gamma$ and h_0 is finer than $\tilde{h}$, where h_0 and $\tilde{h}$ are defined by (3.5.10);*

(ii) *there exists a function $V \in C[R_+ \times R^n \times C(R_+), R_+]$ that is locally Lipschitzian in x, h-positive definite and h_0-decrescent;*

(iii) *$D^+V_{(3.5.1)}(t, x(t), x_t(\cdot)) \le -\gamma(h_0(t, x_t(\cdot)))$ for all $(t, x(t), x_t(\cdot)) \in S(h, \rho) \times S(h_0, \rho_0)$, where $\gamma \in \mathcal{K}$.*

Then the system (3.5.1) is (h_0, h)-uniformly asymptotically stable.

Proof Since $V(t, x, x_t(\cdot))$ is h-positive definite and h_0-decrescent, there exist $\rho, \rho_0 > 0$ and $\alpha, \beta \in \mathcal{K}$ such that

(3.5.11) $\alpha(h(t,x)) \le V(t, x, x_t(\cdot))$ whenever $h(t,x) < \rho$,

(3.5.12) $V(t, x, x_t(\cdot)) \le \beta(h_0(t, x_t(\cdot)))$ whenever $h_0(t, x_t(\cdot)) < \rho_0$.

176   THEORY OF INTEGRO-DIFFERENTIAL EQUATIONS

In view of the assumption (i), we can find a $\lambda > 0$ and a function $\psi \in \mathcal{K}$ such that $h_0(t, x_t(\,\cdot\,)) < \lambda$ implies

(3.5.13) $$\tilde{h}(t, x_t(\,\cdot\,)) \leq \psi(h_0(t, x_t(\,\cdot\,))).$$

Given $0 < \epsilon < \rho$, it follows from the assumptions on β and ψ that there exists $0 < \delta_1 < \rho_0$ and $0 < \delta_2 < \lambda$ such that

(3.5.14) $$\beta(\delta_1) < \alpha(\epsilon) \qquad \psi(\delta_2) < \rho.$$

Choose $\delta = \min(\delta_1, \delta_2)$. We now claim that the system (3.5.1) is (h_0, h)-uniformly stable.

Suppose that the claim is false. Then there is a solution $x(t) = x(t, t_0, x_{t_0}(\,\cdot\,))$ of (3.5.1) such that $h_0(t_0, x_{t_0}(\,\cdot\,)) < \delta$ and for some $t_1 > t_0$, we have

(3.5.15) $$h(t_1, x(t_1)) = \epsilon.$$

The assumption (iii) clearly implies $D^+V_{(3.5.1)}(t, x(t), x_t(\,\cdot\,)) \leq 0$, and hence we obtain from (3.5.11) − (3.5.15) for $t_1 > t_0$ that

$$\begin{aligned}\alpha(\epsilon) &\leq \alpha(h(t_1, x(t_1))) \leq V(t_1, x(t_1), x_{t_1}(\,\cdot\,)) \leq V(t_0, x(t_0), x_{t_0}(\,\cdot\,)) \\ &\leq \beta(h_0(t_0, x_{t_0}(\,\cdot\,))) < \beta(\delta) < a(\epsilon),\end{aligned}$$

which is a contradiction. Hence the system (3.5.1) is (h_0, h)-uniformly stable. To prove the uniform attracting property of (3.5.1), let $0 < \eta < \rho$ be given. Choose $\delta_0 = \min(\delta(\rho), \rho_0)$ such that

(3.5.16) $$\psi(\delta_0) < \rho, \qquad T(\eta) = \frac{\beta(\delta_0)}{\gamma(\delta(\eta))},$$

where $\delta(\eta)$ corresponds to η in uniform stability.

We now claim that every solution $x(t) = x(t, t_0, x_{t_0}(\,\cdot\,))$ of (3.5.1) with $h_0(t_0, x_{t_0}(\,\cdot\,)) < \delta_0$ satisfying the relation

(3.5.17) $$h_0(t_1, x_{t_1}(\,\cdot\,)) < \delta(\eta) \qquad \text{for some } t_1 \in [t_0, t_0 + T],$$

and hence it follows from the (h_0, h)-uniform stability of (3.5.1) that $h(t, x(t)) < \eta$ for all $t \geq t_1$. Suppose (3.5.17) is not true. Then we have

(3.5.18) $$h_0(t, x_t(\,\cdot\,)) \geq \delta(\eta) \qquad \text{for all } t \in [t_0, t_0 + T]$$

and a $t_2 > t_1$, $t_0 \leq t_1 \leq t_0 + T$ such that

(3.5.19) $$h(t_2, x(t_2)) = \eta.$$

From the assumption (iii) we have

(3.5.20) $$V(t, x(t), x_t(\,\cdot\,)) \leq V(t_0, x(t_0), x_{t_0}(\,\cdot\,)) - \int_{t_0}^{t} \gamma(h_0(s, x_s(\,\cdot\,))) \, ds.$$

Thus, for $t_1 = t_0 + T$, $t_2 > t_1$ and $t_2 - t_0 > T$, the relations (3.5.11) – (3.5.13), (3.5.16) and (3.5.18) – (3.5.20) give

$$0 < \alpha(\eta) = \alpha(h(t_2, x(t_2)) \leq V(t_2, x(t_2), x_{t_2}(\,\cdot\,))$$

$$\leq V(t_0, x(t_0), x_{t_0}(\,\cdot\,)) - \int_{t_0}^{t_2} \gamma(h_0(s, x_s(\,\cdot\,))) \, ds$$

$$\leq \beta(h_0(t_0, x_{t_0}(\,\cdot\,))) - \int_{t_0}^{t_0 + T} \gamma(h_0(s, x_s(\,\cdot\,))) \, ds$$

$$< \beta(\delta_0) - \gamma(\delta(\eta)) T = 0,$$

which is a contradiction. Thus at some t_1 such that $t_0 \leq t_1 \leq t_0 + T(\eta)$, we have $h_0(t_1, x_{t_1}(\,\cdot\,)) < \delta(\eta)$, and hence (h_0, h)-uniform stability of (3.5.1) implies $h(t, x(t)) < \eta$ for all $t > t_1$, in particular for $t \geq t_0 + T(\eta)$. Therefore we obtain

$$h(t, x(t)) < \eta \text{ for } t \geq t_0 + T(\eta) \quad \text{whenever } t_0 \geq 0$$

and $h_0(t_0, x_{t_0}(\,\cdot\,)) < \delta_2$. This shows that the system (3.5.1) is (h_0, h)-uniformly attracting, and hence the proof is complete. □

The concept of M_0-stability (see Moore [1]) describes a general type of invariant set and its stability behavior. Moreover, the notion of M_0-stability is a natural generalization of the usual concepts of eventual stability and the stability of asymptotically self-invariant sets. We now introduce a very general type of stability called "(h_0, h, M_0)-stability" by combining the two concepts such as M_0-stability and (h_0, h)-stability, and discuss these new stability properties relative to the system (3.5.1). This new notion of stability allows us to consider the initial values on surfaces that crucially depend on the initial time and also to introduce different topologies in the definition of stability.

Let $M = M[R_+, R^n]$ be the space of all measurable mappings from R_+ to R^n such that $p \in M$ if and only if $p(t)$ is locally integrable and $\sup_{t>0} \int_t^{t+1} |p(s)| \, ds < \infty$. Let $M_0 = M_0[R_+, R^n]$ be a subspace of M consisting of all $p(t)$ such that $\int_t^{t+1} |p(s)| \, ds \to 0$ as $t \to \infty$. The set $S(M_0, \epsilon)$ is the subset of M defined by

$$S(M_0, \epsilon) = \left\{ p \in M : \lim_{t \to \infty} \sup \int_t^{t+1} |p(s)| \, ds \leq \epsilon \right\}.$$

Let $x(t) = x(t, s, \psi(s, x^*))$ be any solution of (3.5.1) with initial values $(s, \psi(s, x^*))$, $x^* \in R^n$, $\psi \in C[R_+ \times R^n, R^n]$ and $x_s(r) = \psi(r, x^*)$, $0 \leq r \leq s$, existing for all $t \geq s \geq 0$.

Definition 3.5.4 Let $A \subset R^n$. A is said to be M_0-invariant relative to the system (3.5.1) if for $x^* \in A$ and $\psi(s, x^*) \in M_0$, $x(\cdot, s, \psi(s, x^*)) \in M_0$.

Definition 3.5.5 Relative to the system (3.5.1), the set A is said to be

(d_1) (h_0, h, M_0)-uniformly stable if for each $\epsilon > 0$, there exist $\tau_1(\epsilon)$, $\tau_1(\epsilon) \to \infty$ as $\epsilon \to 0$ and $\delta_1(\epsilon)$, $\delta_2(\epsilon) > 0$ such that

$$\int_{t_0}^{t_0+1} h(t, x(t, s, \psi(s, x^*))) \, ds < \epsilon \quad \text{for all } t \geq t_0 + 1$$

whenever $x^* \in S(A, \delta_1)$ and $\int_{t_0}^{t_0+1} h_0(s, \psi(s, x^*)) \, ds < \delta_2$, $t_0 \geq \tau_1(\epsilon)$;

(d_2) (h_0, h, M_0)-uniformly attracting if for any $\eta > 0$, there exist positive numbers $\widehat{\delta}_1, \widehat{\delta}_2, \tau_0$ and $T = T(\eta)$ such that

$$\int_{t_0}^{t_0+1} h(t, x(t, s, \psi(s, x^*))) \, ds < \eta \quad \text{for all } t \geq t_0 + T + 1, \quad t_0 \geq \tau_0$$

whenever $x^* \in S(A, \widehat{\delta}_1)$ and $h_0(s, \psi(s, x^*)) \in S(M_0, \widehat{\delta}_2)$;

(d_3) (h_0, h, M_0)-uniformly asymptotically stable if (d_1) and (d_2) hold simultaneously.

We give below a couple of special cases to indicate the generality of the relatively new concept (h_0, h, M_0)-stability:

(i) if $h_0(t,x) = h(t,x) = |x|$ then (d_1) implies M_0-uniform stability of the set A relative to (3.5.1);

(ii) if $h_0(t,x) = |x|$ and $h(t,x) = |x|_k$, $1 < k < n$, then (d_1) implies M_0-partial stability of the set A relative to (3.5.1).

Definition 3.5.6 A function a is said to belong to class $\mathcal{KC}$ if $a \in \mathcal{K}$ and a is convex.

Definition 3.5.7 A function h_0 has the property P if for a given $\epsilon > 0$ there exists a $\delta(\epsilon) > 0$ such that $x^* \in A$ and $\psi(s, x^*) \in S(M_0, \delta)$, then $b(h_0(s, \psi(s, x^*)) \in S(M_0, \epsilon)$ where $b \in \mathcal{K}$.

We need the following well-known result.

Lemma 3.5.1 (Jensen inequality) *If a is a convex function and f is integrable on R_+ then*

$$a\left(\int_0^t f(s)\,ds\right) \leq \int_0^t a(f(s))\,ds.$$

For any $h, h_0 \in C[R_+ \times R^n, R_+]$, let

$$Q(h, \rho) = \{(t, x) \in R_+ \times R^n : h(t, x) < \rho\},$$

$$H(h_0, \rho_0) = \{(t, x_t(s)) \in R_+ \times C(R_+) : \sup_{0 \leq s \leq t} h_0(s, x(s)) < \rho_0\}.$$

We shall now give sufficient conditions for (h_0, h, M_0)-uniform stability of the set A relative to the system (3.5.1) that will indicate the interplay between several concepts introduced. Sufficient conditions for other (h_0, h, M_0)-stability properties can be formulated on the same lines.

Theorem 3.5.4 *Assume that*

(i) $h, h_0 \in \Gamma$, h_0 *is finer than* h *and* h_0 *has the property* P;

(ii) $V \in C[R_+ \times R^n \times C(R_+), R_+]$, V *is locally Lipschitzian in* x, *and satisfies*

(3.5.21) $\quad a(h(t,x)) \leq V(t, x, x_t(\cdot))$, $\quad (t,x) \in Q(h, \rho)$, $\quad a \in \mathcal{KC}$,

(3.5.22) $\quad V(t,x,x_t(\,\cdot\,)) \le b(h_0(t,x)), \quad (t,x) \in Q(h_0,\rho_0), \quad b \in \mathcal{KC},$

where $\rho_0 \in [0,\lambda]$ and $\phi(\rho_0) < \rho$, λ and ϕ being the same as in Definition 3.5.1;

(iii) for $(t,x,x_t(\,\cdot\,)) \in Q(h,\rho) \times H(h_0,\rho_0)$,

$$D^+V_{(3.5.1)}(t,x,x_t(\,\cdot\,)) \le 0.$$

Then the set A is (h_0,h,M_0)-uniformly stable relative to the system (3.5.1).

Proof Let $0 < \epsilon < \rho$ and $t_0 \in R_+$ be given. It follows from the hypothesis (i) that there exists a $\lambda > 0$ and $\phi \in \mathcal{K}$ such that

(3.5.23) $\quad h(t,x) \le \phi(h_0(t,x)) \quad$ whenever $h_0(t,x) < \lambda$.

Since h_0 has the property P, we can find positive numbers $\delta_1(\epsilon), \delta_2(\epsilon) < \rho_0$ and $\tau_1(\epsilon)$ such that

(3.5.24) $\quad \displaystyle\int_{t_0}^{t_0+1} b(h_0(s,\psi(s,x^*)))\,ds < a(\epsilon), \quad a \in \mathcal{K},$

provided $x^* \in S(A,\delta_1)$ and $\displaystyle\int_{t_0}^{t_0+1} h(s,\psi(s,x^*))\,ds < \delta_2$ for $t_0 \ge \tau_1(\epsilon)$.

We claim that the set A is (h_0,h,M_0)-uniformly stable relative to the system (3.5.1). Suppose this is not true. Then for any solution $x(t) = x(t,s,\psi(s,x^*))$ of (3.5.1) with $x^* \in S(A,\delta_1)$ and $\displaystyle\int_{t_0}^{t_0+1} h_0(s,\psi(s,x^*))ds < \delta_2$, there exists a first $t_1 > t_0 + 1$, $t_0 \ge \tau_1(\epsilon)$, such that

(3.5.25) $\quad \begin{cases} \displaystyle\int_{t_0}^{t_0+1} h(t_1,x(t_1,s,\psi(s,x^*)))\,ds = \epsilon \\[6pt] \displaystyle\int_{t_0}^{t_0+1} h(t,x(t,s,\psi(s,x^*)))\,ds < \epsilon \qquad \text{for } t_0+1 \le t < t_1, \end{cases}$

and $t_0 \ge \tau(\epsilon)$. From the assumption (iii), we have

(3.5.26) $\quad D^+V_{(3.5.1)}(t,x(t,s,\psi(s,x^*)),x_t(\,\cdot\,)) \le 0$

and hence from (3.5.21) − (3.5.23) and (3.5.26) we obtain

$$\begin{aligned} a(h(t_1, x(t_1, s, \psi(s, x^*)))) &\leq V(t_1, x(t_1, s, \psi(s, x^*)), x_{t_1}(\cdot)) \\ &\leq V(s, x(s, s, \psi(s, x^*)), x_s(\cdot)) \\ &= V(s, \psi(s, x^*), x_s(\cdot)) \leq b(h_0(s, \psi(s, x^*))). \end{aligned}$$

Now integrating over s between t_0 to $t_0 + 1$ and using Lemma 3.5.1 and the relations (3.5.24) and (3.5.25), we get

$$a(\epsilon) = a\left(\int_{t_0}^{t_0+1} h(t_1, s, \psi(s, x^*))) \, ds\right)$$

$$\leq \int_{t_0}^{t_0+1} a(h(t_1, x(t_1, s, \psi(s, x^*)))) \, ds$$

$$\leq \int_{t_0}^{t_0+1} b(h_0(s, \psi(s, x^*))) \, ds < a(\epsilon),$$

which is a contradiction. Hence the set A relative to (3.5.1) is (h_0, h, M_0)-uniformly stable. This completes the proof. □

3.6 Impulsive Integro-differential Equations

It is now well recognized that the concept of Lyapunov-like functions and the theory of differential and integral inequalities can be employed to study various properties of nonlinear systems. Since many evolution processes are characterized by the fact that at certain moments of time they experience an abrupt change of state, the study of dynamic systems with impulse effects has recently been assuming a greater importance. As we have seen in Section 2.10, impulsive linear integro-differential systems can be discussed by utilizing the variation of parameters technique. In the present section, employing piecewise-Lyapunov functions and the theory of impulsive differential inequalities, we wish to extend Lyapunov's method to nonlinear impulsive integro-differential equations. This important technique, which permits us to reduce the study of impulsive integro-differential equations to the study of impulsive differential equations, depends crucially on choosing suitable minimal sets of functions along which the derivative of the Lyapunov function allows a convenient estimate. We prove a general comparison result that offers a unified choice of minimal sets and also enables us to prove in one set-up various stability criteria.

Consider the integro-differential system

(3.6.1) $$x' = f(t, x, Tx), \quad x(t_0) = x_0, \quad t_0 \geq 0,$$

where $f \in C[R_+ \times S(\rho) \times R^n, R^n]$, $Tx = \int_{t_0}^{t} K(t, s, x(s)) ds$ and $K \in C[R_+^2 \times S(\rho), R^n]$. We need the following well-known comparison result (see Shendge [1]) relative to (3.6.1), which permits us to reduce the study of integro-differential system to the study of a scalar differential equation. As we shall see, this comparison result depends crucially on choosing an appropriate minimal class of functions along which the generalized derivative of the Lyapunov function allows a convenient estimate. To state this comparison theorem, we make the following hypotheses:

(H_1) $g_0, g \in C[R_+^2, R], g_0(t, u) \leq g(t, u), r(t, t_0, u_0)$ is the right maximal solution of

$$u' = g(t, u), \quad u(t_0) = u_0 \geq 0,$$

existing on $[t_0, \infty)$, and $\eta(t, t^0, v_0)$ is the left maximal solution of

$$v' = g_0(t, v), \quad v(t^0) = v_0 \geq 0,$$

existing on $t_0 \leq t \leq t^0$;

(H_2) $V \in C[R_+ \times S(\rho), R_+]$, $V(t, x)$ is locally Lipschitzian in x and for $t \geq t_0$, $x \in \Omega$,

$$D_- V(t, x, Tx) \equiv \varliminf_{h \to 0} \inf \frac{1}{h} \left[V(t+h, x + hf(t, x, Tx)) - V(t, x) \right]$$

$$\leq g(t, V(t, x)),$$

where

$$\Omega_1 = \left\{ x \in C[R_+, R^n] : V(s, x(s)) \leq \eta(s, t, V(t, x(t))), t_0 \leq s \leq t \right\}.$$

Theorem 3.6.1 *Assume that* (H_1) *and* (H_2) *hold. Let* $x(t) = x(t, t_0, x_0)$ *be any solution of* (3.6.1) *existing on* $[t_0, \infty)$ *such that* $V(t_0, x_0) \leq u_0$. *Then*

$$V(t, x(t)) \leq r(t, t_0, u_0), \quad t \geq t_0.$$

Note that Theorem 3.6.1 is contained in Theorem 3.4.6, which we have stated here in a convenient form.

Now we shall consider the impulsive integro-differential system

$$(3.6.2) \quad \begin{cases} x' = f(t, x, Tx), & t \neq t_i, \\ \Delta x|_{t=t_i} = I_i(x(t_i)), & i = 1, 2, 3, \ldots, \\ x(t_0^+) = x_0, & t_0 \geq 0, \end{cases}$$

where

(i) $\Delta x|_{t=t_i} = x(t_i^+) - x(t_i^-)$,
(ii) $0 < t_1 < t_2 < \ldots < t_i < \ldots$ and $\lim_{i \to \infty} t_i = \infty$,
(iii) $f: R_+ \times S(\rho) \times R^n \to R^n$ is continuous on $(t_i, t_{i+1}] \times S(\rho) \times R^n$,
(iv) $Tx = \int_{t_0}^{t} K(t, s, x(s)) \, ds$, where $K: R_+^2 \times S(\rho) \to R^n$ is continuous on $(t_i, t_{i+1}] \times (t_i, t_{i+1}] \times S(\rho)$,
(v) $I_i: S(\rho) \to R^n$.

We shall assume the existence and uniqueness of solutions of (3.6.2), and note that the solutions $x(t) = x(t, t_0, x_0)$ of (3.6.2) are piecewise-continuous functions with points of discontinuity of the first kind at $t = t_i$, at which they are left-continuous. Also, it is understood that when $t_0 \neq t_i$, $x(t_0^+) = x(t_0)$.

Let us define the following classes of functions for convenience. Let PC^+ denote the class of piecewise-continuous functions from R_+ to R_+ with discontinuities of the first kind only at $t = t_i$, $i = 1, 2, \ldots$, and left-continuous at $t = t_i$.

Definition 3.6.1 *A function* $V: R_+ \times S(\rho) \to R_+$ *is said to belong to the class* V_0 *if* $V(t, x)$ *is continuous on* $(t_i, t_{i+1}] \times S(\rho)$ *such that* $\lim_{(t,y) \to (t_i^+, x)} V(t, y) = V(t_i^+, x)$ *exists for* $t > t_i$, $i = 1, 2, \ldots$.

Definition 3.6.2 A function $a \in C[[0,\rho), R_+]$ is said to belong to the class $\mathcal{K}$ if $a(u)$ is strictly increasing and $a(0) = 0$.

Definition 3.6.3 Let $V \in V_0$. Then for any $t \in (t_i, t_{i+1}]$ and $x \in PC^+[R_+, S(\rho)]$, we define

$$D_-V(t,x,Tx) \equiv \lim_{h \to 0^-} \inf \frac{1}{h}\left[V(t+h, x+hf(t,x,Tx)) - V(t,x)\right].$$

Utilizing Theorem 3.6.1, we can now prove the following comparison result which we need to discuss in a unified way stability criteria for the impulsive integro-differential system (3.6.2).

Theorem 3.6.2 Assume that

(i) $g: R_+^2 \to R$ is continuous on $(t_i, t_{i+1}] \times R_+$, $\lim_{\substack{(t,v) \to (t_i^+, u) \\ t \geq t_i}} g(t,v) = g(t_i^+, u)$ exists and $r(t, t_0, u_0)$ is the maximal solution of the impulsive differential equation

(3.6.3)
$$\begin{cases} u' = g(t,u), & t \neq t_i, \\ u(t_i^+) = \psi_i(u(t_i)), \\ u(t_0^+) = u_0 \geq 0, & t_0 \geq 0, \end{cases}$$

existing on $[t_0, \infty)$;

(ii) $g_0 \in C[[r_i, t_{i+1}) \times R_+, R]$ and on each $[t_i, t_{i+1}) \times R_+$, g_0, g satisfy (H_1);

(iii) $V \in V_0$, $V(t,x)$ is locally Lipschitzian in x and for $t > t_0$, $x \in \Omega$ where $\Omega = \{x \in PC^+[R_+, S(\rho)]: V(s, x(s)) \leq \eta(s, t, V(t, x(t)), t_0 \leq s \leq t\}$,

$$D_-V(t,x,Tx) \leq g(t, V(t,x)), \quad t \neq t_i;$$

(iv) $V(t^+, x+I_i(x)) \leq \psi_i(V(t,x))$, $t = t_i$, where $\psi_i: R_+ \to R_+$ is nondecreasing.

Then if $x(t) = x(t, t_0, x_0)$ is any solution of (3.6.2) existing on $[t_0, \infty)$, we have

$$V(t, x(t)) \leq r(t, t_0, u_0), \quad t \geq t_0,$$

provided $V(t_0^+, x_0) \leq u_0$.

Proof Let $t_0 \geq 0$ and $t_0 \in (t_j, t_{j+1}]$ for some $j \geq 1$. Let $x(t) = x(t, t_0, x_0)$ be any solution of (3.6.2) existing on $[t_0, \infty)$ and set $m(t) = V(t, x(t))$. For convenience, we designate $t_i = t_{j+i}$ if $t_0 \neq t_{j+1}$ and $t_i = t_{j+1+i}$ if $t_0 = t_{j+1}$, $i = 1, 2, \ldots$. Then, for $t \in (t_0, t_1]$, Theorem 3.6.1 implies that

$$m(t) \leq r_1(t, t_0, u_0),$$

where $r_1(t, t_0, u_0)$ is the maximal solution of the differential equation

(3.6.4) $$u' = g(t, u),$$

existing on $(t_0, t_1]$ such that $r_1(t_0^+, t_0, u_0) = u_0$. Since $\psi_1(u)$ is nondecreasing in u and $m(t_1) \leq r_1(t_1, t_0, u_0)$, we get from (iv), $m(t_1^+) \leq u_1^+$, where

$$u_1^+ = \psi_1(r(t_1, t_0, u_0)).$$

Using Theorem 3.6.1 again, we obtain

$$m(t) \leq r_2(t, t_1, u_1^+), \quad t \in (t_1, t_2],$$

where $r_2(t, t_1, u_1^+)$ is the maximal solution of (3.6.4) existing on $[t_1, t_2]$ such that

$$r_2(t_1^+, t_1, u_1^+) = u_1^+.$$

We therefore have successively

$$m(t) \leq r_{i+1}(t, t_i, u_i^+), \quad t \in (t_i, t_{i+1}],$$

where $r_{i+1}(t, t_i, u_i^+)$ is the maximal solution of (3.6.4) existing on $(t_i, t_{i+1}]$ such that $r_{i+1}(t_i^+, t_i, u_i^+) = u_i^+$. Thus if we define

$$u(t) = \begin{cases} u_0, & t = t_0 \\ r_1(t, t_0, u_0), & t \in (t_0, t_1], \\ r_2(t, t_1, u_1^+), & t \in (t_1, t_2], \\ \vdots & \\ \vdots & \\ r_{i+1}(t, t_i, u_i^+), & t \in (t_i, t_{i+1}], \\ \vdots & \end{cases}$$

then it is easy to see that $u(t)$ is a solution of (3.6.3) and

$$m(t) \leq u(t), \quad t \geq t_0.$$

Since $r(t, t_0, u_0)$ is the maximal solution of (3.6.3), we get immediately

$$m(t) \leq r(t, t_0, u_0), \quad t \geq t_0$$

and the proof is complete. □

Let us collect several interesting and useful special cases of Theorem 3.6.2 in the following corollary.

Corollary 3.6.1 *If in Theorem 3.6.2, we choose*

(i) $g_0(t, u) = g(t, u) \equiv 0$ *and* $\psi_i(u) = u$ *for all i then $V(t, x(t))$ is nonincreasing in t and*

$$V(t, x(t)) \leq V(t_0^+, x_0), \quad t \geq t_0;$$

(ii) $g_0(t, u) = g(t, u) \equiv 0$ *and* $\psi_i(u) = d_i u$, $d_i \geq 0$ *for all i then*

$$V(t, x(t)) \leq V(t_0^+, x_0) \prod_{t_0 < t_i < t} d_i, \quad t \geq t_0;$$

(iii) $g_0(t, u) \equiv 0$, $g(t, u) = \lambda'(t) u$, *where* $\lambda \in C^1[R_+, R_+], \lambda'(t) \geq 0$ *and* $\psi_i(u) = d_i u$, $d_i \geq 0$ *for all i, then*

$$V(t, x(t)) \leq \left[V(t_0^+, x_0) \prod_{t_0 < t_i < t} d_i \right] \exp[\lambda(t) - \lambda(t_0)], \quad t \geq t_0;$$

(iv) $g_0(t, u) = g(t, u) = -A'(t)/A(t) u$, *where $A(t) > 0$ is continuously differentiable on R_+, and $A(t) \to \infty$, and $\psi_i(u) = d_i u$, $d_i \geq 0$ for all i, then*

$$V(t, x(t)) \leq \left[V(t_0^+, x_0) \prod_{t_0 < t_i < t} d_i \right] \frac{A(t_0)}{A(t)}, \quad t \geq t_0$$

(in particular, $A(t) = e^{\alpha t}$, $\alpha > 0$, is admissible);.

(v) $g_0(t, u) = g(t, u) = -\gamma(u)$, *where $\gamma \in \mathcal{K}$, $\psi_i(u) = u$ for all i, then*

$$V(t, x(t)) \leq J^{-1}[J(V(t_0^+, x_0)) - (t - t_0)], \quad t \geq t_0,$$

where J^{-1} is the inverse function of J and $J'(u) = a/\gamma(u)$.

It is useful to know how the minimal class of functions Ω of the assumption (iii) change depending on the choice of g_0, since the derivative of Lyapunov function has to be estimated along these sets. We shall illustrate this for some important special cases of Corollary 3.6.1 because of their use in the literature.

(a) As in $(i) - (iii)$, if $g_0(t, u) \equiv 0$ then $\eta(s, t^0, v_0) \equiv v_0$, and hence the set

$$\Omega = \left\{x \in PC^+[R_+, S(\rho)]: V(s, x(s)) \leq V(t, x(t)),\ t_0 \leq s \leq t\right\}.$$

(b) As in (iv), if $g_0(t, u) = -[A'(t)/A(t)]u$, then $\eta(s, t^0, v_0) = v_0[A(t^0)/A(s)]$, $t_0 \leq s \leq t^0$, and consequently we have

$$\Omega = \left\{x \in PC^+[R_+, S(\rho)]: V(s, x(s))A(s) \leq V(t, x(t))A(t), t_0 \leq s \leq t\right\}.$$

(c) As in (v), if $g_0(t, u) = -\gamma(u)$, $\gamma \in \mathcal{K}$, then

$$\eta(s, t^0, v_0) = J^{-1}[J(v_0) - (s - t^0)], \quad t_0 \leq s \leq t^0,$$

where J and J^{-1} are the same functions as in (v). Since $\eta(s, t^0, v_0)$ is increasing in s to the left of t^0, fixing an $s_0 < t^0$ and defining $L(u) = \eta(s_0, t^0, u)$, it is clear that $L \in \mathcal{K}$, and as a result

$$\Omega = \left\{x \in PC^+[R_+, S(\rho)]: V(s, x(s)) \leq L(V(t, x(t))), t_0 \leq s \leq t\right\}.$$

We shall now discuss the stability properties of the trivial solution of the impulsive integro-differential system (3.6.2). Our tools are piecewise-Lyapunov functions, theory of impulsive differential inequalities and minimal classes of functions along which we can estimate the derivative of the Lyapunov functions so that the study of impulsive integro-differential systems can be reduced to the study of scalar impulsive differential equations.

Having the general comparison Theorem 3.6.2 at our disposal, it is now easy to give, in a unified way, sufficient conditions for various stability properties to hold. For this purpose, we assume that $f(t, 0, 0) \equiv 0$, $K(t, s, 0) \equiv 0$ and $I_i(0) = 0$ for all i so that we have the trivial solution for (3.6.2).

Theorem 3.6.3 *Suppose that the assumptions of Theorem 3.6.2 hold. Assume further that*

$$b(|x|) \leq V(t,x) \leq a(|x|) \quad \text{on } R_+ \times S(\rho),$$

where $a, b \in \mathcal{K}$, and that there exists a $\rho_0 > 0$ such that $x \in S(\rho_0)$ implies that $x + I_i(x) \in S(\rho)$ for i. Then the stability properties of the trivial solution of the scalar impulsive differential equation (3.6.3) implies the corresponding stability properties of the trivial solution of (3.6.2).

Proof Let $0 < \epsilon < \rho^* = \min(\rho_0, \rho)$ and $t_0 \in R_+$ be given. Suppose that the trivial solution of (3.6.3) is stable. Then, given $b(\epsilon) > 0$ and $t_0 \in R_+$, there exists a $\delta_1(t_0, \epsilon) > 0$ such that

$$0 \leq u_0 < \delta_1 \quad \text{implies} \quad u(t, t_0, u_0) < b(\epsilon), \quad t \geq t_0,$$

where $u(t, t_0, u_0)$ is any solution of (3.6.3). Let $u_0 = a(|x_0|)$ and choose a $\delta_2 = \delta_2(\epsilon)$ such that $a(\delta_2) < b(\epsilon)$. Define $\delta = \min(\delta_1, \delta_2)$. With this δ, we claim that if $|x_0| < \delta$ then $|x(t)| < \epsilon$ for $t \geq t_0$, where $x(t) = x(t, t_0, x_0)$ is any solution of (3.6.2). If this is not true, there would exist a solution $x(t) = x(t, t_0, x_0)$ of (3.6.2) with $|x_0| < \delta$ and a $t^* > t_0$ such that $t_k < t^* \leq t_{k+1}$ for some k, satisfying

$$\epsilon \leq |x(t^*)| \quad \text{and} \quad |x(t)| < \epsilon \quad \text{for } t_0 \leq t \leq t_k.$$

Since $0 < \epsilon < \rho_0$, we obtain $|x_k^+| = |x_k + I_k(x_k)| < \rho$, where $x_k = x(t_k)$ and $|x_k| < \epsilon$. Hence we can find a t^0 such that $t_k < t^0 \leq t^*$, satisfying $\epsilon \leq |x(t^0)| < \rho$. Now setting $m(t) = V(t, x(t))$ for $t_0 \leq t \leq t^0$, we get by Theorem 3.6.2 the estimate

(3.6.5) $\qquad V(t, x(t)) \leq r(t, t_0, a(|x_0|)), \quad t_0 \leq t \leq t^0,$

where $r(t, t_0, u_0)$ is the maximal solution of (3.6.3). We are then lead to the contradiction

$$b(\epsilon) \leq b(|x(t^0)|) \leq V(t^0, x(t^0)) \leq r(t^0, t_0, a(|x_0|)) < b(\epsilon),$$

which proves that $x \equiv 0$ of (3.6.2) is stable.

If we suppose that $u \equiv 0$ of (3.6.3) is uniformly stable then it is clear that δ will be independent of t_0, and thus we get the uniform stability of $x \equiv 0$ of (3.6.2).

Let us suppose that $u \equiv 0$ of (3.6.3) is asymptotically stable, which implies that $x \equiv 0$ of (3.6.2) is stable. Consequently, taking $\epsilon = \rho^*$ and setting $\delta_0^* = \delta(t_0, \rho^*)$, we have

(3.6.6) $\qquad |x_0| < \delta_0^* \quad \text{implies} \quad |x(t)| < \rho, \quad t \geq t_0.$

To prove attractivity, we let $0 < \epsilon < \rho^*$ and $t_0 \in R_+$. Since $u \equiv 0$ of (3.6.3) is attractive, given $b(\epsilon) > 0$ and $t_0 \in R_+$, there exists a $\delta_{10} = \delta_{10}(t_0) > 0$ and a $T = T(t_0, \epsilon) > 0$ such that

$$0 \leq u_0 < \delta_{10} \quad \text{implies} \quad u(t, t_0, u_0) < b(\epsilon), \quad t \geq t_0 + T.$$

We choose $\delta_0 = \min(\delta_0^*, \delta_{10})$ and let $|x_0| < \delta_0$. In view of (3.6.6) the arguments leading to (3.6.5) yield

$$V(t, x(t)) \leq r(t, t_0, a(|x_0|)), \quad t \geq t_0,$$

from which it follows that

$$b(|x(t)|) \leq V(t, x(t)) \leq r(t, t_0, a(|x_0|)) < b(\epsilon), \quad t \geq t_0 + T,$$

which proves that $x \equiv 0$ is attractive. Hence $x \equiv 0$ of (3.6.2) is asymptotically stable.

If we suppose that $u \equiv 0$ of (3.6.3) is uniformly asymptotically stable, it is clear that we find that $x \equiv 0$ of (3.6.2) is also uniformly asymptotically stable, since δ_0 and T will be independent of t_0. Hence the proof of Theorem 3.6.3 is complete. □

Corollary 3.6.2 The functions
(A) $g_0(t, u) = g(t, u) \equiv 0$, $\psi_k(u) = d_k u$, $d_k \geq 0$ for all k, are admissible in Theorem 3.6.3, giving that $x \equiv 0$ of (3.6.2) is uniformly stable provided the infinite product $\prod_{i=1}^{\infty} d_i$ converges. In particular, $d_k = 1$ for all k, is admissible;

(B) $g_0(t, u) \equiv 0$, $g(t, u) = \lambda'(t)u$, $\lambda \in C^1[R_+, R_+]$, $\lambda'(t) \geq 0$, $\psi_k(u) = d_k u$, $d_k \geq 0$ for all k, are admissible in Theorem 3.6.3, impling stability of $x \equiv 0$ of (3.6.2) provided that

(3.6.7) $$\lambda(t_k) + \ln d_k \leq \lambda(t_{k-1}) \quad \text{for all } k;$$

(C) the functions in (B) are also admissible in Theorem 3.6.3, ensuring to assure asymptotic stability of $x \equiv 0$ of (3.6.2) if (3.6.7) is strengthened to

(3.6.8) $$\lambda(t_k) + \ln \alpha d_k \leq \lambda(t_{k-1}) \quad \text{for all } k, \quad \text{where } \alpha > 1.$$

Proof (A) follows directly from Corollary 3.6.1. To prove (B) and (C), we see that any solution $u(t, t_0, u_0)$ of

(3.6.9) $$\begin{cases} u' = \lambda'(t)u, & t \neq t_k, \\ u(t_k^+) = d_k u(t_k), \\ u(t_0^+) = u_0 \geq 0, \end{cases}$$

is given by $u(t, t_0, u_0) = u_0 \prod_{t_0 < t_k < t} d_k \exp[\lambda(t) - \lambda(t_0)]$, $t \geq t_0$. Since $\lambda(t)$ is nondecreasing, it follows from (3.6.7) that

$$u(t, t_0, u_0) \leq u_0 \exp[\lambda(t_1) - \lambda(t_0)], \quad t \geq t_0,$$

provided $0 < t_0 < t_1$. Hence choosing $\delta = \frac{1}{2}\epsilon \exp[\lambda(t_0) - \lambda(t_1)]$, stability of the trivial solution $u = 0$ of (3.6.9) follows. If, on the other hand, (3.6.8) holds then we get $u(t, t_0, u_0) \leq u_0 \exp[\lambda(t_1) - \lambda(t_0)]\alpha^{-k}$, $t_{k-1} < t \leq t_k$, from which $\lim_{t \to \infty} u(t, t_0, u_0) = 0$ follows. Thus Theorem 3.6.3 implies the stated conclusion.□

Finally, we consider a simple situation of (3.6.2) where

$$f(t, x, Tx) = Ax + \int_{t_0}^{t} K(t, s, x(s)) ds,$$

and suppose that

$$|K(t, s, x)| \leq H(t, s)|x| \quad \text{on } R_+ \times S(\rho),$$

other conditions being the same. Then taking $V(t, x) = |x|$ and using the set

$$\Omega = \{x \in PC^+[R_+, S(\rho)] : |x(s)| \leq |x(t)|, t_0 \leq s \leq t\},$$

it is easy to compute

$$D_-V \leq \left[\mu(A) + \int_{t_0}^{t} H(t,s)\,ds\right] V,$$

where $\mu(A)$ is the logarithmic norm of A defined by

$$\mu(A) = \lim_{h \to 0} \frac{1}{h}(|I + hA| - 1),$$

I being the identity matrix. Thus we see that $g_0(t,u) \equiv 0$ and $g(t,u) = \lambda'(t)u$, where $\lambda'(t) = \mu(A) + \int_{t_0}^{t} H(t,s)\,ds$. If $\psi_i(u) = d_i u$, $d_i \geq 0$ for all i, then one can conclude the stability of the trivial solution of (3.6.2) based on Corollary 3.6.2 depending on the choice of $\lambda'(t)$.

It is important to note that impulses do contribute to yield stability properties even when the corresponding integro-differential system without impulses does not enjoy any stability behavior.

3.7 Impulsive Integro-differential Equations (Continued)

In Section 3.6, the Lyapunov method was extended to impulsive integro-differential systems by employing Lyapunov functions and choosing minimal sets of functions so as to reduce the study of impulsive integro-differential systems to integro-differential inequalities. In this section, we shall use Lyapunov functionals to study the stability properties of impulsive integro-differential systems. For this purpose, we shall first investigate linear impulsive integro-differential systems by constructing suitable Lyapunov functionals, and then extend this idea to consider nonlinear impulsive integro-differential systems.

Consider the linear integro-differential system

(3.7.1) $$\begin{cases} x'(t) = A(t)x(t) + \int_{t_0}^{t} K(t,s)x(s)\,ds, & t \neq t_k, \ k = 1,2,\ldots, \\ \Delta x(t) = B_k x(t), & t = t_k, \ x(t_0) = x_0, \ t \in R_+, \end{cases}$$

where $A \in PC^+[R_+, R^{n^2}]$, $K \in PC^+[R_+ \times R_+, R^{n^2}]$, $I_k \in C[R^n, R^n]$ and B_k, $k = 1, 2, \ldots$ are $n \times n$ constant matrices.

Let us also consider the linear impulsive ordinary differential system

(3.7.2)
$$\begin{cases} x' = A(t)x, & t \neq t_k, \\ \Delta x = B_k x, & t = t_k, \quad x(t_0^+) = x_0, \end{cases}$$

where B_k ($k = 1, 2, \ldots$) are $n \times n$ constant matrices such that $\det(I + B_k) \neq 0$, I being the identity matrix.

Let $\phi_k(t, s)$ be a fundamental matrix solution of the linear system

$$x' = A(t)x, \quad t_{k-1} < t \leq t_k.$$

Then the solution of (3.7.2) can be written (see Lakshmikantham, Bainov and Simeonov [1]) in the form

$$x(t, t_0, x_0) = \Psi(t, t_0 + 0)x_0,$$

where

$$\Psi(t, s) = \begin{cases} \phi_k(t, s) & \text{for } t, s \in (t_{k-1}, t_k], \\ \phi_{k+1}(t, t_k)(I + B_k)\phi_k(t_k, s) & \text{for } t_{k-1} < s \leq t_k < t \leq t_{k+1}, \\ \phi_{k+1}(t, t_k)\left[\prod_{j=k}^{i+1}(I + B_j)\phi_j(t_j, t_{j+1})\right](I + B_i)\phi_i(t_i, s) \\ \qquad \text{for } t_{i-1} < s \leq t_i < t_k < t \leq t_{k+1}. \end{cases}$$

Let the matrix $G(t)$ be defined by

$$G(t) = \int_t^\infty \Psi^T(s, t)\Psi(s, t)\,ds, \quad \text{which exists and finite for all } t \in R_+,$$

where Ψ^T is the transpose of Ψ. Clearly $G(t)$ is symmetric.

Theorem 3.7.1 *Assume that the following conditions hold for $|x| < \rho$ and $t \in R_+$:*

(a) $L|x| \leq \langle G(t)x, x \rangle^{1/2} \leq 1/2 \widehat{M}|x|$;
(b) $|G(t)x| \leq \widehat{K}\langle G(t)x, x\rangle^{1/2}$;

(c) $\quad -\widehat{M} + \widehat{\beta}\int\limits_{t}^{\infty} |K(u,t)|\, du \leq 0,\ \widehat{\beta} \geq \widehat{K}$;

(d) $\quad |x| > |x + B_k x|$ and

$$\langle G(t)x, x\rangle^{1/2} > \langle G(t)(x + B_k x), (x + B_k x)\rangle^{1/2}$$

where $L, \widehat{M}, \widehat{K}$ and $\widehat{\beta}$ are positive real numbers.
Then the zero solution of (3.7.1) is uniformly stable.

Proof Let $W(t,x) = \langle G(t)x, x\rangle^{1/2}$. Then we have

(3.7.3) $$W'(t,x) = \frac{\langle G'(t)x, x\rangle}{2\langle G(t)x, x\rangle^{1/2}} + \frac{\langle 2G(t)x', x\rangle}{2\langle G(t)x, x\rangle^{1/2}}$$

for $t \neq t_k$.

From

$$\frac{\partial \Psi}{\partial t}(s,t) = -\Psi(s,t)A(t),$$

$$\frac{\partial \Psi^T}{\partial t}(s,t) = -A^T(t)\psi^T(s,t) \quad \text{for } t \neq t_k,$$

it follows that

$$G'(t) = -I + \int\limits_{t}^{\infty}\left[\frac{\partial \Psi^T}{\partial t}(s,t)\Psi(s,t) + \Psi^T(s,t)\frac{\partial \Psi}{\partial t}(s,t)\right]ds$$

$$= -I - A^T(t)G(t) - G(t)A(t), \quad t \neq t_k.$$

Hence (3.7.3) gives

(3.7.4) $$W'_{(3.7.1)}(t,x) = \frac{-\langle x,x\rangle}{2\langle G(t)x,x\rangle^{1/2}} + \frac{\left\langle G(t)x, \int\limits_{t_0}^{t} K(t,s)x(s)\,ds\right\rangle}{\langle G(t)x,x\rangle^{1/2}}$$

for $t \neq t_k$.

Now we define the Lyapunov functional

$$V(t, x(\,\cdot\,)) = W(t, x(t)) + \widehat{\beta}\int\limits_{t_0}^{t}\int\limits_{t}^{\infty} |K(u,s)|\, du\, |x(s)|\, ds.$$

From (3.7.4) and the assumptions (a) and (b), we obtain

$$V'_{(3.7.1)}(t, x(\,\cdot\,)) \leq -\widehat{M}\,|x(t)| + \widehat{K}\int\limits_{t_0}^{t} |K(t,s)|\,|x(s)|\, ds$$

$$+ \widehat{\beta} \int_t^\infty |K(u,t)|\, du\, |x(t)| - \widehat{\beta} \int_{t_0}^t |K(t,s)|\, |x(s))|\, ds$$

for $t \neq t_k$. Hence, in view of the assumption (c), it follows that

(3.7.5) $\qquad V'_{(3.7.1)}(t, x(\cdot)) \leq 0 \quad \text{for } t \neq t_k.$

Further, from the assumption (d), it is clear that

(3.7.6) $\qquad V(t_k^+, x_k + B_k x_k) \leq V(t_k, x_k), \quad \text{where } x_k = x(t_k).$

Thus, in view of Corollary 3.6.1, (3.7.5) and (3.7.6) give for all $t \geq t_0$ that

$$V(t, x(t)) \leq V(t_0, x_0),$$

where $x(t)$ is any solution of (3.7.1). Thus, together with the assumption (a) and the definition of V, we obtain

$$L\,|x(t)| \leq V(t, x(t)) \leq V(t_0, x_0) \leq \frac{1}{2\widehat{M}} |x_0| \qquad \text{for all } t \geq t_0.$$

This implies that the zero solution of (3.7.1) is uniformly stable, and hence the proof of the theorem is complete. $\qquad\square$

Theorem 3.7.2 *Assume that all the conditions of Theorem 3.7.1 hold except the condition (c) is replaced by*

$(\bar{c})\quad \widehat{\gamma} \leq \widehat{M} - \widehat{\beta}\int_t^\infty |K(u,t)|\, du$ *for some $\widehat{\gamma} > 0$ and $\widehat{\beta} > \widehat{K}$.*

Further, suppose that

$(e)\quad |A(t)| \leq \lambda$ *for all $t \in R_+$.*

If $\widehat{\gamma} > \lambda(\widehat{\beta} - \widehat{k})$ then the zero solution of (3.7.1) is asymptotically stable.

Proof As in the proof of Theorem 3.7.1, we get

(3.7.7) $\quad V'_{(3.7.1)}(t, x(\cdot)) \leq -\widehat{\gamma}\,|x(t)| - (\widehat{\beta} - \widehat{K})\int_{t_0}^t |K_0(t,s)|\, |x(s)|\, ds$

for $t \neq t_k$ and

(3.7.8) $\qquad V(t_k^+, x_k + B_k x_k) \leq V(t_k, x_k)$

where $x_k = x(t_k)$.

With the hypothesis (e), (3.7.1) implies

$$|x'(t)| \leq \lambda\,|x(t)| + \int_{t_0}^t |K(t,s)|\,|x(s)|\, ds \quad \text{for } t \neq t_k.$$

This together with (3.7.7) gives

$$V'_{(3.7.1)}(t, x(\cdot)) \leq -\hat{\gamma} |x(t)| + (\hat{\beta} - \hat{K})[\lambda |x(t)| - |x'(t)|]$$

for $t \neq t_k$. Thus we can find a $\theta > 0$ such that

(3.7.9) $\qquad V'_{(3.7.1)}(t, x(\cdot)) \leq -\theta[|x(t)| + |x'(t)|] \qquad$ for $t \neq t_k$

It follows from (3.7.8) and (3.7.9) that

$$V(t, x(t)) \leq V(t_0, x_0) - \theta \int_{t_0}^{t} [|x(s)| + |x'(s)|] ds$$

for all $t \geq t_0$.

Thus, in view of hypothesis (a), we obtain

$$L|x(t)| \leq V(t_0, x_0) - \theta \int_{t_0}^{t} |x(s)| ds - \theta x[t_0, t]$$

where $x[t_0, t]$ denotes the arc length of $x(t)$ from t_0 to t. Because $|x(t)| \geq 0$ and $\int_{t_0}^{\infty} |x(s)| ds < \infty$, this implies that there is a sequence $\{t_n\} \to \infty$ with $|x(t_n)| \to 0$, $x[t_0, t]$ is bounded. Hence $|x(t)| \to 0$ as $t \to \infty$, and, by Theorem 3.7.1, we have stability. Thus the proof of the theorem is complete. □

Now consider the nonlinear integro-differential system with fixed moments of impulsive effect

(3.7.10) $\qquad \begin{cases} x' = f(t,x) + \int_{t_0}^{t} g(t, s, x(s)) ds, & t \neq t_k, \quad k = 1, 2, \ldots \\ \Delta x = I_k(x), & t = t_k, \quad x(t_0) = x_0, \quad t \in R_+, \end{cases}$

in which $f \in PC^+[R_+ \times R^n, R^n]$, $g \in PC^+[R_+ \times R_+ \times R^n, R^n]$, $g(t, s, 0) \equiv 0$, $I_k \in C^1[R^n, R^n]$, $f_x = \partial f/\partial x$ exists and $f_x \in PC^+[R_+ \times R^n, R^n]$ with $f(t, 0) \equiv 0$ and $I_k(0) = 0$.

Consider the impulsive ordinary differential system

(3.7.11) $\qquad \begin{cases} x' = f(t,x), & t \neq t_k, \quad k = 1, 2, \ldots, \\ \Delta x = I_k(x), & t = t_k, \quad x(t_0) = x_0. \end{cases}$

Let $\Psi(t,s)$ be a fundamental matrix solution of the linear system

$$\begin{cases} y' = f_x(t,0)y, & t \neq t_k, \\ \Delta y = B_k y, & t = t_k, \end{cases}$$

where $B_k = \partial I_k/\partial x\big|_{x=0}$ and $f(t,x) = f_x(t,0)x + F(t,x)$. Let the matrix $\widehat{G}(t)$ be defined by

$$\widehat{G}(t) = \int_t^\infty \widehat{\Psi}^T(s,t)\widehat{\Psi}(s,t)\,ds, \quad \text{which exists and is finite for all } t \in R_+,$$

where $\widehat{\Psi}^T$ is the transpose of $\widehat{\Psi}$, and let the scalar function $W(t,x)$ be defined by

$$W(t,x) = \langle \widehat{G}(t)x, x \rangle^{1/2}.$$

Theorem 3.7.3 Assume that the following conditions hold for $|x| < \rho$ and $t \in R_+$:

(i)   there exists a positive number M such that $|W(t,x) - W(t,y)| \leq M|x-y|$;

(ii) $\alpha|x| \leq \langle \widehat{G}(t)x, x \rangle^{1/2} \leq L|x|$ with $\alpha, L > 0$;

(iii) $|\widehat{G}(t)F(t,x)| \leq |x|/A$ with $A > 0$;

(iv) $|g(u,t,x(t))| \leq R(u,t)|x(t)|$, where

$$\sup_{t \geq t_0} \int_t^\infty R(u,t)\,du < \infty;$$

(v) $\beta - K\int_t^\infty R(u,t)\,du \geq 0$, where $\beta = (A\alpha - 2L)/2Al\alpha > 0$ and $K \geq M$;

(vi) $|x| > |x + I_k(x)|$ and

$$\langle \widehat{G}(t)x, x \rangle^{1/2} > \langle \widehat{G}(t)(x + I_k(x)), x + I_k(x) \rangle^{1/2}.$$

Then the zero solution of (3.7.10) is uniformly stable.

Proof   From the definition $W(t,x) = \langle \widehat{G}(t)x, x \rangle^{1/2}$ and $f(t,x) = f_x(t,0)x + F(t,x)$ it follows for $t \neq t_k$ that

$$W'_{(3.7.11)}(t,x) = \frac{\langle [\widehat{G}'(t) + 2\widehat{G}(t)f_x(t,0)]x, x \rangle}{2\langle \widehat{G}(t)x, x \rangle^{1/2}} + \frac{\langle 2\widehat{G}(t)F(t,x), x \rangle}{2\langle \widehat{G}(t)x, x \rangle^{1/2}}.$$

From
$$\frac{\partial \Psi}{\partial t}(s,t) = -\Psi(s,t)f_x(t,0),$$
$$\frac{\partial \Psi^T}{\partial t}(s,t) = -f_x^T(t,0)\Psi^T(s,t)$$

for $t \neq t_k$, we have
$$\widehat{G}'(t) = -I + \int_t^\infty \left[\frac{\partial \Psi^T}{\partial t}(s,t)\Psi(s,t) + \Psi^T(s,t)\frac{\partial \Psi}{\partial t}(s,t)\right]ds$$

$$= -I - f_x^T(t,0)\widehat{G}(t) - \widehat{G}(t)f_x(t,0).$$

Therefore
$$\langle [\widehat{G}'(t) + 2\widehat{G}(t)f_x(t,0)]x, x \rangle = -\langle x, x \rangle,$$

and hence
$$W'_{(3.7.11)}(t,x) = \frac{-\langle x, x \rangle}{2\langle \widehat{G}(t)x, x\rangle^{1/2}} + \frac{2\langle \widehat{G}(t)F(t,x), x\rangle}{2\langle \widehat{G}(t)x, x\rangle^{1/2}}.$$

This, together with assumptions (ii) and (iii), gives

(3.7.12) $$W'_{(3.7.11)}(t,x) \leq \frac{-|x|}{2L} + \frac{|x|}{A\alpha} = -\beta |x|$$

for $t \neq t_k$.

We now define a Lyapunov functional
$$V(t, x(\,\cdot\,)) = W(t, x(t)) + K \int_{t_0}^{t} \int_t^\infty |g(u, s, x(s))|\, du\, ds.$$

Thus, in view of the assumption (i), we obtain for all $t \neq t_k$ that

$$V'_{(3.7.10)}(t, x(\,\cdot\,)) = W'_{(3.7.11)}(t,x) + M \int_{t_0}^{t} |g(t, s, x(s))|\, ds$$

$$+ K \int_t^\infty |g(u, t, x(t))|\, du$$

$$- K \int_{t_0}^{t} |g(t, s, x(s))|\, ds.$$

Using the hypothesis (iv) and (3.7.12), we get

$(3.7.13) \quad V'_{(3.7.10)}(t, x(\cdot)) \leq -\left[\beta - K \int_t^\infty R(u,t)\,du\right] |x(t)|$

$$+ (M-K) \int_{t_0}^t |g(t,s,x(s))|\,ds$$

for all $t \neq t_k$.

From the assumption (v), it follows that

$$V'_{(3.7.10)}(t, x(\cdot)) \leq 0 \quad \text{for } t \neq t_k.$$

Further, from the hypothesis (vi), we obtain

$$V(t_k^+, x_k + I_k(x_k)) \leq V(t_k, x_k)$$

where $x_k = x(t_k)$. Therefore, for all $t \geq t_0$, we have

$$\alpha |x(t)| \leq V(t, x(t)) \leq V(t_0, x_0) = W(t_0, x_0) \leq L |x_0|,$$

which in turn implies that the zero solution of (3.7.10) is uniformly stable, and hence the proof is complete. □

Theorem 3.7.4 *Suppose all the conditions of Theorem 3.7.3 are satisfied except the condition (v) is replaced by*

(v̄) $\quad \gamma \leq \beta - K \int_t^\infty R(u,t)\,du \quad \text{where} \quad \beta = (A\alpha - 2L)/2AL\alpha > 0, \quad K > M$
and $\gamma > 0$.

Further, assume that

$$|f(t,x)| \leq N \quad \text{for } |x| < \rho \text{ and } t \in R_+.$$

If $\gamma > N(K - M)$ then the zero solution of (3.7.10) is asymptotically stable.

Proof As in the proof of Theorem 3.7.3, it follows by (3.7.13) that

$$V'_{(3.7.10)}(t, x(\cdot)) \leq -\gamma |x(t)| - (K - M) \int_{t_0}^t |g(t,s,x(s))|\,ds.$$

The rest of the proof is exactly similar to that of Theorem 3.7.2, and hence is omitted. □

3.8 Notes and Comments

Theorems 3.1.1 − 3.1.3, Example 3.1.1, and Corollaries 3.1.1 and 3.1.2 are taken from Burton and Mahfoud [1], while Theorems 3.1.4 and 3.1.5, Examples 3.1.2 and 3.1.3, and Theorem 3.1.6 and its special cases are taken from Rama Mohana Rao and Raghavendra [1].

The results of Section 3.2 are contributions of Rama Mohana Rao and Sivasundaram [1]. Further results on equations with unbounded delay are found in Burton and Hering [1], Corduneanu [2], Corduneanu and Lakshmikantham [1], Hale and J. Kato [1], Leitman and Mizel [1], Luca [1], Wu [1] and Wang and Wu [1].

Theorems 3.3.1 and 3.3.2 are due to Hara, Yoneyama and Itoh [1], while Theorem 3.3.3 is taken from Miller [5]. The perturbation Theorems 3.3.4 − 3.4.10 are found in Hara, Yoneyama and Itoh [1]. For further results in this direction, see Becker [1], Burton [2, 4, 5], Brauer [1], Feller [1], Friedman [1], Grossman and Miller [1], Jordan [1], Krisztin [1, 2], Kuen and Ryabakowskii [1], Levin [1, 2], Levin and Shea [1], Rama Mohana Rao and Tsokos [1], Strauss [1], Angelova and Bainov [1] and Staffans [2] for integral and integro-differential equations; Burton [6], Driver [1, 2], Hale [1, 2], Krasovskii [1], Lakshmikantham and Leela [3], Kaplan and Yorke [1], Halanay [1] and Yoshizawa [1] for functional differential equations; and Prakasa Rao and Rama Mohana Rao [1, 2] for stochastic integro-differential equations.

The material covered in Section 3.4 is due to Lakshmikantham and Rama Mohana Rao [1]. See also Lakshmikantham [1, 2], Grimmer and Seifert [1], Haddock and Terjeki [1], J. Kato [1, 2], Rama Mohana Rao and Srinivas [1], Razumikhin [1], Seifert [2, 3] and Shendge [1]. The recent monograph by Lakshmikantham, Leela and Martynyuk [1] discusses various aspects of this topic.

The Lyapunov functions on product spaces discussed in Section 3.5 are due to Rama Mohana Rao and Sanjay [1], and the results on (h_0, h, M)-stability are taken from Leela and Rama Mohana Rao [1]. Similar results for functional differential equations may be found in Lakshmikantham, Leela and

Sivasundaram [1]. For ordinary differential equations, see Burton and Hatvani [1], Hatvani [1] and Moore [1].

The material of Section 3.6 on impulsive systems is taken from Lakshmikantham and Liu [1], while the results of Section 3.7 are due to Rama Mohana Rao, Sanjay and Sivasundaram [1]. Similar results without impulsive effect are found in Elaydi and Sivasundaram [1]. The recent monograph by Lakshmikantham, Bainov and Simeonov [1] is an excellent source of information for impulsive differential systems. For further results on this topic, see Deo and Pandit [1], Lakshmikantham, Liu and Sathananthan [1], Rama Mohana Rao and Raghavendra [2, 3], Rama Mohana Rao and Sree Hari Rao [1, 2], Bainov and Dishliev [1] and Simeonov and Bainov [1, 2].

4 EQUATIONS IN ABSTRACT SPACES

4.0 Introduction

It is well known that the theory of integro-differential equations in abstract spaces has proved in recent years to be a useful tool to study various classes of partial differential equations and integro-partial differential equations. These equations arise in many areas of applied sciences and engineering, such as viscoelasticity, thermodynamics, heat conduction in materials and wave propagation. Several approaches are found in the literature for the discussion of these equations, for example fixed point methods on various function spaces, resolvent operators, semigroups for convolution equations, the theory of evolution operators for problems of existence, uniqueness and well-posedness, and Laplace transform methods and Lyapunov techniques to investigate the stability properties of solutions.

In this chapter, we shall discuss the questions of existence, uniqueness, well-posedness and stability of solutions of Volterra integro-differential equations in a Banach space. Sections 4.1 and 4.2 deal with existence and uniqueness of solutions. In Section 4.3, we introduce some results on well-posedness of solutions of linear equations. A relation between semigroups and the existence of resolvent operators for linear convolution equations is discussed in Section 4.4. The theory of linear evolution operators for linear time-dependent equations and the variation of parameters formula is developed in Section 4.5. The stability and asymptotic behavior of solutions are investigated in Sections 4.6 and 4.7.

4.1 Existence and Uniqueness

In this section, we investigate the initial value problem of first-order nonlinear integro-differential equation of Volterra type, namely

(4.1.1) $$x' = H(t, x, Tx), x(t_0) = x_0$$

in a real Banach space E, where

(4.1.2) $$(Tx)(t) = \int_{t_0}^{t} K(t, s) x(s) \, ds,$$

$K \in C[J \times J, R], \quad |K(t,s)| \leq K_1 \quad \text{on } J \times J,$
$H \in C[J \times \Omega \times \Omega, E],$
$J = [t_0, t_0 + a], \quad t_0 \geq 0,$
$\Omega = B(0, N) = \{|x| < N : x \in E\},$
$x_0 \in \Omega.$

The problem (4.1.1) is equivalent to the following integral equation:

(4.1.3) $$x(t) = x_0 + \int_{t_0}^{t} H(s, x(s), (Tx)(s)) \, ds.$$

We obtain an existence theorem using Darbo's fixed point theorem and an existence and uniqueness theorem by means of the classical method of successive approximations.

Let us begin by prove the following existence result.

Theorem 4.1.1 *Assume that*

(A_1) $K(t,s) \in C(J \times J, R)$ and $|K(t,s)| \leq K_1$ $(K_1 a \geq 1)$ for $(t,s) \in J \times J$;

(A_2) $H(t,x,y) \in C(J \times \Omega \times \Omega_1, E)$, $\Omega_1 = B(0, K_1 a N) \subset E$;

(A_3) $\alpha(H(I \times B_1 \times B_2)) \leq \lambda \; \max[\alpha(B_1), \alpha(B_2)]$ for every bounded subset $B_1 \subset \Omega$, $B_2 \subset \Omega_1$, and every interval $I \subset J$, where λ is some positive number and $\alpha(\cdot)$ is Kuratowski's measure of noncompactness.

Then there exists a $\gamma > 0$ such that (4.1.1) has a solution $x(t)$ for $t \in J_0 = [t_0, t_0 + \gamma]$.

Proof Let $\eta = \sup\{\epsilon\colon B(x_0,\epsilon) \subset \Omega\}$. It follows by the assumption (A_3) that $H(t,x,y)$ maps bounded sets into bounded sets, and hence there exists a constant $M > 0$, such that

$$|H(t,x,y)| \leq M \quad \text{for } t \in J,\ x,y \in B(0, |x_0| + \tfrac{1}{2}\eta) = \bar{\Omega}_0.$$

Let $\gamma = \min(a, \eta/2M, 1/K_1, 1/2\lambda)$. Define the operator $H\colon C(J_0, E) \to C(J_0, E)$ by

$$(H\phi)(t) = x_0 + \int_{t_0}^t H(s, \phi(s), (T\phi)(s))\,ds.$$

Define the set A by

$$A = \{\phi \in C(J_0, \bar{\Omega}_0)\colon \max_{J_0} |\phi(t) - x_0| \leq \tfrac{1}{2}\eta,\ |\phi(t_1) - \phi(t_2)|$$
$$\leq M|t_1 - t_2|, \text{ for } t_1, t_2 \in J_0\}.$$

It is clear that A is closed, bounded and convex. If $\phi \in A$ then, by the assumption (A_1), (A_2) and the choice of γ, we have

$$|(T\phi)(t)| = \left|\int_{t_0}^t K(t,s)\phi(s)\,ds\right| \leq K_1 \gamma \max_{s \in [t_0, t]} |\phi(s)| \leq |x_0| + \tfrac{1}{2}\eta,\ t \in J_0.$$

Hence

$$|(H\phi)(t) - x_0| \leq \int_{t_0}^t |H(s, \phi(s), (T\phi)(s))\,ds \leq M(t - t_0) \leq \tfrac{1}{2}\eta,\ t \in J_0,$$

and

$$|(H\phi)'(t)| = |H(t, \phi(t), (T\phi)(t))| \leq M,\quad t \in J_0.$$

Therefore

$$|(H\phi)(t_1) - (H\phi)(t_2)| \leq M|t_1 - t_2|,\quad t_1, t_2 \in J_0.$$

Thus $HA \subset A$, and clearly $H \in C(A, A)$. Now note that A is a bounded equicontinuous set. Let $B \subset A$; then, using the properties of α, assumption (A_3) and $\gamma\lambda \leq \tfrac{1}{2}$, we have

$$\alpha((HB)(t)) = \alpha\left(\int_{t_0}^t H(s, \phi(s), (T\phi)(s))\,ds + x_0 \mid \phi \in B\right)$$

$$\leq \gamma\alpha(\overline{CO}\{H(s,\phi(s),(T\phi)(s)) + \phi \in B, s \in J_0\})$$

$$\leq \gamma\alpha \left(H(J_0 \times B(J_0) \times (TB)(J_0))\right)$$

$$\leq \gamma\lambda \max[\alpha(B(J_0)), \gamma K_1\alpha(B(J_0))]$$

$$\leq \tfrac{1}{2}\alpha(B),$$

where

$$B(J_0) = \bigcup_{s \in J_0} \{\phi(s), \phi \in B\}$$

and

$$\alpha((TB)(J_0)) = \alpha\left(\bigcup_{t \in J_0} \int_{t_0}^{t} K(t,s)\phi(s)\,ds, \quad \phi \in B\right)$$

$$\leq K_1\gamma\alpha(B(J_0)).$$

Since HB is also a bounded equicontinuous set, we have

$$\alpha(H(B)) = \sup_{J_0} \alpha((HB)(t)) \leq \tfrac{1}{2}\alpha(B).$$

Consequently, by Darbo's fixed point theorem, there exists a fixed point x of H. Clearly such a fixed point is a solution of (4.1.1). The proof is complete.□

Remark 4.1.1 If $E = R^n$ then the assumption (A_3) in Theorem 4.1.1 is superfluous.

We need the following lemmas before we proceed further.

Lemma 4.1.1 *Assume that*

(L_1) $g \in C[R_0, R]$, where $R_0 = [t_0, t_0 + a) \times (-q, q) \times (-K_1 aq, K_1 aq)$, $|g(t,u,v)| \leq M_0$, for $(t,u,v) \in R_0$, and $g(t,u,v)$ is nondecreasing in v for each (t,u);

(L_2) $[t_0, t_0 + a)$ is the largest interval in which the maximal solution $\gamma(t, t_0, u_0)$ of

(4.1.4) $$u' = g(t, u, Su), \quad u(t_0) = u_0$$

exists, where $u_0 \in (-q, q)$ and

$$(Su)(t) = \int_{t_0}^{t} N(t,s)u(s)\,ds$$

$N(t,s) \in C[J \times J, R^+]$ and $N(t,s) \leq K_1$, for $(t,s) \in J \times J$.

Suppose further that $J_1 = [t_0, t_1] \subset [t_0, t_0 + a]$.

Then there is an $\epsilon_0 > 0$ *such that for* $0 < \epsilon < \epsilon_0$, *the maximal solution* $\gamma(t, \epsilon)$ *of*

(4.1.5) $$u' = g(t, u, Su) + \epsilon, \quad u(t_0) = u_0 + \epsilon,$$

exists on J_1 *and*

$$\lim_{\epsilon \to 0} \gamma(t, \epsilon) = \gamma(t, t_0, u_0)$$

uniformly on J_1.

Proof Let Δ be an open interval $\Delta \subset (-q, q)$, and $(t, \gamma(t, t_0, u_0)) \in J \times \Delta$ for $t \in J_1$. We can choose a $b > 0$ such that for $t \in J_1$, the rectangle $R_t^\epsilon = [(s, u): s \in t, t+b]$ and $|u - (\gamma(t) + \epsilon)| \le b$ is included in $J \times \Delta$ for $\epsilon \le \frac{1}{2}b$. Note that

$$|(Su)(t)| \le \int_{t_0}^{t} N(t,s) |u(s)| \, ds \le K_1 |t - t_0| \max_{s \in [t_0, t]} |u(s)| < K_1 a q$$

and $|g(t, u, Su)| \le M_0$ for $(t, u(t), (Su)(t)) \in R_0$. Then it is evident that

$$|g(t, u, Su) + \epsilon) \le M_0 + \tfrac{1}{2}b.$$

Let $\eta = \min(b, 2b/(2M_0 + b), 1/K_1)$. Note that η does not depend upon ϵ. Then using the special case of Theorem 4.1.1 and proceeding as in Theorem 1.3.1 in Lakshmikantham and Leela [1], (4.1.5) has the maximal solution $\gamma(t, \epsilon)$ and

$$\lim_{\epsilon \to 0} \gamma(t, \epsilon) = \gamma(t, t_0, u_0)$$

uniformly on $[t_0, t_0 + \eta]$. The rest of the proof runs along the same lines as Lemma 1.3.1 in Lakshmikantham and Leela [1]. □

Lemma 4.1.2 *Assume that* (L_1) *and* (L_2) *hold, and suppose further that:*

(L_3) $m \in C[[t_0, t_0 + a), R]$ *is such that* $(t, m(t)) \in J \times (-q, q)$ *for* $t \in [t_0, t_0 + a)$, $m(t_0) \le u_0$ *and for a fixed Dini derivative*

$$D_- m(t) \le g(t, m(t), (Sm)(t)), \quad t \in [t_0, t_0 + a)$$

then

(4.1.6) $$m(t) \leq \gamma(t, t_0, u_0), \quad t \in [t_0, t_0 + a).$$

Proof First of all, we prove that $m(t) \leq \gamma(t, \epsilon)$, on $[t_0, t_0 + a)$, where $\gamma(t, \epsilon)$ is the maximal solution of (4.1.5). In fact, if this assertion is false then the set $Z = [t \in [t_0, t_0 + a) : m(t) > \gamma(t, \epsilon)]$ is nonempty. Define $t_1 = \inf Z$. Since $m(t_0) \leq u_0 < \gamma(t_0, \epsilon) + \epsilon$, $t_1 > t_0$. Furthermore $m(t_1) = \gamma(t_1, \epsilon)$ and $m(t) \leq \gamma(t, \epsilon)$ $t \in [t_0, t_1)$. Then

$$D_-m(t_1) \geq D_-\gamma(t_1, \epsilon).$$

On the other hand, since $N(t, s) \geq 0$, $(Sm)(t) \leq (S\gamma)(t)$, $t \in [t_0, t_1)$, and using the monotonicity of g, we obtain

$$\begin{aligned} D_-m(t_1) &\leq g(t_1, m(t_1), (Sm)(t_1)) \\ &\leq g(t_1, \gamma(t_1, \epsilon), (S\gamma)(t_1)) \\ &< D_-\gamma(t_1, \epsilon). \end{aligned}$$

This contradiction proves that Z is empty. Thus

$$m(t) \leq \gamma(t, \epsilon) \quad \text{on } [t_0, t_0 + a).$$

Finally, by Lemma 4.1.1, we have, as $\epsilon \to 0$,

$$m(t) \leq \gamma(t, t_0, u_0), \quad t \in [t_0, t_0 + a).$$

The proof of the lemma is complete. □

We are now in a position to prove our main result.

Theorem 4.1.2 *Assume that*

$(A_1)^*$ $K(t, s) \in C(J \times J, R)$ *and* $|K(t, s)| \leq K$ *for* $(t, s) \in J \times J$;

$(A_2)^*$ $H(t, x, y) \in C[J \times \Omega \times \Omega, E]$ *and* $|H(t, x, y)| \leq M_1$ *for* $(t, x, y) \in J \times \Omega \times \Omega$;

(A_4) $g \in C[J \times [0, 2N] \times [0, 2N_1], R^+]$, $N_1 = KaN$, $g(t, u, v) \leq M_0$ *on* $J \times [0, 2N] \times [0, 2N_1]$, $g(t, 0, 0) \equiv 0$, $g(t, u, v)$ *is nondecreasing in v for fixed t and u, and nondecreasing in u for fixed t and v, and $u \equiv 0$ is the unique solution of the scalar integro-differential equation*

(4.1.7) $$u' = g(t, u, Su), \quad u(t_0) = 0$$

on $[t_0, t_0 + a]$, *where*

$$(Su)(t) = \int_{t_0}^{t} |K(t,s)| u(s) ds;$$

(A_5) $|H(t,x_1,y_1) - H(t,x_2,y_2)| \leq g(t, |x_1-x_2|, |y_1-y_2|)$ on $J \times \Omega \times \Omega$.

Then there exists a $\gamma > 0$ such that the problem (4.1.1) has an unique solution for $t \in [t_0, t_0+\gamma]$.

Proof We shall prove the theorem by the method of successive approximations. Let $\eta = \sup\{\epsilon = B(x_0,\epsilon) \subseteq \Omega\}$, $M = \max(M_0, M_1)$, $\gamma = \min(a, \eta/2M, 1/K_1)$ and $J_0 = [t_0, t_0+\gamma]$. The successive approximations are defined by

(4.1.8) $\begin{cases} x_0(t) = x_0 \\ x_n(t) = x_0 + \int_{t_0}^{t} H(s, x_{n-1}(s), (Tx_{n-1})(s)) ds \quad (n=1,2,\ldots). \end{cases}$

If $x \in C(J_0, \Omega)$ then, by the assumption $(A_1)^*$

$$|(Tx)(t)| \leq \int_{t_0}^{t} |K(t,s)| |x(s)| ds \leq K_1 \gamma \max_{s \in [t_0, t]} |x(s)| < N;$$

that is, T maps $C[J_0\Omega] \to \Omega$. By the assumption $(A_2)^*$ and the choice of γ, we have

$$|x_n(t) - x_0| \leq M|t-t_0| \leq \tfrac{1}{2}\eta \quad \text{on } J_0,$$

and hence

$$|x_n(t)| \leq |x_0| + \tfrac{1}{2}\eta < N.$$

Thus the successive approximations (4.1.8) are well defined and continuous on $[t_0, t_0+\gamma]$.

We also define the successive approximations for the problem (4.1.7) as follows

$$(4.1.9) \quad \begin{cases} u_0(t) = M(t-t_0) \leq \tfrac{1}{2}\eta < N, \\ u_{n+1}(t) = \int_{t_0}^t g(s, u_n(s), (Su_n)(s))\,ds, \quad t \in J_0, \quad n = 1,2,\ldots. \end{cases}$$

Note that

$$u_1(t) = \int_{t_0}^t g(s, u_0(s), (Su_0)(s))\,ds \leq M(t-t_0) = u_0(t)$$

by the assumption (A_4) and $0 \leq |K(t,s)| \leq K_1$ for $(t,s) \in J \times J$. By induction, it is easy to prove that the successive approximations (4.1.9) are well defined and satisfy

$$0 \leq u_{n+1}(t) \leq u_n(t), \quad t \in J_0.$$

Since $|u_n(t)| \leq M$, we conclude by the Ascoli–Arzela theorem and the monotonicity of the sequence $\{u_n(t)\}$ that $\lim_{n \to \infty} u_n(t) = u(t)$ uniformly on J_0. It is also clear that $u(t)$ satisfies (4.1.7). Hence by (A_4), $u(t) \equiv 0$ on J_0.

Now since

$$|x_1(t) - x_0| \leq \int_{t_0}^t |H(s, x_0(s), (Tx_0)(s))|\,ds \leq M(t-t_0) = u_0(t),$$

we assume that $|x_k(t) - x_{k-1}(t)| \leq u_{k-1}(t)$ for a given k; then using the assumption (A_4) and (A_5), we have

$$|x_{k+1}(t) - x_k(t)| \leq \int_{t_0}^t |H(s, x_k(s), (Tx_k)(s)) - H(s, x_{k-1}(s), (Tx_{k-1})(s))|\,ds$$

$$\leq \int_{t_0}^t g(s, |x_k(s) - x_{k-1}(s)|, |(Sx_k)(s) - (Sx_{k-1})(s)|)\,ds$$

$$\leq \int_{t_0}^t g(s, u_{k-1}(s), (Su_{k-1})(s))\,ds = u_k(t).$$

Thus, by induction, the inequality

$$|x_{n+1}(t) - x_n(t)| \leq u_n(t), \quad t \in J_0,$$

is true for all n. Also

$$|x'_{n+1}(t) - x'_n(t)| = |H(t, x_n(t), (Tx_n)(t)) - H(t, x_{n-1}(t), (Tx_{n-1})(t))|$$

$$\leq g(t, |x_n(t) - x_{n-1}(t)|, |(Sx_n)(t) - (Sx_{n-1})(t)|)$$
$$\leq g(t, u_{n-1}(t), (Su_{n-1})(t)).$$

Let $n \leq m$; then we have

$$|x'_n(t) - x'_m(t)| \leq |x'_{n+1}(t) - x'_n(t)| + |x'_{m+1}(t) - x'_m(t)|$$
$$+ |x'_{n+1}(t) - x'_{m+1}(t)|$$
$$\leq g(t, u_{n-1}(t), (Su_{n-1})(t)) + g(t, u_{m-1}(t), (Su_{m-1})(t))$$
$$+ g(t, |x_n(t) - x_m(t)|, |(Sx_n)(t) - (Sx_m)(t)|).$$

Since $u_{n+1}(t) \leq u_n(t)$ for all n, it follows that

$$D^+[|x_n(t) - x_m(t)|] \leq g(t, |x_n(t) - x_m(t)|, |(S(x_n - x_m))(t)|)$$
$$+ 2g(t, u_{n-1}(t), (Su_{n-1})(t)).$$

Using Lemma 4.1.2, we have

$$|x_n(t) - x_m(t)| \leq \gamma_n(t), \quad t \in J_0$$

where $\gamma_n(t)$ is the maximal solution of

$$v' = g(t, v, Sv) + 2g(t, u_{n-1}(t), (Su_{n-1})(t)), \quad v(0) = 0$$

for each n. Since, as $n \to \infty$, $2g(t, u_{n-1}(t), (Su_{n-1})(t)) \to 0$ uniformly on J_0, it follows by Lemma 4.1.2 that $\gamma_n(t) \to 0$ uniformly on J_0. This implies that $x_n(t)$ converges uniformly to $x(t)$, and it is now easy to show that $x(t)$ is a solution of (4.1.1) by standard arguments.

To show that this solution is unique, let $y(t)$ be another solution of (4.1.1) existing on J_0. Define $m(t) = |x(t) - y(t)|$ and note that $m(t_0) = 0$, using the assumption (A_5); then

$$D_-m(t) \leq |x'(t) - y'(t)|$$
$$= |H(t, x(t), (Tx)(t)) - H(T, y(t), (Ty)(t))|$$
$$\leq g(t, m(t), (Sm)(t)).$$

Again applying Lemma 4.1.2, we have

$$m(t) \leq \gamma(t), \quad t \in J_0$$

where $\gamma(t)$ is the maximal solution of (4.1.7). But by the assumption (A_4),

$\gamma(t) \equiv 0$ and this proves that $x(t) \equiv y(t)$. Hence the limit of the successive approximations is the unique solution of (4.1.1). The proof is complete. □

Corollary 4.1.1 *Assume that $(A_1)^*$ and $(A_2)^*$ hold and suppose further that*

(A_6) $|H(t,x_1,y_1) - H(t,x_2,y_2)| \leq L(|x_1 - x_2| + |y_1 - y_2|)$ *for any* $(t,x_1,y_1), (t,x_2,y_2) \in J \times \Omega \times \Omega$, *where L is a Lipschitz constant.*

Then there exists a $\gamma > 0$ such that the problem (4.1.1) has a unique solution on J_0.

4.2 Existence of Maximal and Minimal Solutions

Using the monotone iterative method, we discuss in this section the existence of maximal and minimal solutions of the initial value problem of the nonlinear integro-differential equation

(4.2.1) $$x'(t) = H(t, x, Sx), \quad x(0) = x_0,$$

in a real Banach space E, where $(Sx)(t) = \int_0^t S(t,s)x(s)\,ds$, for $x \in C[I,E]$, $x_0 \in E$, $S(\cdot,\cdot) \in C[I \times I, R_+]$, $S(t,s) \leq S_0$ for $(t,s) \in I \times I$, $H \in C[I \times E \times E, E]$, $I = [0,\tau]$, and $\tau > 0$ is finite.

Let K be a cone of E, then the cone K induces a partial ordering on E defined by $u \leq v$ if $v - u \in K$.

Let E^* denote the set of continuous linear functionals on E. Given a cone K, we let

$$K^* = \{\phi \in E: \phi(u) \geq 0 \text{ for all } u \in K\}.$$

A cone K is said to be normal if there exists a real number $L > 0$ such that $0 \leq u \leq v$ implies $|u| \leq L|v|$, where L is independent of u and v. We shall assume in this section that K is a normal cone.

Definition 4.2.1 *The function $w \in C^1[I,E]$ is called an upper solution of (4.2.1) if*

(4.2.2) $$w'(t) \geq H\left(t, w(t), \int_0^t S(t,s)w(s)\,ds\right), \quad w(0) \geq x_0.$$

Similarly, a function $v \in C^1[I,E]$ is called a lower solution of (4.2.1) if

(4.2.3) $\qquad v'(t) \leq H\left(t, v(t), \int_0^t S(t,s)v(s)\,ds\right), \quad v(0) \leq x_0.$

Definition 4.2.2 The function ρ and $r \in C^1[I, E]$ are called minimal and maximal solutions of (4.2.1) respectively if every solution $x \in C^1[I,E]$ of (4.2.1) satisfies the relation $\rho(t) \leq x(t) \leq r(t)$ for $t \in I$.

We need the following lemmas, which are due to Mönch and Von Harten [1].

Lemma 4.2.1 Let E_1 be a separable Banach space and $\beta(\cdot)$ the Hausdorff measure of noncompactness on E_1. Let $\{x_n\}$ be a sequence of continuous functions from $J = [a,b]$ to E_1 such that there is some function $u \in L^1(a,b)$ with $|x_n(t)| \leq u(t)$ on J. Let $\psi(t) = \beta(\{x_n(t)\}_{n=1}^\infty)$. Then $\psi(t)$ is integrable on J and

$$\beta\left(\left\{\int_a^b x_n(t)\,dt\right\}_{n=1}^\infty\right) \leq \int_a^b \psi(t)\,dt.$$

Lemma 4.2.2 Let $\{x_n\}$ be a sequence of continuously differentiable functions from $J = [a,b]$ to E such that there is some $u \in L^1(a,b)$ with $|x_n(t)| \leq u(t)$ and $|x_n'(t)| \leq u(t)$ on J. Let $\psi(t) = \beta(\{x_n(t)\}_{n=1}^\infty)$. Then ψ is absolutely continuous and

$$\psi'(t) \leq 2\beta(\{x_n'(t)\}_{n=1}^\infty) \quad \text{a.e. on } J.$$

We make the following assumptions for convenience.

(A_1) The functions $v_0, w_0 \in C^1[I, E]$ with $v_0(t) \leq w_0(t)$ on I are lower and upper solutions of (4.2.1) respectively. We define the conical segments

$$[v_0, w_0] = \{u \in E : v_0(t) \leq u \leq w_0(t),\ t \in I\}.$$

Similarly,

$$[Sv_0, Sw_0] = \{u \in E : (Sv_0)(t) \leq u \leq (Sw_0)(t),\ t \in I\}.$$

(A_2) $H(t,x,u) - H(t,\bar{x},\bar{u}) \geq -M(x-\bar{x}) - N(u-\bar{u})$ whenever $x, \bar{x} \in [v_0, w_0]$, $u, \bar{u} \in [Sv_0, Sw_0]$ and $x \geq \bar{x}$, $u \geq \bar{u}$, where $M > 0$, $N \geq 0$ are constants with $NS_0\tau(e^{M\tau} - 1) \leq M$.

(A_3) For any bounded set $B \subset [v_0, w_0]$, $B_1 \subset [Sv_0, Sw_0]$, $t \in I$

$\beta(H(t,B,B)) \leq \lambda \max(\beta(B),\beta(B_1))$ for some $\lambda > 0$, where $\beta(\cdot)$ denotes the Hausdorff measure of noncompactness. Then we have the following main result.

Theorem 4.2.1 *Let the cone K be normal and the assumptions (A_1) – (A_3) hold. Then, for any $x_0 \in [v_0(0), w_0(0)]$, there exist monotone sequences $\{v_n\}$ and $\{w_n\}$ that converge uniformly and monotonically to the minimal and maximal solutions $\rho(t)$ and $r(t)$ respectively of the problem (4.2.1) in $[v_0, w_0]$, and*

$$v_0 \leq v_1 \leq \cdots \leq v_n \leq \cdots \leq \rho \leq u \leq r \leq \cdots \leq w_n \leq \cdots \leq w_1 \leq w_0 \quad \text{on } I.$$

The proof of Theorem 4.2.1 will be completed by the following five lemmas.

Lemma 4.2.3 *Let $y(t) \in C^1[I, R]$ be such that*

$$(4.2.4) \qquad y'(t) \leq -My(t) - N \int_0^t S(t,s) y(s)\,ds, \quad y(0) \leq 0,$$

where $M > 0$, $N \geq 0$, $S(\cdot,\cdot) \in C[I \times I, R^+]$, $S(t,s) \leq S_0$ for $(t,s) \in I \times I$. Suppose further that $NS_0\tau(e^{M\tau} - 1) \leq M$; then $y(t) \leq 0$, $t \in I$.

Proof Set $m(t) = y(t)e^{Mt}$, so that the inequality (4.2.4) reduces to

$$(4.2.5) \qquad m'(t) \leq -N \int_0^t S^*(t,s) m(s)\,ds, \quad m(0) \leq 0,$$

where $S^*(t,s) = S(t,s)e^{M(t-s)}$. It is enough to prove that $m(t) \leq 0$, for $t \in I$. If this is false then there exists $t_1 \in (0, \tau]$, $m(t_1) > 0$. If $m(t) \geq 0$ for $t \in [0, t_1]$ then from (4.2.5) we get $m'(t) \leq 0$, $t \in [0, t_1]$; that is, if $m(t)$ is nonincreasing in $[0, t_1]$ then $m(t_1) \leq m(0) \leq 0$. This is a contradiction. So there exists $t_2 \in [0, t_1]$, $m(t_2) = \min_{t \in [0, t_1]} m(t) = -\mu < 0$. By the mean value theorem on $[t_2, t_1]$; there exists $t_3 \in (t_2, t_1)$ such that

$$m'(t_3) = \frac{m(t_1) + \mu}{t_1 - t_2} > \frac{\mu}{\tau}.$$

On the other hand, it follows from (4.2.5) that

$$m'(t_3) \leq -N \int_0^{t_3} S^*(t_3,s)m(s)\,ds \leq \frac{NS_0\mu}{M}(e^{Mt_3}-1)$$

$$\leq \frac{NS_0\mu}{M}(e^{M\tau}-1).$$

Then from

$$\frac{NS_0\mu}{M}(e^{M\tau}-1) > \frac{\mu}{\tau},$$

we get

$$NS_0\tau(e^{M\tau}-1) > M.$$

This is a contradiction to the assumption $NS_0\tau(e^{M\tau}-1) \leq M$. The proof of Lemma 4.2.3 is complete. □

Lemma 4.2.4 *Let $y(t) \in C[I,R]$, $y(0) \leq 0$, satisfy*

(4.2.6) $$y'(t) \leq A_1 y(t) + A_2 \int_0^t y(s)\,ds \quad \text{a.e. on } I = [0,\tau],$$

where $A_1, A_2 \geq 0$. Then $y(t) \leq 0$ for any $t \in I$.

Proof If the conclusion is false we choose $\eta > 0$ such that $(A_1 + A_2\eta)\eta < 1$ and $n\eta = \tau$ for some integer n. First of all, we prove that $y(t) \leq 0$ for $t \in [0,\eta]$. If this is not true then there exists $t_1 \in [0,\eta]$ such that $y(t_1) = \mu = \max_{t \in [0,\eta]} y(t) > 0$. Define

$$\Sigma = \left\{ t \in (0,t_1] : y'(t) \text{ exists and } y'(t) \leq A_1 y(t) + A_2 \int_0^t y(s)\,ds \right\}.$$

So $\text{mes}(\Sigma) = t_1$. Then it is certain that there exists a $t_2 \in \Sigma$ such that $y'(t_2) \geq \mu/t_1$. Otherwise, if $y'(t) < \mu/t_1$ for all $t \in \Sigma$ then we get

$$\mu \leq \int_0^{t_1} y'(t)\,dt = \int_\Sigma y'(t)\,dt < \mu.$$

This is a contradiction. On the other hand, from (4.2.6), we have

$$\frac{\mu}{t_1} \leq y'(t_2) \leq A_1 y(t_2) + A_2 \int_0^{t_2} y(t)\,dt$$

$$\leq A_1 \mu + A_2 \mu \eta,$$

and hence

$$1 \leq (A_1 + A_2\eta)\eta.$$

This is a contradiction to the choice of η. Thus we have proved that $y(t) \leq 0$, for $t \in [0,\eta]$. We can further prove that $y(t) \leq 0$ for $t \in [\eta, 2\eta]$, by repeating the above process. This argument can be repeated until $n\eta = \tau$. So we get $y(t) \leq 0$, $t \in [0,\tau]$. The proof of Lemma 4.2.4 is complete. □

In order to prove Theorem 4.2.1, let us consider the following linear initial value problem:

(4.2.7) $$x'(t) = -Mx - N\int_0^t S(t,s)x(s)\,ds + \sigma(t), \quad x(0) = x_0$$

where

$$\sigma(t) = H(t,\eta(t),(S\eta)(t)) + M\eta(t) + N(S\eta)(t)$$

for $\eta \in C[I,E]$ such that $\eta(t) \in [v_0, w_0]$ for all $t \in I$ and $x_0 \in E$ such that $v_0(0) \leq x_0 \leq w_0(0)$.

Lemma 4.2.5 *For any $\eta \in C[I,E]$ such that $\eta(t) \in [v_0, w_0]$ for $t \in I$, the problem (4.2.7) possesses an unique solution on I.*

Proof It is easy to check that the problem (4.2.7) is equivalent to the following equation:

(4.2.8) $$x(t) = x_0 e^{-Mt} + \int_0^t \sigma(s)e^{M(s-t)}\,ds$$

$$- Ne^{-Mt}\int_0^t \left[e^{Ms}\int_0^s S(s,\xi)x(\xi)\,d\xi\right]ds.$$

By changing the order of integration, (4.2.8) reduces to the following linear integral equation of Volterra type:

(4.2.9) $$x(t) = x_0 e^{-Mt} + \int_0^t \sigma(s)e^{M(s-t)}\,ds + \int_0^t S^*(t,\xi)x(\xi)\,d\xi,$$

where

$$S^*(t,\xi) = -Ne^{-Mt}\int_\xi^t S(s,\xi)e^{Ms}\,ds \in C[I \times I, R].$$

Now, it is easy to see that (4.2.9) has a unique solution on I. The proof of Lemma 4.2.5 is therefore complete. □

For each $\eta \in C[I, E]$ such that $v_0(t) \leq \eta(t) \leq w_0(t)$ on I, we define the mapping A by $A\eta = x$, where x is the unique solution of (4.2.7) corresponding to η. This mapping will be used to define the sequences that converge to the minimal and maximal solution of (4.2.1).

Lemma 4.2.6 *Assume that the hypotheses (A_1) and (A_2) hold. Then*
(i) $v_0 \leq Av_0$ *and* $w_0 \geq Aw_0$;
(ii) A *is a monotone operator on the segment* $[v_0, w_0]$; *that is, if* $\eta_1(t)$, $\eta_2(t) \in [v_0, w_0]$ *for* $t \in I$ *with* $\eta_1 \leq \eta_2$ *then* $A\eta_1 \leq A\eta_2$ *for* $t \in I$.

Proof To prove (i), set $Av_0 = v_1$, where v_1 is the unique solution of (4.2.7) corresponding to v_0. For any $\phi \in K^*$, set $y(t) = \phi(v_0(t) - v_1(t))$, so that $y(0) \leq 0$, and $y'(t) = \phi(v_0'(t) - v_1'(t))$. Then we have

$$y'(t) \leq \phi\left(H(t, v_0, Sv_0) - H(t, v_0, Sv_0) + M(v_1 - v_0) + N\int_0^t S(t,s)(v_1 - v_0)\,ds\right)$$

$$= -My - N\int_0^t S(t,s)y(s)\,ds.$$

By Lemma 4.2.3, it follows that $y(t) \leq 0$ on I. Since $\phi \in K^*$ is arbitrary, this implies that $v_0(t) \leq v_1(t)$, $t \in I$. Thus we have proved that $v_0 \leq Av_0$ for $t \in I$. Similar argument shows that $w_0 \geq Aw_0$ for $t \in I$, proving (i).

To prove (ii), let $\eta_1, \eta_2 \in C[I, E]$ be such that $\eta_1(t), \eta_2(t) \in [v_0, w_0]$ for $t \in I$ and $\eta_1 < \eta_2$. Suppose that $\xi_1 = A\eta_1$ and $\xi_2 = A\eta_2$. Let $Z(t) = \phi(\xi_1(t) - \xi_2(t))$, so that $Z(0) \leq 0$, where $\phi \in K^*$. Then

$$Z'(t) = \phi(\xi_1'(t) - \xi_2'(t))$$
$$= \phi(H(t, \eta_1(t), (S\eta_1)(t)) - M(\xi_1(t) - \eta_1(t)) - N((S\xi_1)(t) - (S\eta_1)(t))$$
$$- H(t, \eta_2(t), (S\eta_2)(t)) + M(\xi_2(t) - \eta_2(t)) + N((S\xi_2)(t) - (S\eta_2)(t))$$
$$\leq -MZ(t) - N\int_0^t S(t,s)Z(s)\,ds,$$

using (A_2). By Lemma 4.2.3, we get $Z(t) \leq 0$ on I. Since $\phi \in K^*$ is arbitrary, this implies that $\xi_1(t) \leq \xi_2(t)$ on I. That is $A\eta_1 \leq A\eta_2$. The proof of Lemma

4.2.4 is therefore complete. □

We now define the sequences $\{v_n\}$ and $\{w_n\}$ as follows:

$$v_n = Av_{n-1}, \qquad w_n = Aw_{n-1}.$$

It is easy to see from Lemma 4.2.4 that the sequences $\{v_n\}$ and $\{w_n\}$ are monotone nondecreasing and nonincreasing respectively, and $v_n \leq w_n$ and $v_n(t), w_n(t) \in [v_0, w_0]$ for all $t \in I$. Now we show that there exist subsequences of $\{v_n\}$ and $\{w_n\}$ that converge uniformly on I.

Lemma 4.2.7 *Suppose that the assumptions* $(A_1)-(A_3)$ *hold. Then the sequences* $\{v_n\}$ *and* $\{w_n\}$ *are uniformly bounded, equicontinuous and relatively compact on* I.

Proof Since the cone K is assumed to be normal, it follows from $v_n(t), w_n(t) \in [v_0, w_0]$ for all $t \in I$, that $\{v_n\}$ and $\{w_n\}$ are uniformly bounded.

By means of the assumption (A_2) and $v_n(t), w_n(t) \in [v_0, w_0]$, $(Sv_n)(t)$, $(Sw_n)(t) \in [Sv_0, Sw_0]$, for all $t \in I$ and for each n, we have $H(t, w_0, Sw_0) - H(t, v_n, Sv_n) \geq -M(w_0 - v_n) - N(Sw_0 - Sv_n)$. Hence

$$H(t, v_n, Sv_n) \leq H(t, w_0, Sw_0) + M(w_0 - v_0) + N(Sw_0 - Sv_0).$$

Similarly, for each n, we have

$$H(t, v_n, Sv_n) \geq H(t, v_0, Sv_0) - M(w_0 - v_0) - N(Sw_0 - Sv_0).$$

Then it follows that $H(t, v_n, Sv_n)$ is uniformly bounded. Since both $\{v_n\}$ and $\{Sv_n\}$ are uniformly bounded and

$$v'_n = H(t, v_{n-1}, Sv_{n-1}) - M(v_n - v_{n-1}) - N(Sv_n - Sv_{n-1}),$$

this implies the equicontinuity of the sequence $\{v_n\}$. A similar method may be used to prove that the sequence $\{w_n\}$ is equicontinuous.

Now we set $B(t) = \{v_n(t)\}_{n=0}^{\infty}$, $B'(t) = \{v'_n(t)\}_{n=0}^{\infty}$ and $\psi(t) = \beta(B(t))$, and since $\{v_n\}, \{v'_n\}$ are uniformly bounded, it is easy to see that the sequence $\{v_n\}$ satisfies the assumption of Lemma 4.2.1 and Lemma 4.2.2. Note that

$$(4.2.10) \qquad \beta\left(\left\{\int_0^t S(t,s)v_n(s)\,ds\right\}_{n=0}^{\infty}\right)$$

$$\leq \beta_{E_1}\left(\left\{\int_0^t S(t,s)v_n(s)\,ds\right\}_{n=0}^\infty\right)$$

$$\leq S_0\int_0^t \beta_{E_1}\bigl(\{v_n(s)\}_{n=0}^\infty\bigr)ds \leq 2S_0\int_0^t \psi(s)\,ds$$

using the Lemma 4.2.3, where E_1 is the closed subspace of E spanned by $B = (\{v_n(t)\}_{n=0}^\infty, \ t \in I)$. Hence E_1 is separable (β_{E_1} denotes the Hausdorff measure of compactness on E_1). Using (4.2.10), the assumption (A_3) and Lemma 4.2.2, $\psi(t)$ is absolutely continuous and satisfies

(4.2.11) $\quad \psi'(t) \leq 2\beta(B'(t))$

$$= 2\beta\Bigg(\bigg\{H\bigg(t,v_n(t),\int_0^t S(t,s)v_n(s)\,ds\bigg) - M(v_n(t)-v_{n-1}(t))$$

$$- N\bigg[\int_0^t S(t,s)v_n(s)\,ds - \int_0^t S(t,s)v_{n-1}(s)\,ds\bigg]\bigg\}_{n=0}^\infty\Bigg)$$

$$\leq 2\lambda\,\max\bigg[\psi(t),2S_0\int_0^t \psi(s)\,ds\bigg] + 4M\psi(t) + 8NS_0\int_0^t \psi(s)\,ds$$

$$\leq A_1\psi(t) + A_2\int_0^t \psi(s)\,ds,$$

where $A_1 = 2\lambda + 4M > 0$, $A_2 = 4S_0\lambda + 8NS_0 > 0$ and $\psi(0) = 0$. From (4.2.11), applying Lemma 4.2.4 and $\psi(t) \geq 0$, we have $\psi(t) = 0$, $t \in I$. Thus it implies the relative compactness of the sequences $\{v_n(t)\}$ for each $t \in I$. Similarly, $\{w_n(t)\}$ is relatively compact for each $t \in I$. The proof of Lemma 4.2.7 is complete. □

We now apply Ascoli's theorem to the sequences $\{v_n\}$ and $\{w_n\}$ to obtain subsequences $\{v_{n_k}\}$ and $\{w_{n_k}\}$ that converge uniformly on I. Since the sequences $\{v_n\}$ and $\{w_n\}$ are monotone and K is normal, this then shows that the full sequences converge uniformly and monotonically to continuous functions, namely $\lim_{n\to\infty} v_n(t) = \rho(t)$ and $\lim_{n\to\infty} w_n(t) = r(t)$ on I. It then follows easily from (4.2.7) that $\rho(t)$ and $r(t)$ are solutions of the problem (4.2.1) on I.

Finally, we show that $\rho(t), r(t)$ are minimal and maximal solutions of (4.2.1). To this end, let $x(t)$ be any solution of (4.2.1) on I such that $x(t) \in [v_0, w_0]$ for all $t \in I$. Assume that $v_n \leq x \leq w_n$ on I. Set $p(t) = \phi(v_{n+1}(t) - x(t))$. Thus $p(0) = 0$, where $\phi \in K^*$. Then by (A_2) and the assumption that $v_n \leq x$, we have

$$\begin{aligned} p'(t) &= \phi(H(t, v_n(t), (Sv_n)(t)) - M(v_{n+1} - v_n) - N(Sv_{n+1} - Sv_n) \\ &\quad - H(t, x(t), (Sx)(t))) \\ &\leq -Mp(t) - N(Sp)(t). \end{aligned}$$

This implies that $v_{n+1} \leq x$ on I, using Lemma 4.2.3. Similarly, we can show that $x \leq w_{n+1}$ on I. Since $x(t) \in [v_0, w_0]$ for all $t \in I$, we have, by induction, $v_n \leq x \leq u_n$ on I for all n. Thus we obtain, taking the limit as $n \to \infty$, $\rho(t) \leq x(t) \leq r(t)$ on I, proving that $\rho(t)$ and $r(t)$ are minimal and maximal solutions of (4.2.1) on I. The proof is therefore complete. □

4.3 Well Posedness of Linear Equations

Consider the integro-differential equation

$$(4.3.1) \qquad x'(t) = Ax(t) + \int_0^t B(t-s)x(s)\,ds + f(t), \quad x(0) = x_0,$$

in a Banach space E. In this section, we assume that A is an infinitesimal generator of a C_0-semigroup on E satisfying the Hille – Yosida – Phillips [1] conditions

$$|R^n(\lambda; A)| \leq \frac{M}{(\operatorname{Re}\lambda - \omega)^n}, \quad n \geq 1$$

for the resolvent $R(\lambda; A)$ for some $\omega \geq 0$. It is also assumed that f is an element of a Banach space F of E-valued functions defined for $t \geq 0$ where equality in F implies equality a.e., and $B(t)$ is a linear operator on $D(A)$ for $t \geq 0$ so that $B(\cdot)x \in F$ for each fixed $x \in D(A)$. An equation of the form (4.3.1) appears, for example, in the modeling of heat conduction problems in materials with memory, where $A = \Delta$ is the Laplacian and the kernel $B(t)$ is basically of the form $a(t)A$ with $a(t) \in L^1(0,\infty)$. Our approach here is to develop an equivalent ordinary differential equation and investigate sufficient conditions for well-posedness. To this end, let us consider

(4.3.2) $$\frac{d}{dt}z = Cz, \quad z(0) = z_0 \in D(C) \subseteq E \times E \times F,$$

where

$$z \in \begin{bmatrix} w \\ x \\ y \end{bmatrix} \in E \times E \times F \text{ with } |z|^2 = |w|_E^2 + |x|_E^2 + |y|_E^2;$$

(4.3.3) $$C = \begin{bmatrix} 0 & A_0 & 0 \\ 0 & A & \delta_0 \\ 0 & B & D_s \end{bmatrix};$$

B is the linear transformation given by $(Bx)(s) = B(s)x$, A_0 is a closed operator in E with domain $D(A_0) \supseteq D(A)$ with resolvent $R(\lambda; A_0) = (\lambda I - A_0)^{-1}$, δ_0 is the Dirac delta function, and D_s is the differentiation operator on F defined by $D_s f = f'$ on a domain $D(D_s) \subseteq F$, where $f \in D(D_s)$ implies

$$f(s) = \alpha + \int_0^s e(\tau) d\tau$$

for some $e \in F$ and D_s generates the translation semigroup $T(t)$ on F given by $T(t)f(s) = f(t+s)$. The domain $D(C)$ of C, is $E \times D(A) \times D(D_s)$. It is easy to verify that C is a closed operator on $E \times E \times F$. In our analysis, we assume that A_0 is a multiple of either A or some positive fractional power of A if it exists. In the sequel, we use the following notation: let $T(t)$ be the translation semigroup and h be the translated function $T(s)h$, that is, $h_s(u) = h(s+u)$, and let $\mathcal{L}_\lambda$ be the Laplace transformation

$$\mathcal{L}_\lambda(h) = \int_0^\infty e^{-\lambda s} h(s) ds;$$

$*$ denotes the transpose of an element in $E \times E \times F$.

Definition 4.3.1 The equation (4.3.2) is said to be well-posed if for each $z_0 \in D(C)$, there is a unique solution $z(t, z_0)$ of (4.3.2), and for any $T > 0$, there is an $K > 0$ such that

$$|z(t,z_0)| \leq K|z_0|$$

for all $z_0 \in D(C)$, $0 \leq t \leq T$.

Definition 4.3.2 The equation (4.3.1) is said to be well-posed if for each pair (x_0, f) with $(0, x_0, f) \in D(C)$, there is a unique solution $x(t, x_0, f)$ of (4.3.1) and for any $T > 0$, there is an $M > 0$ such that

$$|x(t,x_0,f)|_E \leq M(|x_0|_X + |f|_F), \quad 0 \leq t \leq T.$$

In order to obtain the equivalence relation between the solutions of (4.3.1) and (4.3.2), we need the following hypotheses on B and A_0.

(h_1) $B(\cdot)x(t)$ is continuous as a function of t on R^+ into F whenever $x(t)$ and $A_0x(t)$ are continuous on R^+ into F. In addition, $B(t)R(\lambda; A_0)$ is a bounded operator from E into E that is continuous in t for some $\lambda > 0$.

(h_2) $B(s)x \in D(D_s)$ for each fixed $x \in D(A_0)$.

(h_3) $D_sB(s)x(t)$ is locally integrable as a function of t whenever $x(t)$ and $A_0x(t)$ are continuous.

(h_4) $A_0x(t)$ is continuous as a function of t whenever $Ax(t)$ is continuous as a function of t.

Example 4.3.1 Let $B(t) = a(t)A_0$, where $a(t)$ is a scalar-valued function.

(i)   Let $F = B^2(R^+, E)$. If $a \in L^2(R^+)$ then (h_1) holds.

(ii)   If a is absolutely continuous with $a' \in L^2(0, \infty)$ then $B(s)x = a(s)A_0x$ is in $D(D_s)$ for each fixed $x \in D(A_0)$ and $a'(s)A_0x(t)$ is a continuous function of t into F when $A_0x(t)$ is continuous. Thus (h_2) and (h_3) hold.

(iii)   If $A = -A_0^2$ and A_0 is invertible or if $A = A_0$ then (h_4) is satisfied.

The following result gives an equivalence relation between the solutions of (4.3.1) and (4.3.2).

Theorem 4.3.1 Assume that the hypothesis (h_1) is valid.

(a)   If $(w, x, y)^*$ is a solution of (4.3.2) then $x(t)$ is a solution of (4.3.1) with $f(t) = y(0)(t)$ and $x_0 = x(0)$.

(b)   Let $f \in D(D_s)$ and suppose further that the hypotheses (h_2)-(h_4)

are satisfied. If $x(t)$ is a solution of (4.3.1) then $(w,x,y)^*$ is a solution of (4.3.2) with

$$w(t) = w_0 + \int_0^t A_0 x(s)\,ds,$$

$$y(t)(s) = f(t+s) + \int_0^t B(t-\tau+s)x(\tau)\,d\tau.$$

Proof of (a) First let us assume that (h_1) holds and that $(w,x,y)^*$ is a solution of (4.3.2). Then w, w', x, x', y and y' are all continuous as functions of t from R^+ into either E or F. Since the equation

$$y'(t) = D_s y(t) + B(s)x(t), \quad y(0) = y_0, \quad t \geq 0,$$

has a solution, it follows from Pazy [1] and T. Kato [1] that the solution is given by

$$y(t) = T(t)y_0 + \int_0^t T(t-\tau)B(\cdot)x(\tau)\,d\tau,$$

where $T(t)$ is the semigroup generated by D_s (that is, $T(t)$ is the translation semigroup). Hence if $y(0) = f$, we see that in F,

$$y(t)(s) = f(t+s) + \int_0^t B(t-\tau+s)x(\tau)\,d\tau.$$

As $y \in D(D_s)$, $y(t,\cdot)$ is absolutely continuous and, by the hypothesis (h_1), in x,

$$y(t)(0) = f(t) + \int_0^t B(t-\tau)x(\tau)\,d\tau$$

and is continuous in t. Now $x'(t) = Ax(t) + y(t)(0)$, and hence $Ax(t)$ is continuous. Therefore

$$x'(t) = Ax(t) + \int_0^t B(t-\tau)x(\tau)\,d\tau + f(t)$$

is a solution of (4.3.1).

Proof of (b) Suppose that $(h_1)-(h_4)$ hold and that $x(t)$ is a solution of (4.3.1) with $f \in D(D_s)$. Then $x(t)$ and $Ax(t)$ are continuous. Hence $B(\cdot)x(t) \in D(D_s)$ for each t, continuous in t and $D_s B(s) x(t)$ is locally integrable as a function of t. Thus the equation

$$y'(t) = D_s y(t) + B(s)x(t)$$

has a solution (see Pazy [1]), which is given by

$$y(t)(s) = f(t+s) + \int_0^t B(t-\tau+s)x(\tau)\,d\tau,$$

and in particular

$$y(t)(0) = f(t) + \int_0^t B(t-\tau)x(\tau)\,d\tau.$$

Hence $x(t)$ satisfies

$$x'(t) = Ax(t) + \delta_0 y(t).$$

Given $w_0 \in E$, define

$$w(t) = w_0 + \int_0^t A_0 x(s)\,ds.$$

Then is is clear that $z = (w, x, y)^*$ is a solution of (4.3.2). This completes the proof of Theorem 4.3.1. □

Remark 4.3.1 Under the hypotheses $(h_1)-(h_4)$, if the solutions of (4.3.2) are unique then the solutions of (4.3.1) with $(0, x_0, f) \in D(C)$ are unique when they exist. Similarly, if the solutions of (4.3.1) are unique for $(0, x_0, f) \in D(C)$ then the solutions of (4.3.2) must also be unique. Consequently, if C generates a C_0-semigroup then (4.3.1) is well-posed.

Remark 4.3.2 Suppose that $F = BU(R^+, E)$, the space of bounded uniformly continuous functions from R^+ into E, and $A_0 \equiv A$. Then (4.3.2) can be treated in the simple setting $E \times F$ as in Miller [1].

In view of Remark 4.3.1, to obtain the well-posedness of (4.3.1), we shall now give sufficient conditions ensuring that C generates a C_0-semigroup.

Theorem 4.3.2 *Assume that $F \equiv BU[R^+, E]$ and $A_0 \equiv A$. Suppose that B can be written as $B = FA + K$, where $F: E \to D(D_s)$ and $K: E \to F$ are bounded linear operators. Then C defined by (4.3.3) generates a C_0-semigroup on $E \times E \times F$.*

Remark 4.3.3

(i) Theorem 4.3.2 holds for any F such that δ_0 is a bounded operator from F into E, for example, with $F = H^1[(0,\infty), E]$ and $D(D_s) = H^2[(0,\infty), E]$.

(ii) If C generates a C_0-semigroup and $F_1: E \to D(D_s)$ then A must generate a semigroup.

Proof of Theorem 4.3.2 We first note (see T. Kato [1]) that the operator C_2 on $E \times E \times F$ given by

$$C_2 = \begin{bmatrix} 0 & 0 & 0 \\ 0 & A & 0 \\ 0 & 0 & D_s \end{bmatrix}$$

generates a C_0-semigroup on $E \times E \times F$ and the operator P given by

$$P = \begin{bmatrix} I_E & -I_E & 0 \\ 0 & I_E & 0 \\ 0 & -F_1 & I_F \end{bmatrix}$$

is invertible with inverse

$$P^{-1} = \begin{bmatrix} I_E & I_E & 0 \\ 0 & I_E & 0 \\ 0 & F_1 & I_F \end{bmatrix}.$$

Then for any $(w, x, y) \in P^{-1}(D(C_2)) = P^{-1}(E \times D(A) \times D(D_s))$, we obtain

$$P^{-1}C_2 P(w,x,y) = P^{-1}C_2(w-x, x, -F_1 x + y)$$
$$= P^{-1}(0, Ax, D_s(-F_1 x + y))$$
$$= (Ax, Ax, F_1 Ax - D_s F_1 x + D_s y).$$

Since F_1 maps E into $D(D_s)$, we have

$$P^{-1}(E \times D(A) \times D(D_s)) = E \times D(A) \times D(D_s).$$

Also, $D_s F_1$ is a closed operator from E into F with domain E. Therefore, by the closed graph theorem, $D_s F_1$ is a bounded operator. Hence

$$P^{-1}C_2 P = \begin{bmatrix} 0 & A & 0 \\ 0 & A & 0 \\ 0 & F_1 A - D_s F_1 & D_s \end{bmatrix}.$$

Since
$$L = \begin{bmatrix} 0 & 0 & 0 \\ 0 & 0 & \delta_0 \\ 0 & D_s F_1 + K & 0 \end{bmatrix}$$
is a bounded operator, it follows that
$$P^{-1} C_2 P + L = C$$
is an infinitesimal generator of a C_0-semigroup (see Pazy [1]) with $D(C) = E \times D(A) \times D(D_s)$. This completes the proof. □

Corollary 4.3.1 *Suppose $(h_1) - (h_4)$ are valid with $B(t) = \alpha_1(t)A + \alpha_2(t)I$ where $\alpha_1(t)$, $\alpha_2(t)$, $\alpha_1'(t)$ and $\alpha_2'(t)$ are uniformly bounded continuous scalar functions on $\mathbb{R}^+$. Then the equation (4.3.1) is well-posed for (x_0, f) with $x_0 \in D(A)$ and $f \in D(D_s)$.*

We shall now study the well-posedness of a time-dependent (time-varying) linear integro-differential equation by constructing an equivalent evolution equation (ordinary differential equation). Consider the integro-differential equation

(4.3.4) $$\begin{cases} x'(t) = A(t)x(t) + \int_0^t B(t,s)x(s)\,ds + f(t), & 0 \le t \le T < \infty, \\ x(0) = x_0 \in E \end{cases}$$

in a Banach space E.

We shall assume that for each fixed $t \ge 0$, $A(t)$ is linear and defined in a dense subspace D of E that is independent of t and that $A(t)$ is the generator of a C_0-semigroup on E; the linear operator $B(t,s)$ is defined on a subspace $D_0 \supseteq D$ for each pair (t,s) with $0 \le s \le t < \infty$, and the function f is defined in $\mathbb{R}^+$ with values in E. In particular, let f be an element of a Banach space F of E-valued functions that are defined for $t \ge 0$, and for each $t \ge 0$, let $x \in D_0$, $B_t(\,\cdot\,,t)x \in F$, where

$$(B_t(\,\cdot\,,t)x)(s) = B(t+s,t)x, \quad s \ge 0,$$

belongs to F as a function of s.

Along with (4.3.4), we consider the evolution equation

(4.3.5)
$$\begin{cases} z' = C(t)z, \\ z(0) = z_0 \in D(C) \subseteq E \times E \times F \equiv Z, \end{cases}$$

where Z is a Banach space with norm given by

$$|z| = |w|_E^2 + |x|_E^2 + |y|_F^2$$

for

$$z = \begin{bmatrix} w \\ x \\ y \end{bmatrix} \in Z, \quad C(t) = \begin{bmatrix} 0 & A_0(t) & 0 \\ 0 & A(t) & \delta_0 \\ 0 & B(t) & D_s \end{bmatrix}$$

Here the operator $A_0(t)$ is closed in E and has domain $D_0 \supseteq D$ with resolvent $R(\lambda; A_0(t)) = (\lambda I - A_0(t))^{-1}$ for some $\lambda > 0$ and all $t \geq 0$. If E_1 and E_2 are Banach spaces, let $\mathcal{B}(E_1, E_2)$ denote the space of all bounded linear operators from E_1 to E_2 with the usual norm. If $E_1 = E_2$ then $\mathcal{B}(E_1, E_2) = \mathcal{B}(E_1)$.

We now need the following hypotheses on B and A_0.

($\bar{h}_1$) $B_t(\cdot, t)x(t)$ is continuous as a function of t on R^+ into F whenever $x(t)$ and $A_0(t)x(t)$ are continuous on R^+ into E. Further, for some $\lambda > 0$, let

$$B_t(s, \tau)R(\lambda; A_0(\tau))$$

be a bounded operator on E into E such that for each t, the family

$$\{B_t(s, \tau)R(\lambda, A_0(\tau)): 0 \leq \tau \leq t\}$$

is continuous for $s \in R^+$ into $\mathcal{B}(E)$.

($\bar{h}_2$) $B_t(s, t)x$ is in $D(D_s)$ for each fixed $x \in D_0$ and $t \geq 0$.

($\bar{h}_3$) $D_s B_t(s, t)x(t)$ is locally integrable as a function of t whenever $x(t)$ and $A_0(t)x(t)$ are continuous for $t \geq 0$.

($\bar{h}_4$) $A_0(t)x(t)$ is continuous as a function of t whenever $A(t)x(t)$ is continuous as a function of t.

The following result is similar to Theorem 4.3.1 for time-dependent equations.

Theorem 4.3.3 Assume that $(\bar{h}_1)$ is satisfied.

(i) If $z = (w, x, y)^*$ is a solution of (4.3.5) then $x(t)$ is a solution of (4.3.4) with $f(t) = y(0)(t)$ and $x_0 = x(0)$.

(ii) Suppose that $f \in D(D_s)$ and $(\bar{h}_2) - (\bar{h}_4)$ hold. If $x(t)$ is a solution of (4.3.4), then $(w, x, y)^*$ is a solution of (4.3.5) with $w(t) = w_0 + \int_0^t A_0(s)x(s)ds$ and $y(t)(s) = f(t+s) + \int_0^t B(t+s, \tau)x(\tau)d\tau$.

The proof of this theorem is exactly similar to that of Theorem 4.3.1 and hence is left as an exercise.

Remark 4.3.4 Under the hypotheses $(\bar{h}_1) - (\bar{h}_4)$, it is clear that if the solutions of (4.3.5) are unique then the solutions of (4.3.4) with $(0, x_0, f) \in D(C) = E \times D \times D(D_s)$ are unique when they exists. Similarly, if the solutions of (4.3.4) are unique for $(0, x_0, f) \in D(C)$ then the solutions of (4.3.5) must also be unique. Thus, if $(\bar{h}_1) - (\bar{h}_4)$ are valid, the existence and uniqueness of the solutions of (4.3.5) imply the same for the solutions of (4.3.4) with $(0, x_0, f) \in D(C)$.

Definition 4.3.3 The equation (4.3.4) is said to be well-posed if for each pair (x_0, f) with $(0, x_0, f) \in D(C)$, there is a unique solution $x(t, x_0, f)$ of (4.3.4) existing on $[0, T]$ and there is a constant $M > 0$ such that

$$|x(t, x_0, f)| \leq M(|x_0|_E + |f|_F), \quad 0 \leq t \leq T.$$

Remark 4.3.5 In addition to Definition 4.3.3, if for any $T > 0$, $|x(t, x_0, f)| \to 0$ uniformly on $[0, T]$ as $|(0, x_0, f)| \to 0$ then the equation (4.3.4) is called uniformly well-posed.

The well-posedness of the evolution equation (4.3.5) can be defined in the same way, but, instead, we shall first obtain the existence of a fundamental system for (4.3.5) from which the well-posedness of (4.3.5) will follow.

Definition 4.3.5 A fundamental solution of (4.3.5) is a unique bounded operator-valued function $U(t, s) \in \mathcal{B}(E)$ with $0 \leq s \leq t \leq T$ that satisfies the following conditions:

(i) $U(t, s)$ is strongly continuous in s and t, $U(s, s) = I$ and $|U(t, s)| \leq Me^{\beta(t-s)}$ for some constants M and β;

(ii) $U(t, s) = U(t, \tau)U(\tau, s)$ for $s \leq \tau \leq t$;

(iii) $U(t,s): D(C) \to D(C)$, and for each $v \in D(C)$, $U(t,s)v$ is strongly continuously differentiable in t and s, with

$$\frac{\partial}{\partial t}U(t,s)v = A(t)U(t,s)v,$$

$$\frac{\partial}{\partial s}U(t,s)v = -U(t,s)A(s)v$$

since both sides of these equations are strongly continuous on $0 \leq s \leq t \leq T$.

Remark 4.3.6 A fundamental solution $U(t,s)$ satisfying the conditions $(i)-(iii)$ of Definition 4.3.5 is usually referred to as an evolution operator for (4.3.5).

Remark 4.3.7 If a fundamental solution $U(t,s)$ of (4.3.5) exists then it is easy to verify that for $z_0 \in D(C)$, $z(t) = U(t,0)z_0$ is the unique solution of (4.3.5) with $z(0) = z_0$. Further, it follows from Definition (4.3.5) (i) that the equation (4.3.5) is well-posed. In view of the equivalence theorem, Theorem 4.3.3, we can thus obtain the well-posedness of (4.3.4) through the existence of a fundamental solution (or evolution operator) for (4.3.5) (see Theorem 4.3.4 below).

We need the following definition in our subsequent analysis.

Definition 4.3.6 If $\{A(t)\}$, $0 \leq t \leq T$, is a family of generators of C_0-semigroups then $\{A(t)\}$ is called stable if there are real constants $M \geq 1$ and β such that

$$\left| \prod_{j=1}^{k} (A(t_j) - \lambda I)^{-1} \right| \leq M(\lambda - \beta)^{-k}$$

for all $\lambda > \beta$, $0 \leq t_1 \leq t_2 \leq \ldots \leq t_k \leq T$, $k = 1, 2, \ldots$, where the product is taken as

$$(A(t_k) - \lambda I)^{-1}(A(t_{k-1}) - \lambda I)^{-1}\ldots(A(t_1) - \lambda I)^{-1}.$$

For example, if $A(t)$ is a constant or $A(t)$ generates a semigroup $\{S_t(u)\}$ with $|S_t(s)| \leq e^{\beta s}$ for each $t \geq 0$ then $\{A(t)\}$ is stable.

Theorem 4.3.4 Assume that
(i) $\{A(t)\}$, $0 \leq t \leq T$, is a stable family of generators with common domain D such that $A(t)v$ is strongly continuously differentiable

on $[0,T]$ for $v \in D$;

(ii) $A_0(t) \equiv A(t)$ and $\delta_0 \in \mathcal{B}(F,E)$;

(iii) $B(t)v$ is strongly continuously differentiable on $[0,T]$ for $v \in D$;

(iv) $B(t)$ has the form $B(t) = F_1(t)A(t) + k(t)$, where $F_1: [0,T] \to \mathcal{B}(E,F)$ is such that $F_1(t)$ is of bounded variation on $[0,T]$, $F(t)$ maps E into $D(D_s)$ and $D_s F(t)$ is uniformly bounded on $[0,T]$; $k(t) \in \mathcal{B}(E,F)$ is uniformly bounded on $[0,T]$.

Then the equation (4.3.5) has a fundamental solution $U(t,s)$.

Proof Define the family of generators

$$\{(PC_1 P^{-1})(t)\}$$

where

$$P = \begin{bmatrix} I_E & I_E & 0 \\ 0 & I_E & 0 \\ 0 & F_1(t) & I_F \end{bmatrix}, \quad C_1 = \begin{bmatrix} 0 & 0 & 0 \\ 0 & A(t) & 0 \\ 0 & 0 & D_s \end{bmatrix},$$

$$P^{-1} = \begin{bmatrix} I_E & -I_E & 0 \\ 0 & I_E & 0 \\ 0 & -F_1(t) & I_F \end{bmatrix}.$$

Then it follows from the proof of Theorem 4.3.2 that the family $\{(PC_1 P^{-1})(t)\}$ is a family of generators with domain $D(C) = E \times D \times D(D_s)$.

Let

$$Q_j = (P^{-1}(t_j) - P^{-1}(t_{j-1}))P(t_{j-1}).$$

Then the product

$$\prod_{j=1}^{k} (PC_1 P^{-1}(t_j) - \lambda I)^{-1}$$

is equal to

$$P(t_k)\{(C_1(t_k) - \lambda I)^{-1}(I + Q_k)(C_1(t_{k-1}) - \lambda I)^{-1} \cdots (I + Q_2)(C_1(t_1) - \lambda I)^{-1}\}P^{-1}(t_1)$$

and hence we obtain that

$$\left| \prod_{j=1}^{k} (PC_1 P^{-1}(t_j) - \lambda I)^{-1} \right| \leq \frac{M^2 M_1}{\lambda - \beta_1} \exp(MM_1 V),$$

where M_1 and β_1 are constants corresponding to Definition 4.3.6 for $\{C_1(t)\}$, M is a bound for $|P(t)|$ and $|P^{-1}(t)|$, and V is the total variation of F_1 on $[0, T]$.

Therefore, by Definition 4.3.6, it follows that

$$PC_1 P^{-1}(t) = \begin{bmatrix} 0 & A(t) & 0 \\ 0 & A(t) & 0 \\ 0 & F_1(t)A(t) - D_s F_1(t) & D_s \end{bmatrix}$$

is a stable family of generators.

Further, since $\delta_0 \in \mathcal{B}(F, E)$, $k(t)$ and $D_s F(t)$ are uniformly bounded on $[0, T]$, it follows that $\{C(t)\}$ given by

$$C(t) = PC_1 P^{-1} + L$$

where

$$L = \begin{bmatrix} 0 & 0 & 0 \\ 0 & 0 & \delta_0 \\ 0 & D_s F_1(t) + k(t) & 0 \end{bmatrix}$$

is a stable family of generators and $C(t)z$ is strongly differentiable on $[0, T]$. Hence the equation (4.3.5) has a fundamental solution (see Tanabe [1]). This completes the proof. □

Remark 4.3.8 In view of Remark 4.3.7 and Theorem 4.3.3, the well-posedness of (4.3.4) follows under the hypotheses $(\bar{h}_1) - (\bar{h}_4)$.

4.4 Semigroups and Resolvent Operators

Consider an integro-differential equation

$$(4.4.1) \qquad x'(t) = Ax(t) + \int_0^t B(t-s)x(s)\,ds, \quad t \in R^+,$$

with initial data

$$x(0) = x_0 \in D(A) \subset E_1,$$

where E_1 is a Banach space, under the following hypotheses on A and B.

(h_0) A is the infinitesimal generator of a semigroup of bounded linear operators in E_1. Since A is closed, its domain $D(A)$ becomes a Banach space E_2 with norm $|x|_{E_2} = |x|_{E_1} + |Ax|_{E_1}$.

(h_1) $\{B(t): t \in R^+\}$ is a family of bounded linear operators from E_2 into E_1.

(h_2) For each $x \in E_2$, the function $(Bx)(t) = B(t)x$ is Bochner-measurable (from R^+ to E_1).

(h_3) There exists a constant $\omega \geq 0$ such that

$$\int_0^\infty e^{-\omega t} |B(t)|_{(E_2, E_1)} dt < \infty,$$

where $|B(t)|_{(E_2, E_1)}$ is the usual norm of the (bounded) linear operator $B(t)$ from E_2 into E_1.

We shall now define the resolvent operator corresponding to (4.4.1).

Definition 4.4.1 A family $\{R(t): t \in R^+\}$ of bounded linear operators on E_1 is called a resolvent operator for (4.4.1) if and only if

(a) $R(0) = I$, the identity map on E_1;

(b) for all $x \in E_1$, the map $t \to R(t)$ is a continuous function on R^+;

(c) for all $x \in E_2$, the map $t \to R(t)x$ belongs to $C(R^+, E_2) \cap C^1(R^+, E_1)$ and satisfies

(4.4.2) $$R'(t)x = AR(t)x + \int_0^t B(t-s)R(s)x \, ds,$$

(4.4.3) $$R'(t)x = R(t)Ax + \int_0^t R(t-s)B(s)x \, ds.$$

Remark 4.4.1 The concept of a resolvent operator for (4.4.1) is analogous to the notion of the semigroup associated with the abstract Cauchy problem $x'(t) = Ax(t)$, $x(0) = x_0$. In fact, the function $x(t) = R(t)x_0$ with $x(0) = x_0 \in D(A)$ may be regarded as a generalized solution of the IVP (4.4.1).

We need the following well-known result (see Corduneanu [1]) in our subsequent analysis to have a relation between the semigroup of operators attached to (4.4.1) and the existence of a resolvent operator corresponding to (4.4.1).

Lemma 4.4.1 Let $\{R(t): t \in R^+\}$ be a family of bounded linear operators on E_1 satisfying (a) and (b) of Definition 4.4.1 with

(4.4.4) $$|R(t)| \leq Me^{\omega t} \quad \text{for all } t \in R^+,$$

where M and ω are some constants. Then the following relations are equivalent:

(i) $R(t)$ satisfies (4.4.2);
(ii) $R(t)$ satisfies (4.4.3);
(iii) $\widehat{R}(s) = [sI - A - \widehat{B}(s)]^{-1}$, where $\widehat{R}(s)$ and $\widehat{B}(s)$ are the Laplace transforms of $R(t)$ and $B(t)$ respectively.

We shall now show that the semigroup of operators can be associated with the IVP (4.4.1). This approach will allow us to prove a result on the existence of the resolvent operator for (4.4.1) and also to derive the usual variation of parameters formula for the nonhomogeneous equation

(4.4.5) $$x'(t) = Ax(t) + \int_0^t B(t-s)x(s)\,ds + f(t),$$

where f belongs to a suitable function space $\mathcal{F} = \mathcal{F}(R^+, E_1)$. The pair $(x_0, f) \in E_2 \times \mathcal{F}$ may be considered as the state of the system at time $t = 0$. The following simple calculation indicates how the system changes in time.

From (4.4.5), we have

$$\begin{aligned}
x'(t+s) &= Ax(t+s) + \int_0^{t+s} B(t+s-u)x(u)\,du + f(t+s) \\
&= Ax(t+s) + \int_t^{t+s} B(t+s-u)x(u)\,du \\
&\quad + \int_0^t B(t+s-u)x(u)\,du + f(t+s) \\
&= Ax(t+s) + \int_0^s B(s-u)x(u+t)\,du + g(t,s)
\end{aligned}$$

where

$$g(t,s) = \int_0^t B(t+s-u)x(u)\,du + f(t+s).$$

In particular, for $s = 0$, we obtain

$$x'(t) = Ax(t) + g(t,0).$$

Further, we see that

$$\frac{d}{dt}[g(t,s)] = B(s)x(t) + \int_0^t B'(t+s-u)x(u)\,du + f'(t+s)$$

$$= B(s)x(t) + \frac{d}{ds}\left[\int_0^t B(t+s-u)x(u)\,du + f(t+s)\right]$$

$$= B(s)x(t) + \frac{d}{ds}[g(t,s)].$$

These relations suggest that the solution $x(t)$ of (4.4.5) will be the first component of the solution (x,g) of the IVP

$$(x',g') = C(x,g),$$

$$(x(0),g(0)) = (x_0,f),$$

where

$$C(x,g) = \left(Ax + g(\cdot,0), Bx + \frac{d}{ds}g\right).$$

The above analysis is somewhat informal; however, one could make it more rigorous by selecting a suitable function space $\mathcal{F}$ and assuming sufficient conditions on B that ensure that C is the generator of a semigroup of operators on the product space $E_1 \times \mathcal{F}$. An example is the case where $\mathcal{F}$ is the space $e^{\omega t}L^1(R^+, E_1)$ with norm

$$|x|_{E_1} = \int_0^\infty e^{-\omega t}|x(t)|_{E_2}\,dt,$$

and with B mapping Y into $\mathcal{F}$.

Remark 4.4.2 It is clear that the shift semigroup $T(\cdot)$ given by

$$(T(t)f)(s) = f(t+s)$$

is strongly continuous, and its infinitesimal generator D_u has as domain the Sobolev space $e^{\omega t}W^{1,1}(R^+, E_1)$. This implies that the Dirac operator $\delta: \delta f = f(0)$ is defined on the space $e^{\omega t}W^{1,1}(R^+, E_1)$ and is continuous from $e^{\omega t}W^{1,1}(R^+, E_1) \to E_1$.

Theorem 4.4.1 *Assume that*
(i) *hypotheses* $(h_1) - (h_4)$ *hold;*
(ii) *the operator* $C: E_1 \times e^{\omega t}L^1(R^+, E_1)$ *defined by* $(x, g) = (Ax + \delta g, Bx + D_u g)$ *with* $D(A) = E_2 \times e^{\omega t}W^{1,1}(R^+, E_1) \subset E_1 \times e^{\omega t}L^1(R^+, E_1)$ *generates a semigroup* $S(t)$ *on* $E_1 \times e^{\omega t}L^1(R^+, E_1)$;
(iii) *the conditions of Lemma 4.4.1 are satisfied.*
Then the equation (4.4.1) admits a resolvent $R(t)$ such that $R(t)x$ is the first component of $S(t)(x, 0)$. Moreover, for any $f \in e^{\omega t}L^1(R^+, R)$, the first component of $S(t)(0, f)$ is

$$\int_0^t R(t-s)f(s)\,ds.$$

Proof Let $R(t)x$ be the first component of $S(t)(x, 0)$. Then for sufficiently large $\mathrm{Re}(s)$, $\widehat{R}(s)$ is the first component of

$$(y, g) = (sI - C)^{-1}(x, 0).$$

This implies that

(4.4.6) $$sy - Ay - g(0) = x,$$

(4.4.7) $$sg - D_u g - By = 0.$$

Thus (4.4.7) gives

$$g = (sI - D_u)^{-1}By$$

$$= \int_0^\infty e^{-st}T(t)By\,dt.$$

Hence

$$g(u) = \int_0^\infty e^{-st} B(t+u) y \, dt$$

and, in particular,

$$g(0) = \int_0^\infty e^{-st} B(t) y \, dt = \widehat{B}(s) y.$$

Therefore from (4.4.6), we obtain

$$sy - Ay - \widehat{B}(s)y = x,$$

which shows that

$$y = \left[sI - A - \widehat{B}(s)\right]^{-1} x.$$

Thus the application of Lemma 4.4.1 yields that $R(t)$ is the resolvent operator for (4.4.1). Further, for any $f \in e^{\omega t} L^1(R^+, R)$, let $u(t)$ be the first component of $S(t)(0, f)$. Then $\widehat{u}(s)$ is the first component of

$$(y, g) = (sI - C)^{-1}(0, f).$$

This implies that

(4.4.8)
$$sy - Ay - g(0) = 0$$
$$sg - D_u g - By = f.$$

Thus

$$g = (sI - D_u)^{-1}(By + f).$$

Therefore

$$g(u) = \int_0^\infty e^{-st}\bigl[B(t+u)y + f(t+u)\bigr] dt$$

and, in particular,

$$g(0) = \int_0^\infty e^{-st}\bigl[B(t)y + f(t)\bigr] dt$$
$$= \widehat{B}(s)y + \widehat{f}(s).$$

As a result, (4.4.8) gives

$$sy - Ay = \widehat{B}(s)y - \widehat{f}(s) = 0,$$

and hence

$$y = \widehat{u}(s) = [sI - A - \widehat{B}(s)]^{-1}\widehat{f}(s)$$

$$= \widehat{R}(s)\widehat{f}(s).$$

This is equivalent to

$$u(t) = \int_0^t R(t-\tau)f(\tau)\,d\tau,$$

and this completes the proof. □

Remark 4.4.3 It is well known (see Pazy [1]) that a C_0-semigroup (or strongly continuous semigroup) always produces an exponentially bounded resolvent operator $R(t)$, $t \in R^+$. However, any resolvent that does not satisfy the condition (4.4.4) cannot be obtained by the procedure indicated in Theorem 4.4.1.

In view of Remark 4.4.3, it is clear that the basic requirement of Theorem 4.4.1 is that the operator C must generate a C_0-semigroup. To achieve this, we need the following well-known perturbation result.

Definition 4.4.2 Let $T(t)$, $t \in R^+$, be a linear semigroup of bounded operators on the Banach space E_2. Then the Favard class of $T(t)$ is the set

$$\left\{x \in E_2 : \lim_{t \to 0^+} \sup\left[\frac{|T(t)x - x|}{t}\right] < \infty\right\}.$$

Lemma 4.4.2 (Perturbation Result) *Let E_2 be a Banach space and let $A: D(A) \subset E_2 \times E_2$ be the infinitesimal generator of a C_0-semigroup $T(t)$, $t \in R^+$. Let $K: D(A) \to E_2$ be continuous with respect to the graph norm of A on $D(A)$ such that the range of K belongs to the Favard class of $T(t)$. Then $A + K$ generates a semigroup on E_2.*

We are now in a position to prove the following existence result for the resolvent operator $R(t)$ of (4.4.1).

Theorem 4.4.2 *Assume that*
(i) *the hypotheses $(h_0) - (h_3)$ hold;*
(ii) *the function $e^{-\omega t}B(t)x$ is of bounded variation on R^+ for any $x \in E_2$.*

Then there exists a resolvent operator for the equation (4.4.1).

Proof In view of Theorem 4.4.1, we need only prove that the operator

$$C(x,g) = (Ax + \delta g, Bx + D_u g)$$

generates a semigroup on $E_2 = E_1 \times e^{\omega t} L^1(R^+, E_1)$. We shall first construct the unperturbed semigroup $T(t)$ by letting

(4.4.9) $$T(t)(x,g) = \left(S(t)x + \int_0^t S(t-s)g(s)\,ds, T(t)g \right),$$

where $S(t)$, $t \in R^+$ is the semigroup generated by the operator A on E_1. Then from (4.4.9), we have

$$T(t)T(u)(x,g) = \left(S(t)\left[S(u)x + \int_0^u S(t-\tau)g(\tau)\,d\tau \right] \right.$$

$$\left. + \int_0^t S(t-\tau)(T(u)g)(\tau)\,d\tau, T(t)T(u)g \right)$$

$$= \left(S(t+u)x + \int_0^u S(t+u-\tau)g(\tau)\,d\tau \right.$$

$$\left. + \int_0^t S(t-\tau)g(u+\tau)\,d\tau, T(t+u)g \right)$$

$$= T(t+u)(x,g)$$

and $T(t)$, $t \in R^+$, is a strongly continuous family of bounded operators on E_2. Let $\mathcal{A}$ be the infinitesimal generator of $T(t)$, $t \in R^+$. Then

$$(y, f) = \mathcal{A}(x, g)$$

is equivalent to

(4.4.10) $$y = \lim_{t \to 0^+} \frac{1}{t}\left[S(t)x - x + \int_0^t S(t-\tau)g(\tau)\,d\tau \right],$$

(4.4.11) $$f = \lim_{t \to 0^+}[T(t)g - g] = D_u g.$$

Clearly (4.4.11) implies that g is continuous and hence from (4.4.10), we obtain

$$y = \lim_{t \to 0^+} \tfrac{1}{t}[S(t)x - x] + S(0)g(0)$$
$$= Ax + \delta g.$$

Therefore
$$\mathcal{A}(x,g) = (Ax + \delta g, D_u g),$$

and $\mathcal{A}$ is defined on $E_2 \times D(D_u)$. But $\mathcal{A} = T + L$ with $L(x,g) = (0, Bx)$, and the range of L consists of those elements of the form $(0, f)$, where f is such that $e^{-\omega t}f(t)$ is bounded in R. Thus, to apply Lemma 4.4.2, we need only to show that these elements belong to the Favard class for $T(t)$, $t \in R^+$.

Since f is of bounded variation

(4.4.12) $\qquad \left| \tfrac{1}{t} \int_0^t S(t-\tau)f(\tau)\,d\tau \right| \leq \sup_{0 \leq \tau \leq t} |S(\tau)| \sup_{0 \leq \tau \leq t} |f(\tau)|,$

(4.4.13) $\qquad \tfrac{1}{t}|T(t)f - f| = \tfrac{1}{t}\int_0^\infty e^{-\omega s}|f(t+\tau) - f(\tau)|\,d\tau$

$$\leq \tfrac{1}{t}\int_0^\infty e^{-\omega(t+\tau)}(e^{\omega t} - 1)|f(t+\tau)|\,d\tau$$

$$+ \tfrac{1}{t}\int_0^\infty |e^{-\omega(t+\tau)}f(t+\tau) - e^{-\omega \tau}f(\tau)|\,d\tau.$$

As $t \to 0^+$, the first term on the right-hand side of (4.4.13) is bounded by

$$2\omega \int_0^\infty e^{-\omega \tau}|f(\tau)|\,d\tau,$$

whereas the second term (see Brezis [1]) tends to the variation of $e^{-\omega t}f(t)$ on R^+. Thus
$$\lim_{t \to 0^+} \sup \tfrac{1}{t}|T(t)(0,f) - (0,f)|$$
is finite. This means that the range of L belongs to the Favard class of the semigroup $T(t)$. This completes the proof of the theorem. □

4.5 Evolution Operators and Resolvents

Consider the IVP

(4.5.1)
$$\begin{cases} x'(t) = A(t)x(t) + \int_0^t B(t,s)x(s)\,ds + f(t), & 0 \le s \le t < \infty, \\ x(0) = x_0 \in E_1, \end{cases}$$

in a Banach space E_1.

We shall assume that for each $t \in [0,T]$, $A(t)$ is a closed linear operator with dense domain $D(A) \subset E_1$ that is independent of t, $0 \le t \le T$; for $0 \le s \le t \le T$, $B(t,s)$ is a closed operator with domain at least in $D(A)$. Suppose that E_2 is a Banach space formed from $D(A)$ with graph norm $|y|_{E_2} = |A(0)y| + |y|$, where $|\cdot|$ is the norm on E_1. Let $\mathcal{B}(E_2, E_1)$ be the set of all bounded operators from E_2 to E_1. If $E_2 = E_1$, we shall denote $\mathcal{B}(E_1, E_1) = \mathcal{B}(E_1)$. Since $A(t)$ and $B(t,s)$ are closed linear operators, it is clear that $A(t)$ and $B(t,s)$ belong to $\mathcal{B}(E_2, E_1)$ for $0 \le t \le T$ and $0 \le s \le t \le T$ respectively. Assume further that $A(t)$ and $B(t,s)$ are continuous on $0 \le t \le T$ and $0 \le s \le t \le T$ respectively. As we are interested in our subsequent discussion only in strong solutions, we shall restrict the initial condition for (4.5.1) to $x_0 \in D(A)$ and the forcing function $f \in C[[0,T], E_2]$.

Definition 4.5.1 By a solution $x(t)$ of (4.5.1) on $[0,T]$ with $x(0) = x_0 \in D(A)$, we mean a function $x \in C[[0,T], E_2] \cap C^1[[0,T], E_1]$ such that (4.5.1) is satisfied for all $t \in [0,T]$.

Definition 4.5.2 A resolvent operator for (4.5.1) is a bounded operator-valued function $R(t,s) \in \mathcal{B}(E_2, E_1)$ for $0 \le s \le t \le T$ such that the following conditions hold:

(i) $R(t,s)$ is strongly continuous in t and s, $R(s,s) = I$, $0 \le s \le T$, $|R(t,s)| \le Me^{\beta(t-s)}$ for some constants M and β;

(ii) $R(t,s)E_2 \subset E_2$ is strongly continuous in t and s on E_2;

(iii) for each $x \in D(A)$, $R(t,s)x$ is strongly continuous and differentiable in t and s, and satisfies

(4.5.2)
$$\frac{\partial R}{\partial t}(t,s)x = A(t)R(t,s)x + \int_s^t B(t,\tau)R(\tau,s)x\,d\tau,$$

(4.5.3) $$\frac{\partial R}{\partial s}(t,s)x = -R(t,s)A(s)x - \int_s^t R(t,\tau)B(\tau,s)x\,d\tau,$$

with $\partial R/\partial t(t,s)x$ and $\partial R/\partial s(t,s)x$ strongly continuous on $0 \leq s \leq t \leq T$.

Remark 4.5.1 If $A(t)$ and $B(t,s)$ are defined on R^+ and $R^+ \times R^+$ then we replace $0 \leq t \leq T$ and $0 \leq s \leq t \leq T$ by $0 \leq t < \infty$ and $0 \leq s \leq t < \infty$ respectively in Definitions 4.5.1 and 4.5.2.

The following result shows that there can be at most one resolvent operator if it exists.

Theorem 4.5.1 There exists at most one resolvent operator for (4.5.1).

Proof Suppose that $R(t,s)$ and $S(t,s)$ are resolvent operators for (4.5.1). Let $x \in D(A)$. Then from Definition 4.5.2 (iii), we get

$$S(t,s)x - R(t,s)x = \int_s^t \frac{\partial}{\partial \tau}[R(t,\tau)S(\tau,s)x]d\tau$$

$$= -\int_s^t \int_\tau^t R(t,u)B(u,\tau)S(\tau,s)x\,du\,d\tau$$

$$+ \int_s^t \int_s^\tau R(t,\tau)B(\tau,u)S(u,s)x\,du\,d\tau.$$

Since $x \in D(A)$ and $R(t,\tau)B(\tau,u)S(u,s)$ is continuous in τ and u, it follows from Fubini's theorem that

$$S(t,s)x - R(t,s)x = 0, \quad 0 \leq s \leq t \leq T,$$

and for all $x \in D(A)$. From the facts that $D(A)$ is dense in E_1 and $S(t,s) - R(t,s)$ is a bounded operator, the result follows. This completes the proof. □

Theorem 4.5.2 If $A(t) \equiv A$ and $B(t,s) \equiv B(t-s)$ and there exists a resolvent operator $R(t,s)$ for (4.5.1) then $R(t,s) = R(t-s)$.

Proof Let u $(0 < u < T)$ be fixed. Consider the function $Q(t,s) = R(t+u, s+u)$ on $0 \leq s \leq t \leq T - u$. Then it is clear that $Q(t,s)$ satisfies the conditions (i) and (ii) of Definition 4.5.2. Moreover, if $x \in D(A)$ then $Q(t,s)x$ is strongly continuously differentiable in t and s, $0 \le s \le t \le T$. Further, we have

$$\frac{\partial Q}{\partial t}(t,s)x = AR(t+u, s+u) + \int_{s+u}^{t+u} B(t+u-\tau)R(\tau, s+u)x\, d\tau$$

$$= AQ(t,s) + \int_s^t B(t-\tau)Q(\tau, s)x\, d\tau.$$

This shows that (4.5.2) holds. Similarly, it is easy to verify (4.5.3) for $Q(t,s)$. Thus by uniqueness of the resolvent on $[0, T-u]$, we must have

$$Q(t,s) = R(t,s).$$

If $t = s$ then $R(t,s) = R(t-s, 0)$ is clear, and if $t > s > 0$ then we have

$$R(t,s) = R(t-s, 0)$$

by taking $s = u$.

Theorem 4.5.3 *Suppose that there exists a resolvent operator for (4.5.1). If $x(t)$ is a solution of (4.5.1) then*

(4.5.4) $$x(t) = R(t,0)x_0 + \int_0^t R(t,s)f(s)\, ds.$$

Proof Consider the identity

$$x(t) - R(t,0)x_0 = \int_0^t \frac{\partial}{\partial s}[R(t,s)x(s)]\, ds.$$

Using the fact that $x(t)$ is a solution of (4.5.1) and from the relation (4.5.3), we obtain

$$x(t) - R(t,0)x_0 = \int_0^t \left[-R(t,s)A(s) - \int_s^t R(t,\tau)B(\tau, s)\, d\tau \right] x(s)\, ds$$

$$+ \int_0^t R(t,s)\left[A(s)x(s) + \int_0^s B(s,\tau)x(\tau)\, d\tau + f(s) \right] ds.$$

This implies that

$$(4.5.5) \quad x(t) - R(t,0)x_0 - \int_0^t R(t,s)f(s)\,ds = -\int_0^t \left[\int_s^t R(t,\tau)B(\tau,s)\,d\tau\right]x(s)\,ds$$

$$+ \int_0^t R(t,s)\left[\int_0^s B(s,\tau)x(\tau)\,d\tau\right]ds.$$

From the definition of a solution for (4.5.1), it is clear that $R(t,\tau)B(\tau,s)x(s)$ is continuous. Hence the application of Fubini's theorem yields that the right-hand side of (4.5.5) is equal to zero. This completes the proof. □

Remark 4.5.2 The relation (4.5.4) is the standard linear variation of parameters formula for the integro-differential equation (4.5.1). Since E_1 is infinite-dimensional, the function $x(t)$ satisfying the relation (4.5.4) need not actually be a solution of (4.5.1) as in the case of finite dimensional integro-differential equations. In general, even if $x_0 \in D(A)$, the formula (4.5.4) will not necessarily yield a solution in the usual sense.

Definition 4.5.3 For $x_0 \in D(A)$ and $f \in C[[0,T], E_1]$, a function $x(t)$ satisfying the relation (4.5.4) is called a weak solution of (4.5.1) on $[0,T]$.

A natural question that arises in this connection is to determine when weak solutions are strong solutions. This has been partly discussed in Section 4.3 by constructing an equivalent evolution equation (ordinary differential equation), and this topic will be considered further in this section. However, the following result provides a partial answer.

Theorem 4.5.4 Suppose that a resolvent operator $R(t,s)$ for (4.5.1) exists. If $f \in C[[0,T], Y]$ then

$$(4.5.6) \quad v(t) = \int_0^t R(t,s)f(s)\,ds$$

is a solution of (4.5.1).

Proof From (4.5.6), we have

$$v'(t) = R(t,t)f(t) + \int_0^t \frac{\partial R}{\partial t}(t,s)f(s)\,ds.$$

Using Definition 4.5.2, we obtain

$$v'(t) = f(t) + \int_0^t A(t)R(t,s)f(s)\,ds$$

$$+ \int_0^t \int_s^t B(t,\tau)R(\tau,s)f(s)\,d\tau\,ds.$$

Since $A(t)$ and $B(t,s)$ are closed operators, it follows from Fubini's theorem and (4.5.6) that

$$v'(t) = f(t) + A(t)v(t) + \int_0^t B(t,s)v(s)\,ds,$$

and hence the proof is complete. □

In order to obtain the existence of a resolvent operator for (4.5.1), we shall reformulate the equivalent evolution equation. Then we show that this ordinary differential equation (see (4.5.8)) has an evolution operator $U(t,s)$. From this $U(t,s)$, we shall realize the required resolvent operator. Let BU be the space of bounded uniformly continuous functions on R^+ into E_1, and let $\mathcal{F}$ be a Banach space with a norm stronger than the sup norm on BU. It is further required that for each $t \in R^+$,

$$B(t + \cdot, t)x \in \mathcal{F} \quad \text{for every } x \in D(A),$$

where $(B(t + \cdot, t)x)(s) = B(t + s, t)x$ for $s \geq 0$. This then defines an operator $B(t)$ from E_1 to $\mathcal{F}$ that has the domain $D(A)$. We shall further assume that $\{T(t): T \in R^+\}$ defined by

$$T(t)f(s) = f(t + s)$$

is a C_0-semigroup on $\mathcal{F}$ with generator D_s on domain $D(D_s)$. Let $E = E_1 \times \mathcal{F}$ with the norm

$$|(x,y)| = |x| + |y|_{\mathcal{F}}.$$

We need the following hypotheses in our subsequent discussion.

(H_0) $\{A(t)\}$, $0 \leq t \leq T$, is a stable family of generators (see Definition 4.3.6) such that $A(t)x$ is strongly continuously differentiable on $[0,T]$ for $x \in D(A)$;

(H_1) $B(t)x$ is strongly continuously differentiable on $[0,T]$ for $x \in D(A)$;
(H_2) $B(t)$ is continuous on R^+ into $\mathcal{B}(E_2, \mathcal{F})$;
(H_3) $B(t): E_2 \to D(D_s)$ for all $t \geq 0$;
(H_4) $D_s B(t)$ is continuous on R^+ into $\mathcal{B}(E_2, \mathcal{F})$.

Remark 4.5.3 The hypotheses (H_0) and (H_1) guarantee that $A(t)$ and $B(t)$ are continuous on $[0,T]$ into $\mathcal{B}(E_2, E_1)$ and $\mathcal{B}(E_2, \mathcal{F})$ respectively. In fact, $A(t)$ is Lipschitzian in t, since $A'(t)$ is in $\mathcal{B}(E_2, E_1)$, and the application of the uniform boundedness principle yields that $A'(t)$ is bounded uniformly in t as an element of $\mathcal{B}(E_2, E_1)$. A similar property holds for $B(t)$.

To find a resolvent operator for (4.5.1), it is necessary to establish a relationship between the integro-differential equation

(4.5.7)
$$\begin{cases} x'(t) = A(t)x(t) + \int_{t_0}^{t} B(t,s)x(s)\,ds + f(t), \\ x(t_0) = x_0, \quad 0 \leq t_0 \leq t \leq T, \end{cases}$$

and the evolution equation

(4.5.8) $\qquad z'(t) = C(t)z(t), \quad z(t_0) = z_0, \quad 0 \leq t_0 \leq t \leq T$

on the Banach space E, where

(4.5.9)
$$C(t) = \begin{bmatrix} A(t) & \delta_0 \\ B(t) & D_s \end{bmatrix},$$

$D(C) = D(A) \times D(D_s)$ is the domain of $C(t)$ for $0 \leq t < \infty$ and $\delta_0: \mathcal{F} \to E_1$ is defined by $\delta_0 f = f(0)$.

Definition 4.5.4 Let $E_3 = D(A) \times D(D_s)$ be normed with $|(x,y)|_1 = |x|_{E_2} + |y|_D$ where $|\cdot|_D$ is the graph norm of D_s. By a solution of (4.5.8), we mean a function $z \in C[[0,T], E_3] \cap C^1[[0,T], E]$ such that $z(t_0) = z_0$, and (4.5.8) is satisfied for $0 \leq t_0 < t \leq T$.

Remark 4.5.4 Since $B(0)$ and δ_0 are bounded operators from $E_2 \to E_1$ and $\mathcal{F}$ respectively, repeated application of the triangle inequality shows that $|\cdot|_1$ is equivalent to the graph norm of $C(0)$ on E_3. Also, if (H_0) and (H_1)

hold then $C(t)$ is continuous on $[0,T]$ into $\mathcal{B}(E_3, E)$ and $C(t)$ is strongly continuously differentiable for $z \in E_3$.

Remark 4.5.5 The conditions $(H_0)-(H_4)$ in this section are somewhat much simpler than the corresponding conditions $(\bar{h}_1)-(\bar{h}_4)$ in Section 4.3. This is because of $\mathcal{F}$ consists of continuous functions only. Moreover, the operator $C(t)$ defined in (4.3.5) has been reduced to the form given by (4.5.9), since the problem is slightly reformulated in the section. The arguments in Section 4.3 are equally valid in the case of (4.5.9).

Remark 4.5.6 In the convolution case when $A(t) \equiv A$ and $B(t,s) \equiv B(t-s)$, the operator $B(t) \in \mathcal{B}(E_2, \mathcal{F})$ is constant. Thus (H_0), (H_1) and (H_2) are automatically satisfied if A generates a semigroup. Further, if (H_3) holds, then (H_4) will follow automatically, since D_s is closed.

The following simplified form of Theorem 4.3.3 is of special interest in itself.

Theorem 4.5.5 Suppose that $(H_2)-(H_4)$ are valid and $f \in D(D_s)$. If $x(t)$ is a solution of (4.5.7) then $z(t) = (x(t), y(t))^*$ is a solution of (4.5.8) with $z_0 = (x_0, f_{t_0})^*$ when

$$y(t) = T(t-t_0)f_{t_0} + \int_{t_0}^{t} T(t-s)B(s)x(s)\,ds$$

where $T(t)$ is the translation semigroup generated by D_s and $f_{t_0} = T(t_0)f$. Conversely, if (H_2) holds and $z(t)$ is a solution of (4.5.8) with $z(t_0) = (x_0, f_{t_0})^*$ and $z(t) = (x(t), y(t))^*$, $t_0 \leq t \leq T$, then $x(t)$ is a solution of (4.5.7).

Proof Suppose that $x(t)$ is a solution of (4.5.7). Then by the hypothesis (H_2), $B(t)x(t)$ is continuous on $[t_0, T]$. Further, from the hypotheses (H_3) and (H_4), it follows that $D_s B(t)x(t)$ is continuous, and hence the equation

$$y'(t) = D_s y(t) + B(t)x(t)$$

has a solution

$$y(t) = T(t-t_0)f_{t_0} + \int_{t_0}^{t} T(t-s)B(s)x(s)\,ds.$$

Thus for $\tau \geq 0$, we have

$$(y(t))(\tau) = T(t-t_0)f_{t_0}(\tau) + \int_{t_0}^{t} T(t-s)B(s+\tau,s)x(s)\,ds$$

$$= f(t+\tau) + \int_{t_0}^{t} B(t+\tau,s)x(s)\,ds.$$

Hence

$$\delta_0 y(t) = f(t) + \int_{t_0}^{t} B(t,s)x(s)\,ds,$$

and $z(t) = (x(t), y(t))^*$ is a solution of (4.5.8) with the prescribed initial data. Conversely, suppose that $z(t)$ is a solution of (4.5.8) on $[t_0, T]$ with $z(t_0) = (x_0, f_{t_0})^*$ for $t_0 \leq t \leq T$. Then with $z(t) = (x(t), y(t))^*$, we get

$$x'(t) = A(t)x(t) + \delta_0 y(t),$$

$$y'(t) = D_s y(t) + B(t)x(t).$$

Since $z \in C[[t_0, T], E_3] \cap C^1[[t_0, T], E]$, it follows that

$$x \in C[[t_0, T], E_2] \cap C^1[[t_0, T], E_1].$$

From the hypothesis (H_2), $B(t)x(t)$ is a continuous function on $[t_0, T]$ into $\mathcal{F}$, and $y(t)$ can be written as

$$y(t) = T(t-t_0)f_{t_0} + \int_{t_0}^{t} T(t-s)B(s)x(s)\,ds$$

for $t_0 \leq t \leq T$. From the fact that

$$\delta_0 y(t) = f(t) + \int_{t_0}^{t} B(t,s)x(s)\,ds,$$

it is clear that $x(t)$ is a solution of (4.5.7). This completes the proof. □

It is also interesting in itself to relate (4.5.7) to the evolution equation

(4.5.10) $\qquad z'(t) = C(t)x(t) + g(t), \quad z(t_0) = z_0, \quad t_0 \leq t \leq T,$

where $g(t) = f(t), 0)^*$.

Corollary 4.5.1 *Suppose that $(H_2) - (H_4)$ hold and f is continuous from $[t_0, T]$ into E_1. If $x(t)$ is a solution of (4.5.7) then $z(t) = (x(t), y(t))^*$ is a solution of (4.5.10) with $z(t_0) = (x_0, 0)^*$, where*

$$y(t) = \int_{t_0}^{t} T(t-s)B(s)x(s)\,ds.$$

Conversely, suppose that (H_2) holds and $f(t)$ is continuous from $[t_0, T]$ into E_1. If $z(t)$ is a solution of (4.5.10) with $z(t_0) = (x_0, 0)^*$ and $z(t) = (x(t), y(t))^*$, $t_0 \leq t \leq T$, then $x(t)$ is a solution of (4.5.7).

The proof of this corollary is exactly similar to that of Theorem 4.5.5, and hence is left as an exercise.

The following modified form of the definition (see Definition 4.3.5) of an evolution operator (or fundamental solution) is useful in our subsequent discussion.

Definition 4.5.5 An evolution operator (or fundamental solution) for (4.5.8) is a bounded operator valued function $V(t,s) \in \mathcal{B}(E)$ with $0 \leq s \leq t \leq T$ that satisfies the following conditions:

(i) $V(t,s)$ is strongly continuous in s and t, $V(s,s) = I$ and $|V(t,s)| \leq Me^{\beta(t-s)}$ for some constants M and β;

(ii) $V(t,s) = V(t,\tau)V(\tau,s)$, $0 \leq s \leq \tau \leq t \leq T$;

(iii) $V(t,s)E_3 \subset E_3$, $V(t,s)$ is strongly continuous in s and t on E_3;

(iv) for each $z \in E_3$, $V(t,s)$ is strongly continuously differentiable in t and s with

$$\frac{\partial}{\partial t}V(t,s)z = C(t)V(t,s)z,$$

$$\frac{\partial}{\partial s}V(t,s)z = -V(t,s)C(s)z.$$

Theorem 4.5.6 *Suppose that $(H_0) - (H_4)$ hold. Then there exists a constant $\gamma > 0$ such that*

(4.5.11) $$|B(t)x| \leq \gamma(|x| + |A(t)x|)$$

for all $x \in D(A)$ and $t \in [0,T]$. Furthermore, (4.5.8) with $t_0 = 0$ has an evolution operator.

Proof It follows from the hypothesis (H_1) that there is a constant $\gamma_1 > 0$ such that

(4.5.12) $$|B(t)x| \leq \gamma_1(|x| + |A(0)x|)$$

for all $x \in D(A)$. Since $A(t)$ is continuous on $[0,T]$ into $\mathcal{B}(E_2, E_1)$, if $t - s$ is small, then, using a Neuman series expansion, it follows from the identity

$$A(t) - \lambda I \equiv \left[I + (A(t) - A(s))(A(s) - \lambda I)^{-1}\right]\left[A(s) - \lambda I\right]$$

that

$$(A(t) - \lambda I)^{-1} = (A(s) - \lambda I)^{-1} \sum_{n=0}^{\infty} (-1)^n \left[(A(t) - A(s))(A(s) - \lambda I)^{-1}\right]^n.$$

Thus $[A(t) - \lambda I]^{-1}$ is continuous from $[0,T]$ into $\mathcal{B}(E_1, E_2)$. This implies that $A(0)(A(t) - \lambda I)^{-1}$ is uniformly bounded on $[0,T]$. Therefore, from (4.5.12), we obtain

$$|B(t)x| \leq \gamma_1\left[|x| + |A(0)(A(t) - \lambda I)^{-1}| \, |(A(t) - \lambda I)x|\right]$$

$$\leq \gamma_1 |x| + M|A(t)x| + M\lambda |x|,$$

where

$$M = \max_{t \in [0,T]} |A(0)[A(t) - \lambda I]^{-1}|.$$

Choose

$$\gamma = \max[\gamma_1 + M\lambda, M],$$

so that the inequality (4.5.11) follows. The rest of the proof is similar to that of Theorem 3.3 in Chen and Grimmer [1]. □

Theorem 4.5.7 *Let $t_0 = 0$. Assume that the hypotheses $(H_0) - (H_4)$ are valid. Then the integro-differential equation (4.5.7) with $x_0 \in D(A)$ has a resolvent operator.*

Proof Let $V(t,s)$ be the evolution operator for (4.5.8), which is guaranteed by Theorem 4.5.6. Write $V(t,s)$ as

$$V(t,s) = \begin{bmatrix} V_{11}(t,s) & V_{12}(t,s) \\ V_{21}(t,s) & V_{22}(t,s) \end{bmatrix}.$$

We shall now show that $V_{11}(t,s)$ is a resolvent operator for (4.5.1) with $x_0 \in D(A)$. It is clear that $V_1(t,s) \in \mathcal{B}(E_1, E_1)$, $0 \leq s \leq t \leq T$. Since $V(s,s) = I$, it follows that

$$V_{11}(s,s) = I_E.$$

As $V(t,s)$ is strongly continuous in t and s, $V_{11}(t,s)$ is also strongly continuous. Further, we have

$$|V_{11}(t,s)| \leq |V(t,s)| \leq Me^{\beta(t-s)}.$$

If $x \in E_2$, $(x,0)^* \in E_3$ and $V(t,s)(x,0)^* \in E_2$ then it is clear that $V_{11}(t,s) \in E_2$. As $V(t,s)$ is strongly continuous in t and s on E_3, the same is true for $V_{11}(t,s)$ on E_2. Thus $V_{11}(t,s)$ satisfies the conditions (i) and (ii) of Definition 4.5.2. To see that $V_{11}(t,s)$ also satisfies the condition (iii), let $x_0 \in E_2$. Then $(x_0,0) \in E_3$ and $V(t,s)(x_0,0)^*$ is a solution of (4.5.8). Hence by Theorem 4.5.5., $V_{11}(t,s)x_0$ is a solution of

$$x'(t) = A(t)x(t) + \int_s^t B(t,\tau)x(\tau)\,d\tau, \qquad x(s) = x_0.$$

This implies for $x_0 \in E_2$ that

$$\frac{\partial}{\partial t}V_{11}(t,s) = A(t)V_{11}(t,s)x_0 + \int_s^t B(t,\tau)V_{11}(\tau,s)x_0\,d\tau.$$

Since $\partial/\partial s\, V(t,s)(x_0,0)^* = -V(t,s)C(s)(x_0,0)$, it follows that

(4.5.13) $$\frac{\partial}{\partial s}V_{11}(t,s)x_0 = -V_{11}(t,s)A(s)x_0 - V_{12}(t,s)B(s)x_0.$$

Now we must determine $V_{12}(t,s)$. Let $(0,f)^* \in E_3$. Then $(0,f_s)^* \in E$, and $V(t,s)(0,f_s)^*$ is a solution of (4.5.8). Hence by Theorem 4.5.5, $V_{11}(t,s)f_s$ is a solution (4.5.7). However, by Corollary 4.5.1, the nonhomogeneous equation (4.5.10) has a solution with initial condition $(0,0)^*$, which must be

$$\int_s^t V(t,\tau)(f(\tau),0)\,d\tau,$$

since $V(t,s)$ is an evolution operator for (4.5.8). Thus we have

$$x(t) = \int_s^t V(t,\tau)f(\tau)\,d\tau.$$

Therefore for $f \in D(D_s)$, it follows that

$$V_{12}(t,s)f_s = \int_s^t V_{11}(t,\tau)f(\tau)\,d\tau.$$

This implies

$$V_{12}(t,s)f = \int_s^t V_{11}(t,s)f(\tau-s)\,d\tau.$$

As $D(D_s)$ is dense in $\mathcal{F}$, we may extend the above relation for all $\mathcal{F}$. Recalling that $B(s)x_0$ is a function of τ and is given by

$$(B(s)x_0)(\tau) = B(\tau+s)x_0,$$

it follows from (4.5.13) that

$$\frac{\partial}{\partial s}V_{11}(t,s)x_0 = -V_{11}(t,s)A(s)x_0 - \int_s^t V_{11}(t,\tau)B(\tau,s)x_0\,ds.$$

This implies that $V_{11}(t,s)$ satisfies (4.5.3). Similarly, it is easy to verify that $V_{11}(t,s)$ also satisfies (4.5.2). Hence $R(t,s) \equiv V_{11}(t,s)$ is a resolvent operator for (4.5.7). This completes the proof. □

Corollary 4.5.2 *Suppose that the hypotheses $(H_0)-(H_4)$ are valid. If $x_0 \in D(A)$ and $f \in C^1[[0,T],E]$ then the solution $x(t)$ of (4.5.7) with $x(0) = x_0$ is given by*

$$x(t) = R(t,0)x_0 + \int_0^t R(t,s)f(s)\,ds.$$

Proof From Theorem 4.5.6 and the proof of Theorem 3.2 in Chen and Grimmer [1], it follows that $\{C(t)\}$ is a stable family of generators with a common domain such that $C(t)z$ is strongly differentiable for $z \in E_3$. Thus the hypotheses of Theorem 4.5.3 in Tanabe [1] are verified with $g(t) = (f(t),0)^*$. Hence the equation (4.5.8) has a solution

$$z(t) = V(t,0)(x_0,0)^* + \int_0^t V(t,s)(f(s),0)^*\,ds.$$

This implies that $x(t)$ satisfies the equation

$$x(t) = V_{11}(t,s)x_0 + \int_0^t V_{11}(t,s)f(s)\,ds.$$

Since by Theorem 4.5.7, $R(t,s) \equiv V_{11}(t,s)$, the result follows. Thus the proof is complete. □

4.6 Asymptotic Behavior and Perturbations

As in the theory of integro-differential equations in finite-dimensional spaces, it is extremely useful to have some knowledge of the structure of the resolvent operator to investigate sufficient conditions for asymptotic behavior of Volterra integro-differential equations in abstract spaces.

For simplicity, let us consider the convolution equation

(4.6.1)
$$\begin{cases} x'(t) = Ax(t) + \int_0^t B(t-s)x(s)\,ds + f(t), & t \in R^+, \\ x(0) = x_0 \in D(A) \subset E_1, \end{cases}$$

under the same hypotheses on A and $B(t)$ as in Section 4.3.

The main technique is to use known properties of the resolvent operator $R(t)$ corresponding to (4.6.1) along with the variation of constants formula

(4.6.2)
$$x(t) = R(t)x_0 + \int_0^t R(t-s)f(s)\,ds.$$

If A generates a semigroup $T_1(t)$ with

$$|T_1(t)| \leq Me^{\omega t}$$

then we say that $A \in G(M,\omega)$. In particular, if $A \in G(M,-\alpha)$, $\alpha > 0$, then A^{-1} exists as a bounded operator and $B(t) = B(t)A^{-1}A$. Thus we may write

$$B(t) = F_1(t)A,$$

where $F_1(t)$ is a bounded operator. Let $BU = BU(R^+,E_1)$ be the space of bounded uniformly continuous functions defined from R^+ into E_1. Define the space $\mathfrak{F}_\alpha^a, a > 0$ by

$$\mathfrak{F}_\alpha^a = \{f \in BU : e^{\alpha t}f(t) \in BU\}$$

with norm

$$|f| = \sup\{|ae^{\alpha t}f(t)| : t \geq 0\}.$$

If $T(t)$ is the translation semigroup generated by D_s, it is clear that on $\mathfrak{F}_\alpha^a$

$$|T(t)f| = \sup\{|ae^{\alpha t}f(t+s)| : s \geq 0\}$$
$$\leq e^{-\alpha t}|f|.$$

EQUATIONS IN ABSTRACT SPACES

Theorem 4.6.1 Suppose that $A \in G(M, -\alpha)$ and $B(t) = F_1(t)A$, with $F_1(t)x \in D(D_s) \subset \mathcal{F}_\beta^a$ for some $\beta > 0$ and for each $x \in E_1$. Further, assume that

$$|ae^{\beta t}F(t)| \leq 1,$$

$$|a^2 e^{\beta t} D_s F(t)| \leq 1,$$

with $M/a < \min(\alpha, \beta) = \gamma$. Then

$$|R(t)| \leq M e^{(-\gamma + M/a)t} \quad \text{for all } t \geq 0.$$

Proof On $E = E_1 \times \mathcal{F}_\beta^a$ with norm

$$|(x, y)| = \max(|x|, |y|),$$

it is clear that

$$C_1 = \begin{bmatrix} A & 0 \\ 0 & D_s \end{bmatrix} \in G(M, -\alpha).$$

Let $P = \begin{bmatrix} I & 0 \\ F_1 & I \end{bmatrix}$ and define a new norm on E by

$$|||z||| = |P^{-1}z|.$$

If $S(t)$ is generated by C_1 then we have

$$|PS(t)P^{-1}z| = |S(t)P^{-1}z|$$

$$\leq Me^{-\gamma t}|||z|||.$$

Since

$$L = \begin{bmatrix} 0 & \delta_0 \\ D_s F_1 & 0 \end{bmatrix}$$

is a bounded operator, it follows that

$$PC_1 P^{-1} + L \in G(M, -\gamma + M|||L|||)$$

where

$$|||L||| = |P^{-1}L| \leq \max\{|\delta_0|, |D_s F_1|, |F_1 \delta_0|\}.$$

From the fact that $\delta_0: \mathcal{F}_\beta^a \to E_1$ has norm $1/a$, it is clear that

$$|||L||| \leq \frac{1}{a}.$$

If $S_1(t)$ is the semigroup generated by $PC_1P^{-1}+L$, we see that

$$|R(t)| \leq |S_1(t)|.$$

In particular,

$$|||S_1(t)(x_0,0)^*||| \leq \max\{|R(t)x_0|, |S_{21}(t)x_0 - F_1R(t)x_0|\},$$

with $S_{21}(t) = \prod_2 S(t)\prod_1$, where $\prod_1$ is the projection from E to E_1 and $\prod_2$ is the projection from E to $\mathfrak{F}_\beta^a$. The result now follows immediately, and hence the proof is complete.

Remark 4.6.1 To obtain the asymptotic behavior of solutions of (4.6.1), we note that if the hypotheses of Theorem 4.6.1 are satisfied, then

$$R(t)x_0 \to 0 \quad \text{as } t \to \infty$$

for each $x_0 \in E_1$. Further, if ρ is the operator defined by

$$\rho(f)(t) = \int_0^t R(t-s)f(s)\,ds,$$

then ρ maps each of the following spaces into itself (see Miller [1]):

(i) $BC_0 = \{f \in BU: f(t) \to 0 \text{ as } t \to \infty\}$;
(ii) $BC_q = \{f \in BU: \lim_{t \to \infty} f(t) \to q \text{ exists}\}$;
(iii) $A_w = \{f \in BU: f = g + h, g(t+w) = g(t), h \in BC_0\}$;
(iv) L^p, $p \geq 1$;
(v) $L^p \cap BC_0$.

Example 4.6.1 Consider the equation

(4.6.3)
$$\begin{cases} Ku_{tt} + B(0)u_t = \alpha(0)\nabla^2 u - \int_{-\infty}^t \beta'(t-s)u_s(x,s)\,ds \\ \quad + \int_{-\infty}^t \alpha'(t-s)\nabla^2 u(x,s)\,ds + r_t(x,t) \end{cases}$$

where a prime indicates differentiation with respect to t. This equation arises in the study of heat conduction in materials with memory (see Miller and Wheeler [1]). Let us assume that Ω is a bounded open connected subset of R^3 with C^∞ boundary. Also, assume that K is a positive constant and, α and β are in $C^2[R^+, R]$ with $\alpha(0)$ and $\beta(0)$ positive. If we suppress the dependence of u and r on $x \in \Omega$ and also if we assume that u is zero for $t \in R^-$, then the equation (4.6.3) takes the form

(4.6.4)
$$\begin{cases} Ku'' + \beta(0)u' = \alpha(0)\nabla^2 u - \int_0^t \beta'(t-s)u'(s)\,ds \\ \qquad + \int_0^t \alpha'(t-s)\nabla^2 u(s)\,ds + f(t) \end{cases}$$

where ∇^2 is the Laplacian on Ω with boundary condition $u|_\Gamma = 0$ and $f(t) = r'(t)$. Further, we assume the following initial conditions:

$$u(x,0) = u_0(x) \in H^2(\Omega) \cap H_0^1(\Omega),$$
$$u_t(x,0) = u_1(x) \in H_0^1(\Omega).$$

Then (4.6.4) can be written as

$$\begin{bmatrix} X \\ Y \end{bmatrix}' = \begin{bmatrix} 0, & I \\ \alpha_1(0)\nabla^2, & -\beta(0)I \end{bmatrix} \begin{bmatrix} X \\ Y \end{bmatrix}$$
$$+ \int_0^t \begin{bmatrix} 0, & 0 \\ \alpha_1'(t-\tau)\nabla^2, & -\beta_1'(t-\tau)I \end{bmatrix} \begin{bmatrix} X(\tau) \\ Y(\tau) \end{bmatrix} d\tau + \begin{bmatrix} 0 \\ f(t) \end{bmatrix}, \quad \begin{bmatrix} X(0) \\ Y(0) \end{bmatrix} = \begin{bmatrix} u_0 \\ u_1 \end{bmatrix},$$

where $\alpha_1 = \alpha/K$ and $\beta_1 = \beta/K$. Let $z = (X,Y)^*$; then this equation takes the form

$$z'(t) = Az(t) + \int_0^t F(t-s)Az(s)\,ds + G(t),$$

$$z(0) \in D(A) \subset H,$$

where H is the space $H_0^1(\Omega) \oplus H^0(\Omega)$ with inner product

$$\langle (x_1,y_1),(x_2,y_2) \rangle = \int_\Omega (\operatorname{grad} x_1 \operatorname{grad} x_2 + y_1 y_2)\,dx$$

and A has the domain $D(A) = (H_2^2(\Omega) \cap H_0^1(\Omega) \oplus H_0^1(\Omega))$, and $F(t) = (F_{ij})$, $F_{11}(t) \equiv F_{12}(t) \equiv 0$, $F_{22}(t) = \alpha_1'(t)/\alpha_1(0)I$ and $F_{21}(t) = -\beta_1'(t)I + \beta_1(0)F_{22}(t)$. Assume that $\alpha_1'(t)e^{\gamma t}$, $\alpha_1''(t)e^{\gamma t}$, $\beta_1'(t)e^{\gamma t}$ and $\beta_1''(t)e^{\gamma t}$ are bounded and uniformly continuous. Then it follows from Chen [1] that A generates a semigroup $\{T(t)\}$ on H with

$$|T(t)| \le M e^{-\gamma t}, \quad \gamma > 0.$$

Moreover, by letting $\beta_0 = \beta_1(0)$ and denoting γ_0 by the largest eigenvalue of $\alpha(0) \nabla^2$ in $H_0^1(\Omega)$, it follows from Pritchard and Zabczyk [1] that

$$-\gamma \le -2\beta_0|\gamma_0|\left[4|\gamma_0| + \beta_0\left(\beta_0 + \beta_0^2 + 4|\gamma_0|\right)^{1/2}\right]^{-1}.$$

Define $F_1(t)$ and $F_2(t)$ by

$$F_1(t) = \max(|F_{22}(t)|, |F_{21}(t)|)$$

and

$$F_2(t) = \max(|F'_{22}(t)|, |F'_{21}(t)|).$$

If $F_1(t) \le e^{-\gamma t}/a$ and $F_2(t) \le e^{-\gamma t}/a^2$ then it follows from Theorem 4.6.1 that

$$|R(t)| \le M e^{-\eta t}$$

where $\eta = \gamma - M/a$.

4.7 Stability of Solutions

In this section, we shall investigate sufficient condition for stability of solutions of integro-differential equations in abstract spaces. The main tools are Laplace transforms and Lyapunov functions.

Consider the IVP

$$(4.7.1) \quad \begin{cases} x'(t) = Kx(t) + \int_0^t [B(t-\tau)Ax(\tau) + G(t-\tau)x(\tau)]d\tau + F(t), \quad t \in R^+, \\ x(0) = x_0 \in E, \end{cases}$$

in a Hilbert space E, where K is a real constant, $B(t)$ and $G(t)$ are real-valued functions in $C^2(R^+)$ with $B(0) > 0$, and $F \in C^2(R^+, E)$. An equation of the form (4.7.1) arises in a model of heat flow in materials with memory proposed by Gurtrim and Pipkin [1]. This model exhibits certain behavior of a hyperbolic nature.

We need the following assumptions in our subsequent discussion.

(H_1) A is a self-adjoint, negative definite, closed, densely defined linear operator. Moreover, the resolvent operator $R(\lambda, A) = (A - \lambda I)^{-1}$

is compact when it exists.

(H_2) The scalar functions B and G are of the form

$$B(t) = 1 + \int_0^t b(\tau)d\tau, \quad G(t) = \gamma + \int_0^t g(\sigma)d\sigma,$$

with b and g in $C^1(R^+) \cap L^1(R^+)$ and γ a real constant.

Remark 4.7.1 It is well known (see Taylor [1], p. 343) that if (H_1) holds then there exist eigenvalues $\{\mu_n\}$ satisfying $-\infty < \ldots \leq \mu_2 \leq \mu_1 < 0$ and a complete orthonormal set of corresponding eigenvectors $\{\phi_n\}$ associated with A. Moreover, a complex number μ is in the spectrum of A (i.e. $\mu \in \sigma(A)$) if and only if $\mu = \mu_n$ for some n, and in this case $A\phi_n = \mu_n\phi_n$.

In view of the assumption (H_2), (4.7.1) can be written as

(4.7.1) $$\begin{cases} x''(t) = Kx'(t) + (A + \gamma I)x(t) + \int_0^t [b(t-\tau)Ax(\tau) + g(t-\tau)x(\tau)]d\tau + f(t), \\ x(0) = x_0, \quad x'(0) = v_0, \end{cases}$$

where $f = F'$ and $v_0 = Kx_0 + F(0)$.

Let $L^p(R^+, E)$, $1 \leq p < \infty$, denote the space of measurable functions $f(t)$ from R^+ into E with $|f(t)| \in L^p(R^+)$. Let $\widehat{b}(\lambda)$ be the Laplace transform of $b(t)$.

Definition 4.7.1 The IVP (4.7.2) is called L^2-stable if for any $x_0 \in E$, $v_0 \in E$ and $f \in L^1(R^+, E)$, the unique solution $x(t) = x(t, x_0, v_0, f)$ of (4.7.2) belongs to $L^2(R^+, E)$.

The following Laplace transform condition is necessary for L^2-stability of (4.7.2).

Theorem 4.7.1 Assume that (H_1) and (H_2) hold. If (4.7.2) is L^2-stable then $1 + \widehat{b}(\lambda) \neq 0$ when $\text{Re}\,\lambda > 0$.

Proof Suppose $1 + \widehat{b}(\lambda_0) = 0$ for some λ_0 with $\text{Re}\,\lambda_0 > 0$. Then either

$$\lambda_0^2 - K\lambda_0 - \gamma - \widehat{g}(\lambda_0) = 0$$

or

$$\frac{\lambda^2 - K\lambda - \gamma - \widehat{g}(\lambda)}{1 + \widehat{b}(\lambda)}$$

has a pole at $\lambda = \lambda_0$.

Since $\mu_n \to -\infty$ as $n \to \infty$, it follows that for all sufficiently large n, we can find solutions $\lambda = \lambda_n$ of the equation

$$\lambda_n^2 - K\lambda_n - \gamma - \widehat{g}(\lambda_n) - \mu_n\left[1 + \widehat{b}(\lambda_n)\right] = 0$$

with $\operatorname{Re}\lambda_n > 0$.

It is easy to verify that the function $x_n(t) = e^{\lambda_n t}\phi_n$ is a solution of (4.7.2) with $x_0 = \phi_n$, $v_0 = \lambda_n\phi_n$ and

$$f(t) = \int_t^\infty \exp(\lambda_n(t-\tau))\left[\mu_n b(\tau) + g(\tau)\right]d\tau\,\phi_n.$$

Using the fact that $\operatorname{Re}\lambda_n > 0$, it is easy to show that

$$|f(t)| \in L^1(R^+).$$

Finally, since $\operatorname{Re}\lambda_n > 0$, $|x_n(t)| \to \infty$, and hence it follows that (4.7.2) is not L_2-stable. This completes the proof. □

In view of Theorem 4.7.1, we further assume the following Laplace transform conditions to obtain L^2-stability of (4.7.2).

(T_1) $1 + \widehat{b}(\lambda) \neq 0$ for $\operatorname{Re}\lambda \geq 0$;
(T_2) $\lambda^2 - K\lambda - \gamma - \widehat{g}(\lambda) - \mu_n(1 + \widehat{b}(\lambda)) \neq 0$, for $n = 1, 2, \ldots$ and $\operatorname{Re}\lambda \geq 0$.

Remark 4.7.2 Theorem 4.7.1 and its proof show that a necessary condition for (4.7.2) to be L^2-stable is that (T_1) and (T_2) hold for λ in the open half-plane $\operatorname{Re}\lambda > 0$. But, in practice, if L^2-stability is to be guaranteed, we should generally have that (T_1) and (T_2) hold on $\operatorname{Re}\lambda = 0$ as well as in the open half-plane; otherwise if $1 + \widehat{b}(\lambda) = 0$ had a solution on $\operatorname{Re}\lambda = 0$ then a small perturbation could shift this root into the open right half-plane. Thus discussion of the cases where $1 + \widehat{b}(\lambda) \neq 0$ for $\operatorname{Re}\lambda \geq 0$, $\lambda \neq 0$ and $1 + \widehat{b}(0) = 0$ separately is of special interest (see Miller and Wheeler [1]).

Let $B(E)$ denote the space of bounded linear operators on E. Also, let $H^2(0, E)$ denote the Hardy class consisting of all functions $h(\lambda)$ from $\operatorname{Re}\lambda > 0$ into E which are analytic on $\operatorname{Re}\lambda > 0$ and satisfy:

(i) $\sup\left\{\int_{-\infty}^{\infty} |h(\sigma + i\tau)|^2 d\tau : \sigma > 0\right\} < \infty$;

(ii) $h(i\tau) = \lim_{\sigma \to 0^+} h(\sigma + i\tau)$ exists a.e. and belongs to $L^2(R, E)$.

The following well-known results (see Miller and Wheeler [1]) on Laplace transforms will be used in the proof of our main result, giving conditions that are sufficient to guarantee that (4.7.2) is L^2-stable.

Lemma 4.7.1 *Assume that (H_1) holds. Let $p(t)$ and $v(t)$ be real valued functions in $L^1(R^+)$. Suppose that*

(4.7.3) $$\lambda^2 - \alpha\lambda - \lambda\widehat{p}(\lambda) - \gamma - \widehat{q}(\lambda) - \mu_n \neq 0$$

for $n = 1, 2, \ldots$ and $\operatorname{Re}\lambda \geq 0$, where $\alpha < 0$ and γ are real constants. Then the $B(E)$-valued function

(4.7.4) $$\widehat{T}(\lambda) = \left[\lambda^2 - \alpha\lambda - \lambda\widehat{p}(\lambda) - \gamma - \widehat{q}(\lambda) - A\right]^{-1}$$

is defined for $\operatorname{Re}\lambda \geq 0$ and satisfies

$$\sup\left\{\int_{-\infty}^{\infty} |\widehat{T}(\sigma + i\tau)|^2 \, d\tau : \sigma \geq 0\right\} < \infty.$$

Lemma 4.7.2 *Suppose that the hypotheses of Lemma 4.7.1 hold. Let the $B(E)$-real-valued function $\widehat{R}(\lambda) = \lambda\widehat{T}(\lambda)$ for $\operatorname{Re}\lambda \geq 0$ where $\widehat{T}(\lambda)$ given by (4.7.4). Then for each $x \in E$,*

$$\sup\left\{\int_{-\infty}^{\infty} |\widehat{R}(\sigma + i\tau)x|^2 \, d\tau : \sigma \geq 0\right\} < \infty.$$

We shall now give sufficient conditions for L^2-stability of the integro-differential equation (4.7.2).

Theorem 4.7.2 *Assume that (H_1), (H_2), (T_1) and (T_2) hold. Let $b'(t) \in L^1(R^+)$ and $b(0) + K < 0$. Then the equation (4.7.2) is L^2-stable.*

Proof Define the scalar function $\rho(t)$ by

(4.7.5) $$\rho(t) = b(t) - (b*\rho)(t),$$

where

$$(b*\rho)(t) = \int_0^t b(t-\tau)\rho(\tau)\,d\tau$$

is the convolution product.

It is well known (see Miller [3]) that $\rho \in C(R^+)$. Since (T_1) holds, the classical result of Paley and Weiner [1] yields $\rho \in L^1(R^+)$. By differentiating (4.7.5), it is clear that we also have

$$\rho'(t) \in C(R^+) \cap L^1(R^+).$$

Fix x_0 and v_0 in $D(A)$, $f \in C^1(R^+, E) \cap L^1(R^+, E)$ and let $x(t) = x(t, x_0, v_0, f)$ be the unique solution of (4.7.2) existing for all $t \in R^+$. Then integration by parts gives

(4.7.6) $\qquad (\rho * x'')(t) = (\rho' * x)(t) + b(0)x'(t) - \rho(t)v_0,$

where $b(0) = \rho(0)$.

Thus, if we convolute both sides of (4.7.2), and use (4.7.5) and (4.7.6), we find that $x(t)$ satisfies

(4.7.7) $\quad x''(t) = \alpha x'(t) + (p * x')(t) + [A + \gamma I]x(t) + (q * x)(t) + h(t), \quad t \in R^+,$

where $\alpha = b(0) + K < 0$, $p(t) = \rho'(t) - K\rho(t)$, $q(t) = g(t) - (\rho * g)(t) - \gamma \rho(t)$ and

$$h(t) = f(t) - (\rho * f)(t) - \rho(t)v_0.$$

It is clear that the real-valued functions $p(t)$ and $q(t)$ lie in $C(R^+) \cap L^1(R^+)$ and $h \in C^1(R^+, E) \cap L^1(R^+, E)$. An elementary calculation yields

$$\lambda^2 - \alpha\lambda - \lambda \widehat{p}(\lambda) - \gamma - \widehat{q}(\lambda) = \frac{\lambda^2 - K\lambda - \gamma - \widehat{g}(\lambda)}{1 + \widehat{b}(\lambda)}.$$

Hence, using (T_2), it follows that (4.7.3) holds. Therefore we can define the E-valued function $\widehat{z}(\lambda)$ on $\operatorname{Re}\lambda \geq 0$ by

$$\widehat{z}(\lambda) = \widehat{R}(\lambda)x_0 + \widehat{T}(\lambda)\big[-\alpha x_0 - \widehat{p}(\lambda)x_0 + v_0 + \widehat{h}(\lambda)\big],$$

where $\widehat{R}(\lambda)$ and $\widehat{T}(\lambda)$ are the $B(E)$-valued functions defined in Lemmas 4.7.1 and 4.7.2 respectively.

It is clear that $\widehat{z}(\lambda)$ is continuous on $\operatorname{Re}\lambda \geq 0$ and analytic on $\operatorname{Re}\lambda > 0$. Since $|\widehat{p}(\lambda)| \leq |p|_1$ and $|\widehat{h}(\lambda)| \leq |h|_1$, we obtain

$$|\widehat{z}(\lambda)| \leq |\widehat{R}(\lambda)x_0| + |\widehat{T}(\lambda)|[(|\alpha| + |p|_1)|x_0| + |v_0| + |h|_1].$$

Thus, by Lemmas 4.7.1 and 4.7.2, it follows that

$$\sup\left\{\int_{-\infty}^{\infty}|\widehat{z}(\sigma+i\tau)|^2 d\tau : \sigma \geq 0\right\} < \infty.$$

Therefore $\widehat{z}(\lambda)$ lies in the Hardy space $H^2(0,E)$. Since E is a Hilbert space, there exists a function $z \in L^2(R^+, E)$ (see Friedman and Shinbrot [1]) whose Laplace transform is $\widehat{z}(\lambda)$ for all $\operatorname{Re}\lambda \geq 0$. Further, by a remark in Miller and Wheeler [1], it follows that

$$|x(t)|, \ |x'(t)| \ \text{and} \ |Ax(t)|$$

are all of exponential order whenever both x_0 and $v_0 \in D(A)$, and $f \in C^1(R^+, E) \cap L^1(R^+, E)$ with $|f'(t)|$ is of exponential order. Thus there exists a σ_0 such that we may take Laplace transforms in (4.7.7) when $\operatorname{Re}\lambda \geq \sigma_0$. Using the definitions of $\widehat{T}(\lambda)$, $\widehat{R}(\lambda)$ and $\widehat{z}(t)$, a simple computation gives

$$\widehat{x}(\lambda) = \widehat{z}(\lambda)$$

for $\operatorname{Re}\lambda \geq \sigma_0$.

Therefore the uniqueness of Laplace transforms implies that
$$x(t) = z(t)$$
for almost all $t \in R^+$, and hence $x \in L^2(R^+, E)$.

Since the solutions of (4.7.2) depend continuously on initial data, the elementary density argument may be used to show that the solution $x(t)$ of (4.7.2) belongs to $L^2(R^+, E)$ for all x_0, $v_0 \in E$ and $f \in L^1(R^+, E)$. This completes the proof. □

Remark 4.7.3 If $\alpha = b(0) + K > 0$ in Theorem 4.7.2 then the equation (4.7.2) is L^2-unstable. To see this, we write (4.7.2) in the form of (4.7.7) and observe that when $\alpha > 0$, $\lambda^2 - \alpha\lambda - \gamma - \mu_n$ has roots at

$$\lambda_n = \tfrac{1}{2}\left\{\alpha \pm \left[\alpha^2 + 4(\gamma + \mu_n)\right]^{1/2}\right\},$$

$\operatorname{Re}\lambda_n = \tfrac{1}{2}\alpha > 0$ for all large n. Therefore on a circle of radius $\tfrac{1}{2}\alpha$ about λ_n, the quadratic form $\lambda^2 - \alpha\lambda - \gamma - \mu_n$ has modulus approximately equal to $\alpha|\lambda_n|$.

Since

$$|\lambda \widehat{p}(\lambda) + \widehat{q}(\lambda)| = O(|\lambda|) \quad \text{as } \lambda \to \infty$$

in $\operatorname{Re}\lambda\lambda \geq 0$, Rouche's theorem implies that for all sufficiently large n, the relation (4.7.3) fails with λ in the open half-plane $\operatorname{Re}\lambda > 0$. Thus the condition (T_2) is not satisfied for some λ in $\operatorname{Re}\lambda > 0$, and hence the proof of Theorem 4.7.1 shows that the equation (4.7.2) is L^2-unstable. The case where $\alpha = 0$ appears to be somewhat difficult to analyze.

In the next few results, we shall employ Lyapunov functions to investigate sufficient conditions for asymptotic stability properties of solutions of equations with unbounded delay in Hilbert spaces.

Let E_0 be a real separable Hilbert space with inner product $(\cdot, \cdot)$ and norm $|\cdot|_0$. In particular, we shall study the asymptotic behavior of functions $u(t)$ on $(-\infty, \infty)$ with specified "history"

(4.7.8) $$u(t) = v(t), \quad t \in (-\infty, 0]$$

that satisfy on $R^+ = [0, \infty)$ the integro-differential equation

(4.7.9) $$\frac{d}{dt}[Au'(t)] + Cu(t) + \int_{-\infty}^{t} G(t-\tau)u(\tau)\,d\tau = 0,$$

in which A is a bounded self-adjoint operator on E_0 while C and $G(t)$ are unbounded self-adjoint operators with domains $D(C)$ and $D(G(t))$, $D(C) \subset D(G(t))$, $t \in R^+$, dense in E_0. Further, we assume that

(4.7.10) $\quad (A\omega, \omega) \geq \rho_0 |\omega|_0^2, \quad \rho_0 > 0, \quad$ for all $\omega \in E_0$;

(4.7.11) $\quad (C\omega, \omega) \geq \rho_1 |\omega|_0^2, \quad \rho_1 > 0, \quad$ for all $\omega \in D(C)$;

(4.7.12) $\quad (G(t)\omega, \omega) \leq 0 \quad$ for all $\omega \in D(G(t))$, $t \in R^+$.

The domain $D(C)$ of C equipped with the norm

$$|\omega|_2 \equiv |C\omega|_0$$

induces a Hilbert space that will be denoted by E_2. Similarly, we define the Hilbert space E_1 as the domain $D(C^{1/2})$ of the operator $C^{1/2}$ equipped with the norm

$$|\omega|_1 \equiv |C^{1/2}\omega|_0.$$

Finally, by E_{-1} we shall denote the dual of E_1 via $(\cdot,\cdot)$. Thus E_{-1} will be the completion of E_0 under the norm

$$|\omega|_{-1} \equiv \sup_{v \in E_1} \frac{|(\omega,v)|}{|v|_1}.$$

It is clear that $E_2 \subset E_1 \subset E_0 \subset E_{-1}$ algebraically and topologically. We shall further assume that the injection of E_2 into E_0 is compact. This implies that the injection of E_i into E_{i-1} is compact, $i = 0,1,2$.

Note that for every $\omega \in E_2$,

$$|C\omega|_{-1} \equiv \sup_{v \in E_1} \frac{|(C\omega,v)|}{|v|_1} = \sup_{v \in E_1} \frac{|(C^{1/2}\omega, C^{1/2}v)|}{|v|_1} = |\omega|_1,$$

which shows that C can be extended into $\mathcal{L}(E_1, E_{-1})$.

It is also assumed that $G(t)$ satisfies the following hypothesis

(H_3) (i) $G(t) \in C^0(R^+, \mathcal{L}(E_2, E_0)) \cap L^1(R^+, \mathcal{L}(E_2, E_0))$
      (ii) $G(t), G'(t) \in C^0(R^+, \mathcal{L}(E_1, E_{-1})) \cap L^1(R^+, \mathcal{L}(E_1, E_{-1}))$.

We need the following basic result in our subsequent discussion.

Lemma 4.7.3   Let $f \in L^1(R^+)$. Then there exists an increasing function $p \in C^0(R^+)$ with $p(0) = 1$, $p(t) \to \infty$ as $t \to \infty$, such that $f \cdot p \in L^1(R^+)$.

Proof   Let $I \equiv \int_0^\infty |f(t)| \, dt$. We construct a strictly increasing sequence $\{t_n\}$ with $t_1 \equiv 0$, $t_n \to \infty$ as $n \to \infty$, such that

$$\int_{t_n}^\infty |f(t)| \, dt \le \frac{1}{n^3} I,$$

and define a function $p(t)$ by

$$p(t) \equiv n + \frac{t - t_n}{t_{n+1} - t_n}, \quad t_n \le t \le t_{n+1},$$

$n = 1,2,\ldots$. Then for any $T > 0$, we have

$$\int_0^T |f(t) p(t)| \, dt \le \sum_{n=1}^\infty \int_{t_n}^{t_{n+1}} \left(n + \frac{t - t_n}{t_{n+1} - t_n}\right) |f(t)| \, dt$$

$$\le I \sum_{n=1}^\infty \frac{n+1}{n^3} < \infty.$$

This completes the proof. □

Remark 4.7.4 The hypothesis $(H_3)(ii)$ implies that $|G(t)|_{\mathcal{L}(E_1, E_{-1})} \to 0$ as $t \to \infty$. Moreover, from the hypothesis (H_3) and Lemma 4.7.3, it follows that there exists a decreasing function $h \in C^0(R^+)$ with $h(0) = 1$, $h(t) \to 0$ as $t \to \infty$, such that

$$(4.7.13) \quad \int_0^\infty \left[|G(t)|_{\mathcal{L}(E_1, E_{-1})} + |G'(t)|_{\mathcal{L}(E_1, E_{-1})} \right] h^{-2}(t)\, dt < \infty.$$

We now fix $h(t)$ in (4.7.13) and define two families of Banach spaces that will be involved directly in our discussion of asymptotic stability property of solutions of (4.7.8) and (4.7.9).

Definition 4.7.2 By $\mathcal{C}_K$, $K = 0, 1, \ldots$, we denote the Banach space of functions $\omega \in C^K\!\left[(-\infty, 0], E_1\right] \cap C^{K+1}\!\left[(-\infty, 0], E_0\right]$ such that

$$(4.7.14) \quad |||\omega|||_{\mathcal{C}_K} \equiv \sum_{i=0}^{K} \sup_{(-\infty,0]} \left[h(-\tau) |\omega^{(i)}(\tau)|_1 \right]$$

$$+ \sum_{i=0}^{K+1} \sup_{(-\infty,0]} \left[h(-\tau) |\omega^{(i)}(\tau)|_0 \right] < \infty,$$

where $\omega^{(K)}(t)$ is the Kth derivative of $\omega(t)$.

Definition 4.7.3 By $\mathcal{B}_K$, $K = 0, 1, \ldots$, we denote the Banach space of functions $\omega \in C^K\!\left[(-\infty, 0], E_2\right] \cap C^{K+1}\!\left[(-\infty, 0], E_1\right] \cap C^{K+2}\!\left[(-\infty, 0], E_0\right]$ such that

$$(4.7.15) \quad |||\omega|||_{\mathcal{B}_K} \equiv \sum_{i=0}^{K} \sup_{(-\infty,0]} |\omega^{(i)}(\tau)|_2 + \sum_{i=0}^{K+1} \sup_{(-\infty,0]} |\omega^{(i)}(\tau)|_1$$

$$+ \sum_{i=0}^{K+2} \sup_{(-\infty,0]} |\omega^{(i)}(\tau)|_0 < \infty$$

where $\omega^{(K)}(t)$ is the Kth derivative of $\omega(t)$. It is clear that $\mathcal{B}_K \subset \mathcal{C}_{K+1} \subset \mathcal{C}_K$, $K = 0, 1, \ldots$ algebraically and topologically. The norm of $\mathcal{C}_K$ is usually referred to as "fading memory" type. The following result will find application in the discussion on asymptotic stability.

Lemma 4.7.4 *The injection of $\mathcal{B}_K$ into $\mathcal{C}_K$ is compact.*

Proof Let $\{\omega_n(\tau)\}$ be a sequence in $\mathcal{B}_K$ such that $|||\omega_n|||_{\mathcal{B}_K} < 1$, $n = 1, 2, \ldots$. From (4.7.15), it follows that

$$\{\omega_n(\tau)\}, \{\omega_n'(\tau)\}, \ldots, \{\omega_n^{(K)}(\tau)\}$$

is a uniformly bounded and equicontinuous sequence of functions from $(-\infty, 0]$ to E_0. Furthermore, for any fixed $\tau \in (-\infty, 0]$, the sets

$$\{\omega_n(\tau)\}, \ldots, \{\omega_n^{(K)}(\tau)\}$$

are bounded in E_2, and hence precompact in E_1. Finally, $\{\omega_n^{(K+1)}(\tau)\}$ is precompact in E_0. Therefore, by the Ascoli – Arzelà theorem, it is clear that there exists a function

$$\omega \in C^K[(-\infty, 0], E_1] \cap C^{K+1}[(-\infty, 0], E_0]$$

with

$$\sum_{i=0}^{K} \sup_{(-\infty, 0]} |\omega^{(i)}(\tau)|_1 + \sum_{i=0}^{K+1} \sup_{(-\infty, 0]} |\omega^{(i)}(\tau)|_0 < 1$$

and a subsequence $\{\omega_{n_j}(\tau)\}$ of $\{\omega_n(\tau)\}$ such that

(4.7.16)
$$\begin{cases} \omega_{n_j}^{(i)}(\tau) \xrightarrow{E_1} \omega(\tau), & i = 0, \ldots, K, \\ \omega_{n_j}^{(K+1)}(\tau) \xrightarrow{E_0} \omega(\tau), \end{cases}$$

as $j \to \infty$, $\tau \in (-\infty, 0]$, and this convergence is uniform on every compact subinterval of $(-\infty, 0]$. Thus the proof is complete if we prove that

$$\omega_{n_j} \xrightarrow{\mathcal{C}_K} \omega \quad \text{as } j \to \infty.$$

To do this, given $\epsilon > 0$, let $T > 0$ be such that $h(t) < \tfrac{1}{4}\epsilon$ if $t > T$, where $h(t)$ is the same as given in Definition (4.7.2). For $j = 1, 2, \ldots$, we have

(4.7.17)
$$\sum_{i=0}^{K} \sup_{(-\infty, -T]} \left[h(-\tau) |\omega_{n_j}^{(i)}(\tau) - \omega^{(i)}(\tau)|_1 \right]$$

$$+ \sum_{i=0}^{K+1} \sup_{(-\infty, -T]} \left[h(-\tau) |\omega_{n_j}^{(i)}(\tau) - \omega^{(i)}(\tau)|_0 \right] < \tfrac{1}{2}\epsilon.$$

Since the convergence (4.7.16) is uniform on $[-T, 0]$, there exists a J_ϵ such that

(4.7.18) $$\sum_{i=0}^{K} \sup_{[-T,0]} |\omega_{n_j}^{(i)}(\tau) - \omega^{(i)}(\tau)|_1$$

$$+ \sum_{i=0}^{K+1} \sup_{[-T,0]} |\omega_{n_j}^{(i)}(\tau) - \omega^{(i)}(\tau)|_0 < \tfrac{1}{2}\epsilon$$

if $j \geq J_\epsilon$. Thus by combining (4.7.17) with (4.7.18), we obtain for $j \geq J_\epsilon$ that

$$|||\omega_{n_j} - \omega|||_{\mathcal{C}_K} < \epsilon.$$

This completes the proof. □

Now the following two results (see Theorems 2.1 and 2.2 in Dafermos [1]) will give sufficient conditions for the existence and boundedness of solutions of (4.7.8) and (4.7.9) in $\mathcal{C}_K$ and $\mathcal{B}_K$.

Theorem 4.7.3 *For $v \in \mathcal{C}_K$ and $T > 0$, there exists a unique $u \in C^K[(-\infty,T], E_1] \cap C^{K+1}[(-\infty,T], E_0]$ that satisfies (4.7.8) on $(-\infty,0]$ and (4.7.9) on $[0,T]$. Furthermore,*

(4.7.19) $$\sum_{i=0}^{K} \sup_{[0,T]} |u^{(i)}(t)|_1 + \sum_{i=0}^{K+1} \sup_{[0,T]} |u^{(i)}(t)|_0 \leq K_1 |||v|||_{\mathcal{C}_K}$$

where K_1 is independent of v.

Theorem 4.7.4 *For $v \in \mathcal{B}_K$ and $T > 0$, there exists a unique $u \in C^K[(-\infty,T], E_2] \cap C^{K+1}[(-\infty,T], E_1] \cap C^{K+2}[(-\infty,T], E_0]$ that satisfies (4.7.8) on $(-\infty,0]$ and (4.7.9) on $[0,T]$. Furthermore,*

(4.7.20) $$\sum_{i=0}^{K} \sup_{[0,T]} |u^{(i)}(t)|_2 + \sum_{i=0}^{K+1} \sup_{[0,T]} |u^{(i)}(t)|_1 + \sum_{i=0}^{K+1} \sup_{[0,T]} |u^{(i)}(t)|_0$$

$$\leq K_1 |||v|||_{\mathcal{B}_K},$$

where K_1 is independent of v.

Finally, we need the following assumptions to obtain the asymptotic stability properties of solutions of (4.7.8) and (4.7.9):

(H_4) (i) $(G'(t)\omega,\omega) \geq 0$ for all $\omega \in E_1$, $t \in R^+$;

(ii) $\int_0^\infty |G(t)|_{\mathcal{L}(E_2,E_0)} dt \equiv M_\infty < 1$;

(H_5) $(\rho\omega,\omega) \geq a_0 |\omega|_1^2$, $a_0 > 0$ for all $\omega \in E_1$, where

$$\rho \equiv C + \int_0^\infty G(t)\,dt.$$

We now fix some integer K and consider the mapping $\omega: \mathcal{C}_K \times R^+ \to \mathcal{C}_K$ that maps the pair $(v, \mathcal{S})$, $v \in \mathcal{C}_K$, $\mathcal{S} \in R^+$, onto $\omega(v, \mathcal{S}) \in \mathcal{C}_K$ by the relation

$$\omega(v, \mathcal{S})(\tau) \equiv u(\mathcal{S} + \tau), \quad \tau \in (-\infty, 0],$$

where $u(t)$ satisfies (4.7.8) on $(-\infty, 0]$ and (4.7.9) on $[0, \mathcal{S}]$. It is clear from Theorem 4.7.3 that ω is well-defined. Further, by Theorem 4.7.4, we have

$$\omega: \mathcal{B}_K \times R^+ \to \mathcal{B}_K.$$

Clearly,
$$\omega(v, 0) = v \quad \text{for all } v \in \mathcal{C}_K,$$

$$\omega(v, \mathcal{S}_1 + \mathcal{S}_2) = \omega(\omega(v, \mathcal{S}_1), \mathcal{S}_2) \quad \text{for all } \mathcal{S}_1, \mathcal{S}_2 \in R^+, v \in \mathcal{C}_K.$$

From (4.7.19) and (4.7.20), it follows that for fixed t, ω is continuous on $\mathcal{C}_K$ and $\mathcal{B}_K$.

We are now in a position to prove the following main result on asymptotic stability.

Theorem 4.7.5 *Assume that the hypotheses* (H_3), (H_4) *and* (H_5) *hold. Suppose further that for every eigensolution* ω_n *of the eigenvalue problem*

(4.7.21) $$C\omega - \lambda A\omega = 0,$$

there is at least one $\mathcal{S}_n \in R^+$ *such that*

(4.7.22) $$G'(\mathcal{S}_n)\omega_n \neq 0.$$

Let $u(t)$ be the solution of (4.7.8) and (4.7.9) with $v \in \mathcal{C}_m$, $m = 0, 1, \ldots$. Then the following properties are true:

(a) $u^{(i)}(t) \xrightarrow{E_1} 0$ as $t \to \infty$, $i = 0, 1, \ldots, m$,

(b) $u^{(m+1)}(t) \xrightarrow{E_0} 0$ as $t \to \infty$.

Proof Define a Lyapunov functional $V_K(v)$ on $\mathcal{C}_K$ by

(4.7.23) $V_K(v)(t) = \frac{1}{2}(Au^{(K+1)}(t), u^{(K+1)}(t)) + \frac{1}{2}(Au^{(K)}(t), u^{(K)}(t))$

$$-\frac{1}{2}\int_{-\infty}^{t} \Big(G(t-\tau)[u^{(K)}(t) - u^{(K)}(\tau)], u^{(K)}(t) - u^{(K)}(\tau)\Big) d\tau \geq 0,$$

where $u(t)$ is a solution of (4.7.8) and (4.7.9). It is clear from (4.7.13) that $V_K(v)$ is continuous on $\mathcal{C}_K \times R^+$. Differentiating (4.7.9) i times, $i = 0, 1, \ldots, K$, we obtain

$$(4.7.24) \quad \frac{d}{dt}\bigl[Au^{(i+1)}(t)\bigr] + Cu^{(i)}(t) + \int_{-\infty}^{t} G(t-\tau)u^{(i)}(\tau)\,d\tau = 0.$$

Let $v \in \mathcal{C}_K$. Then the time derivative of $V_K(v)(t)$ along the solutions of (4.7.8) and (4.7.9) together with (4.7.24) for $i = K$ is given by

$$(4.7.25) \quad \frac{d}{dt}[V_K(v)(t)]$$

$$= -\frac{1}{2}\int_{-\infty}^{t} \Bigl(G'(t-\tau)[u^{(K)}(t) - u^{(K)}(\tau)], u^{(K)}(t) - u^{(K)}(\tau)\Bigr)\,d\tau \le 0.$$

We shall first prove the theorem for $v \in \mathcal{B}_K$, and then extend it to $v \in \mathcal{C}_K$. In view of (4.7.23), (4.7.25), (H_5) and (4.7.10), it follows that

$$(4.7.26) \quad \sup_{[0,\infty)} |u^{(K)}(t)|_1 \le \Bigl[\tfrac{2}{a_0}V_K(v)(0)\Bigr]^{1/2}, \quad K = 0, \ldots, m+1,$$

$$(4.7.27) \quad \sup_{[0,\infty)} |u^{(K+1)}(t)|_0 \le \Bigl[\tfrac{2}{\rho_0}V_K(v)(0)\Bigr]^{1/2}, \quad K = 0, \ldots, m+1.$$

From (4.7.27) and (H_4), we can conclude that $|u^{(K)}(t)|_2$ is uniformly bounded on R^+, $K = 0, \ldots, m$. Thus the orbit ω originating at v is then uniformly continuous on R^+, and its range is bounded in $\mathcal{B}_m$ and precompact in $\mathcal{C}_m$. Applying Propositions 3.2, 3.3 and 5.1 of Dafermos [2], one can show that

$$\omega(v, \mathcal{S}) \xrightarrow{\mathcal{C}_m} N \quad \text{as } \mathcal{S} \to \infty,$$

where N is the largest invariant set of ω whose elements satisfy

$$(4.7.28) \quad \frac{d}{dt}[V_K(v)(t)] \equiv 0.$$

Thus in view of LaSalle's invariance principle, it is enough if we show that $N = \{0\}$. Let $\bar{u}(t)$ be the solution of (4.7.8) and (4.7.9) with $v \in N$. From (4.7.28), (4.7.25) and (H_4), we get

$$(4.7.29) \quad G'(t-\tau)[\bar{u}^{(m)}(t) - \bar{u}^{(m)}(\tau)] = 0 \quad \text{for } t \in (-\infty, \infty), \quad \tau \in (-\infty, 0].$$

Integrating (4.7.29) with respect to τ over $(-\infty, t)$ and using integration by parts, we obtain

$$\int_{-\infty}^{t} G(t-\tau)\bar{u}^{(m+1)}(\tau)\,d\tau = 0, \quad t \in (-\infty, \infty).$$

Hence, using (4.7.24) for $i = m+1$, we get

(4.7.30) $\quad \dfrac{d}{dt}[A\bar{u}^{(m+2)}(t)] + C\bar{u}^{(m+1)}(t) = 0 \quad \text{in } E_{-1}.$

Hence

(4.7.31) $\quad \bar{u}^{(m+1)}(t) = \operatorname{Re}\sum_{n=1}^{\infty} \alpha_n \exp(i(\lambda_n^{1/2} t))\, \omega_n,$

where $\{\lambda_n\}$ is the sequence of eigenvalues and $\{\omega_n\}$ the sequence of eigensolutions of (4.7.21). Substituting from (4.7.31) into (4.7.29) and putting $t - \tau = \mathcal{S}$, we get

(4.7.32) $\quad \operatorname{Im} \sum_{n=1}^{\infty} \dfrac{\alpha_n}{\lambda_n^{1/2}}\Big[1 - \exp(-i\lambda_n^{1/2}\mathcal{S})\Big]\exp(i\lambda_n^{1/2} t)\, G'(\mathcal{S})\omega_n = 0,$

for $t \in R$, $\mathcal{S} \in R^+$. Using the elementary properties of almost periodic functions, we obtain

(4.7.33) $\quad \alpha_n\Big[1 - \exp(-i\lambda_n^{1/2}\mathcal{S})\Big] G'(\mathcal{S})\omega_n = 0$

for $\mathcal{S} \in R^+$, $n = 1, 2, \ldots$.

Hence in view of the assumption (4.7.22), it is clear from (4.7.33) that
$$\alpha_n = 0 \quad \text{for } n = 1, 2, \ldots.$$

Therefore $\bar{u}^{(m+1)}(t)$ vanishes identically. By the same argument, we can show that $\bar{u}^{(K)}(t) \equiv 0$ for all $K = 0, 1, \ldots, m$. This establishes the required properties (a) and (b) of Theorem 4.7.5 for $v \in \mathcal{B}_m$.

To see that these properties also hold for $v \in \mathcal{C}_m$, it is clear from Lemma 4.7.3 and the relation (4.7.13) that there exists a decreasing function $\bar{h} \in C^0(R^+)$ with $\bar{h}(0) = 1$ and $\bar{h}(t) \to 0$ as $t \to \infty$, such that

(4.7.34) $\quad \displaystyle\int_0^{\infty} \Big[|G(t)|_{\mathcal{L}(E_1, E_{-1})} + |G'(t)|_{\mathcal{L}(E_1, E_{-1})}\Big] h^{-2}(t)\bar{h}^{-2}(t)\,dt < \infty.$

For $\omega \in \mathcal{C}_K$, we get

$$(4.7.35) \quad |||\omega|||_{\overline{\mathcal{C}}_K} \equiv \sum_{i=0}^{K} \sup_{(-\infty, 0]} \left[h(-\tau)\overline{h}(-\tau) | \omega^{(i)}(\tau)|_1 \right]$$

$$+ \sum_{i=0}^{K+1} \sup_{(-\infty, 0]} \left[h(-\tau)\overline{h}(-\tau) | \omega^{(i)}(\tau)|_0 \right].$$

From (4.7.23), (4.7.34) and (4.7.35), it follows that

$$(4.7.36) \quad V_K(v_1 - v_2)(0) \leq c |||v_1 - v_2|||_{\overline{\mathcal{C}}_K}^2,$$

for all $v_1, v_2 \in \mathcal{C}_K$ and some positive constant c.

Therefore from (4.7.26), (4.7.27) and (4.7.36), we can conclude that the validity of (a) and (b) is extended to the completion of $\mathcal{B}_m$ under the norm $||| \cdot |||_{\overline{\mathcal{C}}_m}$. An argument similar to that used in the proof of Lemma 4.7.4 yields that $\mathcal{C}_m$ is contained in the completion of $\mathcal{B}_m$ under $||| \cdot |||_{\overline{\mathcal{C}}_m}$, and hence (a) and (b) of Theorem 4.7.5 hold for $v \in \mathcal{C}_m$. This completes the proof. □

4.8 Notes and Comments

The existence and uniqueness results discussed in Section 4.1 are taken from Zhuang [1], and are direct extensions of the corresponding results for ordinary differential equations (see Lakshmikantham and Leela [1, 2]). Lemmas 4.2.1 and 4.2.2 are due to Mönch and Von Harten [1]. The proof of Theorem 4.2.1 is based on the monotone iterative technique, and is formulated through various Lemmas 4.2.3 – 4.2.7, which are the contributions of Chen and Zhuang [1]. For a detailed discussion of the monotone iterative method, see Ladde, Lakshmikantham and Vatsala [1]. For further results on this topic, see Brezis [1], Khavanin and Lakshmikantham [1] and Hu, Zhuang and Khavanin [1]. For basic material on abstract spaces, see Ambrosetti and Prodi [1], Barbu [1], Lax [1] and Friedman and Shinbrot [1].

Section 4.3 contains material adapted from the work of Chen and Grimmer [2]. See also Grimmer and Miller [1], Miller [1, 2], Crandall,

London and Nohel [1], DaPrato and Iannelli [1] and Crandall and Nohel [1] for further investigations in this direction.

The theory of semigroups and the existence of resolvents considered in Section 4.4 are taken from Desch and Schappacher [1]. The findings of Desch and Grimmer [1, 3], Grimmer and Miller [1, 2], Grimmer and Pritchard [1], Grimmer and Kappel [1] and Grimmer and Press [1] are a few more new additions in this area. For basic material on evolution equations and semigroups, see T. Kato [1], Pazy [1], Tanabe [1] and Hille and Phillips [1].

Theorems 4.5.1 – 4.5.4 are taken from Grimmer [1]. The relationship between time-dependent linear integro-differential equations and evolution equations (Theorem 4.5.5) is due to Chen and Grimmer [1]. For related work, see Grimmer and Miller [2]. The results on asymptotic behavior of solutions discussed in Section 4.6 are due to Grimmer [1]. See also Hannsgen [1], Hannsgen and Wheeler [1] and Lunardi and DaPrato [1].

The stability analysis described in Section 4.7 using Laplace transform methods is taken from Miller [2], while that using the Lyapunov technique is adapted from Dafermos [1]. Additional material on stability may be found in Miller and Wheeler [1].

5 APPLICATIONS

5.0 Introduction

In this chapter, we shall discuss some important and interesting integro-differential equations which arise in many applications such as physical, biological and engineering sciences.

Section 5.1 is devoted to biological population models based on Volterra equations with unbounded delay. A system of nonlinear integro-differential equations with unbounded delay for grazing a grassland by a cattle population is described in Section 5.2. In Section 5.3, we shall apply the theory of semigroups developed in Chapter 4 to discuss necessary and sufficient conditions for a nonlinear partial integro-differential equation to be hyperbolic in the sense that solutions propagate with finite speed. Section 5.4 deals with stability problems for a class of integro-differential equations that occur in the theory of nuclear reactors. By using the method of energy estimates, the results on boundedness and asymptotic behavior of solutions are presented in Section 5.5 for a mathematical model that represents the motion of an unbounded one-dimensional nonlinear viscoelastic body. Finally, in Section 5.6, we shall introduce large-scale systems (also called composite systems, interconnected systems and multiloop systems) and obtain sufficient conditions for L_2-stability, L_2-instability and asymptotic stability in the sense of Lyapunov for these systems.

5.1 Biological Population

In his study of the growth of biological populations, Volterra [1, 2] introduced the mathematical models for a single population as

$$(5.1.1) \qquad N'(t) = N(t)\left[a - bN(t) - \int_0^t k(t-s)N(s)\,ds\right], \quad t \in R^+,$$

and

$$(5.1.2) \qquad N'(t) = N(t)\left[a - bN(t) - \int_{-\infty}^t k(t-s)N(s)\,ds\right], \quad t \in R,$$

in which $N(t)$ is the population size at time t, a and b are positive rate constants, and $k(t)$ is the "hereditary" influence.

In his original work, Volterra [1] assumed that $k(t) \geq 0$ for all $t \geq 0$. If $k(t) \equiv 0$, then (5.1.1) and (5.1.2) reduce to the law of Verhulst and Pearl (cf. Pearl [1]), and the solution is the logistic curve

$$(5.1.3) \qquad N(t) = \frac{aN(0)}{bN(0) + [a - bN(0)]e^{-at}}.$$

Thus if $k(t)$ is sufficiently small then it is clear that the asymptotic behavior of solutions of (5.1.1) or (5.1.2) is the same as that of (5.1.3). However, (5.1.3) is monotone while the solutions of either (5.1.1) or (5.1.2) need not be monotone. Equations (5.1.1) and (5.1.2) represent rather simple growth models. The following well-known results (see Goh [1]) are useful in our discussion.

If $x(t)$ is defined for all $t \in R^+$ then its positive limit (or ω-limit) set $\Omega(x(t))$ consists of all points q such that there is a sequence $\{t_k\}$, $t_k \to \infty$ as $k \to \infty$ with

$$x(t_k) \to q \quad \text{as } t \to \infty.$$

Lemma 5.1.1 *If $x(t)$ is continuous and bounded for $t \in R^+$ then the positive limit set $\Omega(x(t))$ is nonempty, compact and*

$$\lim_{t \to \infty} d[x(t), \Omega(x(t))] = 0,$$

where $d(y, M)$ is the distance between the point and the set M.

Consider the integro-differential equation

(5.1.4) $$x'(t) = f(x(t)) + \int_\alpha^t B(t-s)G(x(s))\,ds,$$

where $f, G \in C[R]$, $B \in C[R^+]$, and $\alpha = 0$ or $\alpha = -\infty$.

Lemma 5.1.2 *Let $B \in L^1(R^+)$. If $x(t)$ is a bounded solution of (5.1.4) for all $t \in R^+$ then $\Omega(x(t))$ is an invariant set with respect to the equation*

(5.1.5) $$y'(t) = f(y(t)) + \int_{-\infty}^t B(t-s)G(y(s))\,ds.$$

Proof Given z in $\Omega(x(t))$, we prove that there is a sequence $\{t_m\}$, $t_m \to \infty$ as $m \to \infty$ and there is a solution $y(t)$ of (5.1.5) such that $y(0) = z$, $y(t)$ exists and is bounded for $t \in R$, and

$$\lim_{m \to \infty} x(t + t_m) = y(t)$$

uniformly on compact subsets of $-\infty < t < \infty$. This will show that $\Omega(x(t))$ is the union of solutions of (5.1.5).

Given z, from the definition of $\Omega(x(t))$, we select $\{t_m\}$ such that

(5.1.6) $$|x(t_m) - z| < \tfrac{1}{m}, \qquad t_m > m.$$

Define $x_m(t) = x(t + t_m)$ for all $t \geq -t_m$. Since $x(t)$ is a solution of (5.1.4), we have

(5.1.7) $$x_m(t) = x(t_m) + \int_0^t f(x_m(s))\,ds$$

$$+ \int_0^t \int_{\alpha - t_m}^u B(u-s)G(x_m(s))\,ds\,du,$$

for all $t > -t_m$ and $m = 1, 2, 3, \ldots$. Since $x(t)$ is bounded and solves (5.1.4), it is uniformly continuous on R^+. Hence for any compact subset of $-\infty < t < \infty$, there is a positive number M such that $\{x_m(t): m \geq M\}$ is defined, equicontinuous and uniformly bounded on this set. Thus from Ascoli's theorem, there is a subsequence $\{t_{m_k}\}$ and a function $y(t)$ such that

(5.1.8) $$|x_{m_k}(t) - y(t)| < \omega\left(\frac{1}{m}\right)$$

uniformly for $-m_k \leq t \leq m_k$. The function $\omega(\epsilon)$ is the modulus of continuity of $G(x)$ in the region $|x| \leq \rho$, and ρ is a bound for $|x(t)|$ when $t \in R^+$. For any $t \in R$, we take $m_k > |t|$. Then

(5.1.9) $$\int_0^t \int_{-\infty}^u B(u-s)G(y(s))\,ds\,du - \int_0^t \int_{\alpha-t_{m_k}}^u B(u-s)G(x_{m_k}(s))\,ds\,du$$

$$= \int_0^t \int_{-m_k}^u B(u-s)\big[G(y(s)) - G(x_{m_k}(s))\big]\,ds\,du$$

$$+ \int_0^t \int_{-\infty}^{-m_k} B(u-s)G(y(s))\,ds\,du$$

$$- \int_0^t \int_{\alpha-t_{m_k}}^{-m_k} B(u-s)G(x_{m_k}(s))\,ds\,du =: I_1 + I_2 - I_3.$$

It follows from (5.1.8) that

$$|I_1| \leq \int_0^{|t|} \int_{-\infty}^u \frac{|B(u-s)|}{m_k}\,ds\,du$$

$$\leq \frac{|t|}{m_k} \int_0^\infty |B(s)|\,ds.$$

Let M be the bound for $|G(x)|$ on $|x| \leq \rho$. Then

$$|I_2| \leq \int_0^{|t|} \int_{-\infty}^{-m_k} |B(u-s)|\,M\,ds\,du$$

$$\leq |t|\,M \int_{m_k-|t|}^\infty |B(s)|\,ds.$$

A similar estimate can be made for I_3. Thus $|I_1| + |I_2| + |I_3| \to 0$ as $k \to \infty$ for any fixed $t \in R$. This together with (5.1.6) – (5.1.8) gives

$$y(t) = z + \int_0^t f(y(s))\,ds + \int_0^t \int_{-\infty}^u b(u-s)G(y(s))\,ds\,du.$$

Differentiating this relation with respect to t, we obtain (5.1.5). Thus, in view of (5.1.6) and the definition of $x_m(t)$, the result follows. This completes the proof of Lemma 5.1.2. □

We now make the following hypotheses:

(H_1) $k(t) \not\equiv 0$, $k \in C(R^+)$, $k \in L^1(R^+)$ and $b - \int_0^\infty |k(s)|\,ds > 0$;

(H_2) for any positive, continuous, bounded function $g(t)$ on $(-\infty, 0]$, there is a unique solution $N(t)$ of (5.1.2) for $t \in R^+$ such that

$$N(t) = g(t) \quad \text{for } t \in (-\infty, 0].$$

Remark 5.1.1 It follows from the form of (5.1.2) and the hypotheses (H_1) and (H_2) that $N(t) > 0$ for all $t > 0$ if and only if $N(0) > 0$.

Define

(5.1.10) $$N^* = a\left[b + \int_0^\infty k(t)\,dt\right]^{-1}.$$

It is clear from the hypothesis (H_1) that N^* is a positive equilibrium of (5.1.2). Set $x(t) = \log(N(t)/N^*)$ where $N(t)$ is a solution of (5.1.2) for $t \in R^+$ and N^* is defined by (5.1.10). In view of Remark 5.1.1, the function $x(t)$ is well defined. Let $G(x) = N^*(e^x - 1)$. Then the equation

$$N'(t) = N(t)\left[-b(N(t) - N^*) - \int_{-\infty}^t k(t-s)(N(s) - N^*)\,ds\right]$$

takes the form

(5.1.11) $$x'(t) = -bG(x(t)) - \int_{-\infty}^t k(t-s)G(x(s))\,ds.$$

Lemma 5.1.3 *Suppose that $G(x)$ is continuous and strictly increasing with $G(0) = 0$. Let $k \in L^1(R^+)$, $k(t) \not\equiv 0$, and*

$$(5.1.12) \qquad b - \int_0^\infty |k(t)|\, dt \geq 0.$$

For any t_0, if $|x(t)| \leq \rho$ for all $t \leq t_0$ and $x(t)$ solves (5.1.11) for $t \geq t_0$ then, for $t > t_0$,

$$|G(x(t))| \leq \max[G(\rho), -G(\rho)] = G_0.$$

Proof Suppose for some $T > t_0$, we have

$$G(x(T)) > |G(x(t))|$$

for all $t < T$. Then from (5.1.11) and (5.1.12), we obtain

$$x'(T) < -bG(x(T)) + \int_{-\infty}^T |k(T-s)|\, G(x(T))\, ds$$

$$\leq 0.$$

Thus $G(x(t))$ is strictly decreasing in a neighborhood of T, which contradicts the definition of T. Similarly, if $-G(x(T)) > |G(x(t))|$ for all $t < T$, we obtain the contradiction $x'(T) > 0$.

If the lemma is not true then there exists a $t_1 > t_0$ with $|G(x(t_1))| > G_0$. Further, the set

$$D = \left\{ t: t_0 < t \leq t_1 \text{ and } |G(x(t))| = \max_{t_0 < s \leq t_1} |G(x(s))| \right\}$$

is compact. Let $T = \inf D > t_0$. Therefore there is a $T_1 < T$ with

$$|G(x(T_1))| \geq |G(x(T))| > G_0.$$

This contradictions the definition of T, and hence the lemma is proved. □

Remark 5.1.2 It is clear from Lemma 5.1.3 that the zero solution of (5.1.11) is uniformly stable.

Theorem 5.1.1 Assume that the hypotheses (H_1) and (H_2) hold. Then the solution $N(t)$ of (5.1.2) exists for $t \in R^+$ and satisfies

$$(5.1.13) \qquad \lim_{t \to \infty} N(t) = N^*,$$

where N^* is the positive equilibrium defined by (5.1.10).

Proof The existence and boundedness of solutions of (5.1.2) follow by the standard argument and the application of Lemma 5.1.3. The same argument also yields the boundedness of $x(t) = \log(N(t)/N^*)$. Further, it follows from Lemmas 5.1.1 and 5.1.2 that $x(t)$ tends to a compact invariant set of (5.1.11).

Let $y(t)$ be any fixed solution of (5.1.11) that exists and bounded for $t \in R$. Define $z(t) = y(-t)$. Then from (5.1.11), we get

$$(5.1.14) \qquad z'(t) = bG(z(t)) + \int_t^\infty k(s-t)G(z(s))\,ds.$$

We claim that $z(t) \to 0$ as $t \to \infty$. Suppose that this is not true. Let

$$z_0 = \limsup_{t\to\infty} z(t),$$

$$z_1 = \liminf_{t\to\infty} z(t).$$

Without loss of generality, we assume that $z_0 > 0$ and $G(z_0) \geq -G(z_1)$. (Otherwise replace z by $-z$ and $G(z)$ by $-G(-z)$). Suppose that there is a sequence $\{t_m\}$, $t_m \to \infty$ as $m \to \infty$, with

$$z'(t_m) = 0 \quad \text{and} \quad z(t_m) \to z_0.$$

Define $G_m = \sup[|G(z(t))| : t \geq t_m]$. Then, for any m, from (5.1.14), we obtain

$$0 = z'(t_m) \geq bG(z(t_m)) - \int_0^\infty |k(s)|G_m\,ds.$$

Taking the limit as $m \to \infty$ and using the hypothesis (H_1), we get

$$0 \geq \left[b - \int_0^\infty |k(s)|\,ds\right] G_m > 0,$$

which is a contradiction. Hence no such sequence $\{t_m\}$ exists. This means that $z(t)$ tends monotonically to z_0 for sufficiently large t.

Further, if $z'(t) \geq 0$ for all sufficiently large t then from (5.1.14) and the hypothesis (H_1), it follows that

$$z'(t) \geq bG(z(t)) - \int_0^\infty |k(s)|G(z_0)\,ds$$

$$\geq \left[b - \int_0^\infty |k(s)|\,ds \right] \frac{G(z_0)}{2} > 0.$$

Thus $z(t) \to +\infty$ as $t \to \infty$. This contradicts the fact that $z(t)$ is bounded. Similarly, if $z'(t) \leq 0$ for all sufficiently large t then we again obtain

$$0 \geq z'(t) \quad \geq bG(z(t)) - \int_0^\infty |k(s)|\,G(z(t))\,ds$$

$$\geq \left[b - \int_0^\infty |k(s)|\,ds \right] G(z(t)) > 0.$$

Hence

$$y(t) = z(-t) \to 0 \quad \text{as } t \to \infty.$$

Using Lemma 5.1.3 and the fact that $y(t) \to 0$ as $t \to -\infty$, we conclude that $y(t) \equiv 0$ for all t. This shows that the only compact invariant set of (5.1.11) is the single point $y = 0$. Thus the application of Lemmas 5.1.1 and 5.1.2 yields that all the solutions of (5.1.11) tend to zero as $t \to \infty$. Since $x(t) = \log(N(t)/N^*)$, the assertion of the theorem follows, and this completes the proof. □

Theorem 5.1.2 *Suppose that the hypothesis (H_1) holds. Then for any $N_0 > 0$, there exists a unique, positive solution $N(t)$ of (5.1.1) with $N(0) = N_0$. This solution satisfies (5.1.13).*

Proof The local existence and uniqueness of solutions can be proved by the standard arguments. Let $N(t)$ be a local solution of (5.1.1). Then, as long as $N(t)$ exists, we have

(5.1.15) $$\frac{N'(t)}{N(t)} = -b[N(t) - N^*] - \int_0^t k(t-s)[N(s) - N^*]\,ds + h(t),$$

where

(5.1.16) $$h(t) = N^* \int_t^\infty k(s)\,ds \to 0 \quad \text{as } t \to \infty.$$

Define

$$h_0 = \max\,[2\,|h(t)|] \quad \text{for } t \geq 0,$$

$$h_1 = b - \int_0^\infty |k(s)|\, ds,$$

$$h_2 = \max\left[N_0, N^* + \frac{h_0}{h_1}\right].$$

We now claim that $N(t) < h_2$ as long as it exists. Suppose that this is not true. Then there exists a time t_1 such that $N(t_1) = h_2$. At $t = t_1$, it follows from (5.1.15) that

$$\frac{N'(t_1)}{N(t_1)} = -b[N(t_1) - N^*] - \int_0^{t_1} k(t_1 - s)[N(s) - N^*]ds + h(t_1)$$

$$< -b[N(t_1) - N^*] + \int_0^{t_1} |k(t_1 - s)|[N(s) - N^*]ds + h_0$$

$$< -b(h_2 - N^*) + \int_0^{t_1} |k(s)|(h_2 - N^*)ds + h_0$$

$$\leq -h_1(h_2 - N^*) + h_0$$

$$\leq 0.$$

This is a contradiction. Hence $N(t) < h_2$ as long as it exists. By standard arguments, it is easy to show that $N(t)$ exists and bounded for all $t \geq 0$. If zero is not in $\Omega(N(t))$ then it is clear by Theorem 5.1.1 that

$$\Omega(N(t)) = \{N^*\}.$$

Therefore it is enough if we show that zero is not in $\Omega(N(t))$.

Suppose that $N(t) \to 0$ as $t \to \infty$. Then from the continuity of $N(t)$ and the fact that $k \in L^1(R^+)$, it follows that for a given $\epsilon > 0$ there exists a $T > 0$ such that, for all $t \geq T$,

$$N(t) < \epsilon \quad \text{and} \quad \int_t^\infty |k(s)|\, ds < \epsilon.$$

Hence for $t \geq T$, (5.1.1) gives

$$\frac{N'(t)}{N(t)} \geq a - b\epsilon - \int_0^T |k(t-s)|\epsilon\, ds - \int_T^t |k(t-s)|h_2$$

$$> a - \epsilon\left[b + \int_0^\infty |k(s)|\, ds + h_2\right].$$

Taking ϵ small, we obtain $N'(t) > 0$ for all sufficiently large t. Since $N(t) > 0$ and $N(t) \to 0$ as $t \to \infty$, this is a contradiction. On the other hand, suppose that

$$\liminf_{t \to \infty} N(t) = 0$$

while

$$\limsup_{t \to \infty} N(t) > 0.$$

Let $x(t)$ and $G(x)$ be as defined in (5.1.11). Then (5.1.15) and (5.1.16) lead to

(5.1.17) $$x'(t) = -bG(x(t)) - \int_0^t k(t-s)G(x(s))\, ds + h(t).$$

Since $x(t)$ is bounded above and $G(x) > -1$ for $x < 0$, there exists a positive number M such that

$$|G(x(t))| < M \quad \text{for } t \in R^+.$$

Therefore $x(t)$ is uniformly continuous on R^+. Let $x_0 = \limsup_{t \to \infty} x(t)$. Fix T_0 and T_1 such that $T_1 > T_0$,

$$|x(T_0) - x_0| < 1$$

and

$$x(T_1) < -m - 1 - |x_0|$$

for some positive real number m.

We may assume (by possibly reducing the interval) assume that the endpoints are the maximum and minimum values of $x(t)$ in the interval $T_0 \leq t \leq T_1$. For $\epsilon = 1$, fix δ by using the uniform continuity of $x(t)$. Then $T_1 - T_0 \geq \delta_m$. Therefore for $t = T_1$, we obtain from (5.1.17) that

$$x'(T_1) > -bG(x(T_1)) + \int_{T_1-\delta_m}^{T_1} |k(T_1-s)| G(x(t)) \, ds$$

$$- \int_0^{T_1-\delta_m} |k(T_1-s)| M \, ds - h(T_1)$$

$$> \left[\int_0^\infty |k(s)| \, ds - b \right] G(-m) - M \int_{\delta_m}^\infty |k(s)| \, ds - h(T_1).$$

Taking T_0 and m sufficiently large,

$$x'(T_1) > 0.$$

This contradicts the fact that $x(T_1)$ is the minimum value of $x(t)$ on $T_0 \leq t \leq T_1$. Therefore $x(t)$ is bounded below. Further, from the definition of $x(t)$, it is clear that $N(t)$ is uniformly bounded away from zero. Therefore

$$\Omega(N(t)) = \{N^*\},$$

and thus, in view of Lemma 5.1.1, the conclusion of (5.1.13) follows. This completes the proof. □

Remark 5.1.3 In the terminology of microbiology, equations (5.1.1) and (5.1.2) represent a batch process (see Goh [1]). For a continuous process, one must replace the constant a by $a - D$, where D is the dilution rate. The case $a - D > 0$ is covered in Theorems 5.1.1 and 5.1.2. If $a - D \leq 0$ and $k(t) \geq 0$, it is easy to show that all solutions of (5.1.1) and (5.1.2) tend to positive equilibrium N^* as $t \to \infty$.

5.2 Grazing Systems

The degradation of a grass biomass in a forested grassland due to overgrazing by ungulates, cattle migrating from plains into the grassland during the draught period, and cutting of grass for food, fodder and fuel has been stressed in the literature in recent years, and the stability analysis of a grazing system using the concept of prey − predator models has been discussed. Since the availability of grass biomass at a given instant inherently depends upon its growth rate and the grazing rate, both in the present and in the past, it is quite natural to consider some kind of delay effect of grass biomass in the model. Although several investigations related to ecological models with delay can be found in the literature, none is applicable to the dynamics of grassland grazing, since this requires greater consideration of growth and depletion rates. With this in mind, a general model for grazing grassland on the pattern of a prey – predator system has been suggested in Rama Mohana Rao and Pal [1] by considering the effect of delay in the growth rate of the cattle population.

Let $G(t)$ and $U(t)$ be continuously differentiable positive functions representing the densities of grass biomass and cattle population respectively in the grassland. Motivated by the characteristic nature of prey – predator systems with large delay, we shall consider in this section the dynamics of grass biomass and cattle population by the following system of nonlinear integro-differential equations:

(5.2.1)
$$\begin{cases} \frac{dG}{dt} = Gg(G,U), \\ \frac{dU}{dt} = U\left[-\gamma - f(U) + \beta \int_{-\infty}^{t} w(t-s)h(G(s))\,ds\right], \end{cases}$$

in which $g(G,U)$ denotes the specific growth rate of grass biomass, $\gamma > 0$ is the natural death rate of the cattle population, $f(U)$ is the self-inhibition rate of the cattle population; $h(G)$ is the grass response to cattle, $w(t)$ is a nonnegative continuous function called the weight function and $\beta > 0$ is the delay effect.

We make the following hypotheses on $g(G,U), f(U), w(t)$ and $h(G)$.

(H_0) For any positive, continuous, and bounded initial functions $(\phi, \psi) = (\phi(t), \psi(t))$ on $-\infty < t \leq 0$, there exists a unique positive solution $(G(t), U(t)) \equiv (G(t, \phi), U(t, \psi))$ of (5.2.1) for all $t \in R^+$ such that $(G(t), U(t)) = (\phi(t), \psi(t))$ on $-\infty < t \leq 0$.

(H_1) The functions g, f and h are continuously differentiable in the set Ω defined by

$$\Omega = \{G, U \in R^2 : 0 < G < \infty, \; 0 < U < \infty\}.$$

(H_2) $g(0,0) > 0$, $g(k,0) = 0$, $\partial g/\partial G < 0$ and $\partial g/\partial U < 0$, where $k > 0$ is the carrying capacity of the grass biomass in the forested grassland.

(H_3) $h(0) = 0$ and $h'(G) > 0$ for $0 < G < \infty$.

(H_4) $f(0) = 0$ and $f'(U) > 0$ for $0 < U < \infty$.

(H_5) $\int_{-\infty}^{t} \omega(t-s)\,ds = \omega_0$, where ω_0 is a positive constant.

For example, the following weight (or memory) functions are admissible in our subsequent discussion:

(i) $\omega(t) = (1/r)\exp(-t/r)$, $t \in R^+$;
(ii) $\omega(t) = (\pi/2r)\sin(\pi t/r)$, $t \in [0,r]$;
(iii) $\omega(t) = \frac{1}{2}\delta(0) + \frac{1}{2}\delta(t)$.

Where r is a small positive constant and $\delta(t)$ is the Dirac–delta function, defined by

$$\delta(t) = \begin{cases} 0 & \text{if } t < r, \\ \frac{1}{\epsilon} & \text{if } r \leq t \leq r + \epsilon, \\ 0 & \text{if } t > r + \epsilon, \end{cases}$$

ϵ being a sufficiently small positive constant.

In view of the hypotheses (H_0)–(H_5), using the method of isoclines, it is easy to verify the existence of a positive equilibrium (G^*, U^*) given by

$$g(G, U) = 0,$$

and

(5.2.2) $$\gamma + f(U) = \beta \int_{-\infty}^{t} \omega(t-s) h(G(s))\,ds,$$

provided that

(5.2.3) $$\gamma < \beta \omega_0 h(G^*).$$

For a proof, see Freedman and Sree Hari Rao [1] and Stephan [1].

In this section, we shall give sufficient conditions for the asymptotic stability of the positive equilibrium (G^*, U^*) of (5.2.1) under the hypotheses (H_0)–(H_5).

Linear System

Consider the transformation

$$G(t) = G^* + g_1(t), \quad U(t) = U^* + u_1(t).$$

Then the linearized system corresponding to (5.2.1) is given by

(5.2.4)
$$\begin{cases} \dfrac{dg_1}{dt} = G^*\left[\dfrac{\partial g}{\partial G}(G^*,U^*)g_1(t) + \dfrac{\partial g}{\partial U}(G^*,U^*)u_1(t)\right], \\ \dfrac{du_1}{dt} = U^*\left[-f'(U^*)u_1(t) + \beta h'(G^*)\displaystyle\int_{-\infty}^{t} \omega(t-s)g_1(s)\,ds\right]. \end{cases}$$

Equations (5.2.4) can be written in vector − matrix form as

(5.2.5)
$$\dfrac{dx}{dt} = Ax + \int_{-\infty}^{t} \omega(t-s)\widehat{M}x(s)\,ds$$

where

$$x(t) = \begin{bmatrix} g_1(t) \\ u_1(t) \end{bmatrix},$$

$$A = \begin{bmatrix} G^*\dfrac{\partial g}{\partial G}(G^*,U^*) & G^*\dfrac{\partial g}{\partial U}(G^*,U^*) \\ 0 & -U^*f'(U^*) \end{bmatrix}$$

$$\widehat{M} = \begin{bmatrix} 0 & 0 \\ \beta U^*h'(G^*) & 0 \end{bmatrix}.$$

Remark 5.2.1 From the hypotheses $(H_2)-(H_4)$, it is clear that

$$G^*\dfrac{\partial g}{\partial G}(G^*,U^*) < 0,$$

$$G^*\dfrac{\partial g}{\partial U}(G^*,U^*) < 0,$$

$$U^*f'(U^*) > 0,$$

$$\beta U^*h'(G^*) > 0.$$

(A) *Lyapunov-Razumikhin Technique*

To study the asymptotic stability of the linear system (5.2.5), we shall employ the Lyapunov — Razumikhin technique and introduce the class of functions $x \in R^2$ such that

(5.2.6) $$|x(s)| \leq c|x(t)| \quad \text{for all} \; -\infty < s \leq t < \infty,$$

where $c \geq 1$ is a real constant. Let

(5.2.7) $$V(x) = x^T B x,$$

where B is a 2×2 real symmetric constant matrix. Since A is a stable matrix, it is well known that the matrix equation

(5.2.8) $$A^T B + BA = -I,$$

where I is an identity matrix, has a positive definite matrix solution B. Thus the function $V(x)$ defined by (5.2.7) is positive definite in R^2. Further, the time derivative of $V(x)$ along the solutions of (5.2.5) together with (5.2.8) yields

$$\frac{dV}{dt} = -x^T(t)x(t) + \left[\int_{-\infty}^{t} \omega(t-s) x^T(s) \widehat{M}^T ds \right] Bx(t)$$

$$+ x^T(t) B \left[\int_{-\infty}^{t} \omega(t-s) \widehat{M} x(s) ds \right]$$

$$\leq -|x(t)|^2 + 2|B||\widehat{M}||x(t)| \left[\int_{-\infty}^{t} \omega(t) |x(s)| ds \right].$$

Therefore from (5.2.6) and the hypothesis (H_5), we obtain

$$\frac{dV}{dt} \leq -|x(t)|^2 + 2c|\widehat{M}||B||x(t)|^2 \int_{-\infty}^{t} |\omega(t-s)| ds$$

$$= -\left(1 - 2c\omega_0 |\widehat{M}||B|\right)|x(t)|^2.$$

Choose ω_0 such that

$$\omega_0 < \frac{1}{2c|B||\widehat{M}|}.$$

Then dV/dt is negative definite. Hence the linear system (5.2.5) is asymptotically stable.

(B) *Construction of Lyapunov functionals*

We introduce a functional

(5.2.9) $\quad W(g_1(\cdot), u_1(\cdot)) = |g_1(t)| + |u_1(t)|$

$$+ m_{21} \int_{-\infty}^{t} \left[\int_{t}^{\infty} \omega(\tau - s) d\tau |g_1(s)| ds \right],$$

where

$$m_{21} = \beta U^* h'(G^*) > 0.$$

It is clear that W is a positive definite functional on R^2. Further, the functional W has continuous first-order partial derivatives with respect to all variables $((g_1, u_1) \neq (0,0))$. Computing the time derivative of (5.2.9) along a solution $((g_1, u_1) \neq (0,0))$ of (5.2.4), we obtain

$$\frac{dW}{dt} = \frac{d}{dt}|g_1(t)| + \frac{d}{dt}|u_1(t)| + m_{21} \int_{t}^{\infty} \omega(\tau - t) |g_1(t)| d\tau$$

$$- m_{21} \int_{-\infty}^{t} \omega(t-s) |g_1(s)|) ds$$

$$= G^* \frac{\partial g}{\partial G}(G^*, U^*) |g_1(t)| + G^* \frac{\partial g}{\partial U}(G^*, U^*) |u_1(t)|$$

$$- U^* f'(U^*) |u_1(t)| + \beta U^* h'(G) \int_{-\infty}^{t} \omega(t-s) |g_1(s)| ds$$

$$+ m_{21} \int_{t}^{\infty} |\omega(\tau - t)| |g_1(t)| d\tau$$

$$- m_{21} \int_{-\infty}^{t} |\omega(t-s)| |g_1(s)| ds$$

$$= \left[G^* \frac{\partial g}{\partial G}(G^*, U^*) + m_{21} \int_{t}^{\infty} \omega(\tau - t) d\tau \right] |g_1(t)|$$

$$+\left[G^*\frac{\partial g}{\partial U}(G^*,U^*) - U^*f'(U^*)\right]|u_1(t)|.$$

Since $\int_t^\infty |\omega(\tau-t)|\,d\tau = \omega_0$, it follows from the fact that $m_{21} = \beta U^* h'(G^*) > 0$ and Remark 5.2.1 that the linear system (5.2.4) is asymptotically stable if

$$G^*\frac{\partial g}{\partial U}(G^*,U^*) + \beta U^*\omega_0 h'(G^*) < 0.$$

Nonlinear System

We shall now discuss the asymptotic stability of the positive equilibrium (G^*, U^*) of the nonlinear system (5.2.1).

In addition to the hypotheses $(H_0) - (H_5)$, we also assume, following Cushing [2],

(H_6) all the positive solutions $(G(t), U(t))$ of (5.2.1) exist and are bounded for all $t \in R^+$ and are such that

$$K_1 \leq G(t) \leq K_2,$$
$$L_1 \leq U(t) \leq L_2,$$

where K_1, K_2, L_1 and L_2 are positive real numbers.

We now consider the transformation

$$g_1 = \log\left(\frac{G}{G^*}\right), \quad u_1 = \log\left(\frac{U}{U^*}\right),$$

and reduce the system (5.2.1) to

(5.2.10)
$$\begin{cases} \frac{dg_1}{dt} = g(G^*e^{g_1}, U^*e^{u_1}) \equiv F_1(g_1, u_1), \\ \frac{du_1}{dt} = -\gamma - f(U^*e^{u_1}) + \beta\int_{-\infty}^t \omega(t-s)h(G^*e^{g_1(s)})\,ds \\ \equiv F_2(t_1, u_1, t), \end{cases}$$

for all $(t_1, u_1, t) \in \Omega_1 \times (-\infty, \infty)$ where

$$\Omega_1 = \left\{(g_1, u_1) \in R^2 : \log\left(\frac{K_1}{G^*}\right) \leq g_1 \leq \log\left(\frac{K_2}{G^*}\right), \log\left(\frac{L_1}{U^*}\right) \leq u_1 \leq \log\left(\frac{L_2}{U^*}\right)\right\}.$$

The two-dimensional system (5.2.10) can be expressed in vector form as

(5.2.11) $$\frac{dx}{dt} = \widehat{f}(x, t),$$

where

$$x = \begin{bmatrix} g_1 \\ u_1 \end{bmatrix}, \quad \widehat{f}(x, t) = \begin{bmatrix} F_1(g_1, u_1) \\ F_2(g_1, u_1, t) \end{bmatrix}.$$

It is clear that $\widehat{f}(0, t) \equiv 0$ for all $t \in R$. Define a matrix $Q = Q(x, t)$ such that

(5.2.12) $$Q = \tfrac{1}{2}(J^T + J),$$

where J is the Jacobian of $\widehat{f}(x, t)$. Choose a positive definite function $V = x^T x$. Then the time derivative of V along the solutions of (5.2.11) is given by

(5.2.13) $$\frac{dV}{dt} = \widehat{f}^T(x, t)x + x^T \widehat{f}(x, t).$$

Let $Y = (Y_1, Y_2)$ and $Z = (Z_1, Z_2)$ be vectors. Consider the function

(5.2.14) $$q(Y) = \sum_{i,j=1}^{2} F_i(Y, t)\delta_{ij} Z_j$$

as a function of Y only, where δ_{ij} is the Kronecker delta. Then application of mean value theorem for

$$q(Y) = q(Y) - q(0) + q(0)$$

gives

(5.2.15) $$q(Y) = \sum_{i,j,k=1}^{2} \left[\frac{\partial F_i}{\partial Y_k}(\theta Y, t) Y_k \delta_{ij} Z_j\right] + q(0), \quad 0 < \theta < 1.$$

Setting $Y = x$ and $Z = x$, we obtain from (5.2.14) and (5.2.15) that

$$\widehat{f}^T(x, t)x = x^T J^T(\theta x, t)x + \widehat{f}^T(0, t)x.$$

In view of the fact that $\widehat{f}(0, t) \equiv 0$, it follows that

$$\widehat{f}^T(x, t)x = x^T J^T(\theta x, t)x.$$

Similarly

$$x^T \widehat{f}(x,t) = x^T J(\theta x, t) x.$$

Thus from (5.2.12) and (5.2.13), we get

(5.2.16) $$\frac{dV}{dt} = 2x^T Q(x,t) x$$

for all $(x,t) \in \Omega_1 \times (-\infty, \infty)$. Since

$$Q(x,t) = \begin{bmatrix} \dfrac{\partial F_1}{\partial g_1} & \dfrac{1}{2}\left(\dfrac{\partial F_1}{\partial u_1} + \dfrac{\partial F_2}{\partial g_1}\right) \\ \dfrac{1}{2}\left(\dfrac{\partial F_1}{\partial u_1} + \dfrac{\partial F_2}{\partial g_1}\right) & \dfrac{\partial F_2}{\partial u_1} \end{bmatrix},$$

it is clear from (5.2.16) that dV/dt is negative definite if both

(5.2.17) $$\begin{cases} (i) \ \dfrac{\partial F_1}{\partial g_1} < 0, \\ (ii) \ \dfrac{\partial F_1}{\partial g_1} \dfrac{\partial F_2}{\partial u_1} > \dfrac{1}{4}\left(\dfrac{\partial F_1}{\partial u_1} + \dfrac{\partial F_2}{\partial g_1}\right)^2 \end{cases}$$

hold for all $(g_1, u_1, t) \in \Omega_1 \times (-\infty, \infty)$. Thus an application of Krasovskii's theorem (see Hahn [1], Theorem 55.5) yields that the positive equilibrium (G^*, U^*) of (5.2.1) is asymptotically stable provided the conditions (i) and (ii) of (5.2.17) hold.

Remark 5.2.2 Krasovskii's method can be adapted for linear stability analysis without assuming the hypothesis (H_6), and the corresponding stability criterion can be obtained.

Remark 5.2.3 Krasovskii's method guarantees the asymptotic stability of (G^*, U^*) for the system (5.2.1) if $Q(x,t)$ is negative definite, but does not lead to any answer when $Q(x,t)$ is not negative definite. Since the negative definiteness of $Q(x,t)$ requires that the matrix $Q(x,t)$ have nonzero elements on its main diagonal, Krasovskii's method cannot be used if $f \equiv 0$ in (5.2.1).

Example 5.2.1 (Delay model with functional response) As a special case of (5.2.1), we consider the following system of equations:

(5.2.18)
$$\begin{cases} \dfrac{dG}{dt} = G\left[r_G(1 - \dfrac{G}{K}) - \dfrac{\alpha U}{1 + \alpha_0 U}\right], \\ \dfrac{dU}{dt} = U\left[-\gamma - \delta U + \beta \displaystyle\int_{-\infty}^{t} \omega(t-s) G(s)\, ds\right], \end{cases}$$

where r_G, α, α_0 and δ are positive real numbers. The positive equilibrium point (G^*, U^*) of (5.2.18) is given by (see (5.2.2))

$$G^* = \frac{\gamma + \delta U^*}{\beta \omega_0},$$

$$U^* = \frac{-q + (q^2 - 4pr)^{1/2}}{2p},$$

where $p = \delta r_G \alpha_0$, $q = \beta k \omega_0(\alpha - \alpha_0 r_G) + (\delta \alpha_0 \gamma) r G$ and $r = r_G(\gamma - k\beta \omega_0)$, with $\gamma < \beta G^* \omega_0$ ($< \beta k \omega_0$), ω_0 being the same as in (H_5).

The transformation

(5.2.19) $$g_1 = \log\left(\frac{G}{G^*}\right), \qquad u_1 = \log\left(\frac{U}{U^*}\right)$$

reduces (5.2.18) to

(5.2.20)
$$\begin{cases} \dfrac{dg_1}{dt} = r_G(1 - \dfrac{G^* e^g}{k}) - \dfrac{\alpha U^* e^{u_1}}{(1 + \alpha_0 U^* e^{u_1})} \equiv F_1(g_1, u_1), \\ \dfrac{du_1}{dt} = -\gamma - \delta U^* e^{u_1} + \beta \displaystyle\int_{-\infty}^{t} \omega(t-s) e^{g_1(s)}\, ds \equiv F_2(g_1, u_1, t). \end{cases}$$

Then we have

$$\frac{\partial F_1}{\partial g_1} = -r_G \frac{G^* e^{g_1}}{k},$$

$$\frac{\partial F_1}{\partial u_1} = -\frac{\partial U^* e^{u_1}}{(1 + \alpha_0 U^* e^{u_1})^2},$$

$$\frac{\partial F_2}{\partial g_1} = \beta G^* \int_{-\infty}^{t} \omega(t-s) e^{g_1(s)} ds,$$

$$\frac{\partial F_2}{\partial u_1} = -\delta U^* e^{u_1}.$$

Thus, using Krasovskii's method, the conditions (5.2.17) for asymptotic stability of $(0,0)$ for the system (5.2.20) now take the form

$$r_G \frac{\delta G^* U^*}{k} e^{(g_1 + u_1)} > \frac{1}{4} \left[\beta G^* \int_{-\infty}^{t} \omega(t-s) e^{g_1(s)} ds - \frac{\alpha U^* e^{u_1}}{(1+\alpha_0 U^* e^{u_1})^2} \right]^2.$$

This implies, in view of (5.2.19), that

(5.2.21) $$r_G \frac{\delta GU}{k} > \frac{1}{4} \left[\beta \int_{-\infty}^{t} \omega(t-s) G(s) ds - \frac{\alpha U}{(1+\alpha_0 U)^2} \right]^2.$$

Therefore, from the hypotheses (H_5) and (H_6), it is clear that the positive equilibrium (G^*, U^*) of (5.2.18) is asymptotically stable if

(5.2.22) $$r_G \frac{\delta k_1 L_1}{k} + \frac{1}{2} \frac{\alpha \beta k_1 L_1 \omega_0}{(1+\alpha_0 L_2)^2} > \frac{1}{4} \left[\beta^2 k_2^2 \omega_0^2 + \frac{\alpha^2 L_2^2}{(1+\alpha_0 L_1)^2} \right].$$

Remark 5.2.4 It is easy to verify that the condition (5.2.22) implies (5.2.21). However, the converse is not true.

Example 5.2.2 (Delay model without functional response) Consider the two-dimensional system

(5.2.23) $$\begin{cases} \frac{dG}{dt} = G \left[r_G \left(1 - \frac{G}{k} \right) - \alpha U \right], \\ \frac{dU}{dt} = U \left[-\gamma - \delta U + \beta \int_{-\infty}^{t} \omega(t-s) G(s) ds \right]. \end{cases}$$

The positive equilibrium point (G^*, U^*) of (5.2.23) is given by

$$G^* = \frac{k(\alpha\gamma + \delta r_G)}{\alpha\beta k\omega_0 + \delta r_G},$$

$$U^* = \frac{r_G(\beta k\omega_0 - \gamma)}{\alpha\beta k\omega_0 + \delta r_G},$$

with $\gamma < \beta k\omega_0$, ω_0 being the same as in (H_5). Then

$$F_1(g_1, u_1) = r_G\left(1 - \frac{G^* e^{g_1}}{k}\right) - \alpha U^* e^{u_1},$$

$$F_2(g_1, u_1, t) = -\gamma - \delta U^* e^{u_1} + \int_{-\infty}^{t} w(t-s) e^{g_1(s)} ds.$$

Hence the positive equilibrium (G^*, U^*) of (5.2.23) is asymptotically stable if

$$r_G \frac{\delta k_1 L_1}{k} + \frac{1}{2}\alpha\beta k_1 L_1 \omega_0 > \frac{1}{4}\left(\beta^2 k^2 \gamma \omega_0^2 + \alpha^2 L_2^2\right).$$

Example 5.2.3 (*Systems with no delay and without functional response*) Take $w(t)$ in (5.2.23) as a Dirac delta function defined by

$$w(t) = \delta_\epsilon(t) = \begin{cases} \frac{1}{\epsilon} & \text{for } 0 \leq t < \epsilon, \\ 0 & \text{otherwise}, \end{cases}$$

where ϵ is a sufficiently small positive constant. Then the system (5.2.23) (see MacDonald [1]) reduces to

(5.2.24)
$$\begin{cases} \frac{dG}{dt} = G\left(r_G(1 - \frac{G}{k}) - \alpha U\right), \\ \frac{dU}{dt} = U(-\gamma - \delta U + \beta G) \end{cases}$$

in the limiting case as $\epsilon \to 0$.

The positive equilibrium (G^*, U^*) of (5.2.24) can be obtained as

$$G^* = \frac{k(\alpha\gamma + \delta r_G)}{\alpha\beta k + \delta r_G},$$

$$U^* = \frac{r_G(\beta k - \gamma)}{\alpha\beta k + \delta r_G},$$

with $\gamma < \beta k$.

Following the same procedure as in Example 5.2.1, it is easy to verify that (G^*, U^*) of (5.2.24) is asymptotically stable provided that

$$r_G \frac{\delta k_1 L_1}{k} + \frac{1}{2}\alpha\beta k_1 L_1 > \frac{1}{4}\left(\beta^2 k_2^2 + \alpha^2 L_2^2\right).$$

Remark 5.2.5 It is well known that increasing delay usually destabilizes (see Stephan [1]) the system. This fact is quite evident from the condition (5.2.22) for $\beta > 1$. However, increasing r_G (the logistic growth rate) may stabilize the system, since the choices of δ and k are somewhat limited.

5.3 Wave Propagation

Consider the general problem of an integro-differential equation

(5.3.1)
$$\begin{cases} x'(t) = Ax(t) + \int_0^t B(t-s)x(s)\,ds, & t \in R^+ \\ x(0) = x_0 \in E_1 \end{cases}$$

in a Banach space E_1 with norm $|\cdot|$, where A is the generator of a C_0-semigroup $T(t)$ on E_1 and $B(t)$ is closed and defined on the domain $D(A)$ of A with $B(t)x \in C^1(R^+, E_1)$ for each $x \in D(A)$.

In this section, we shall give necessary and sufficient conditions for (5.3.1) is "hyperbolic" in the sense that its solutions propagate with finite speed. Equations of the form (5.3.1) occur in the theory of heat conduction for materials with memory. Many more interesting results about hyperbolic equations can be found in the monograph of Bloom [1]. The main feature of most partial integro-differential equations is that they result in equations of the form (5.3.1), where the equation

(5.3.2) $$x'(t) = Ax(t), \quad t \in R^+,$$

is hyperbolic and is known to have the property of propagating waves with finite speed.

From the fact that A generates a C_0-semigroup, it is clear that $A - \lambda I$ is invertible for sufficiently large λ. Noting that

$$B(t) = B(t)(A - \lambda I)^{-1}(A - \lambda I)$$
$$\equiv F(t)A + K(t),$$

where $F(t) = B(t)(A - \lambda I)^{-1}$ and $K(t) = -\lambda F(t)$, it follows, by the closed graph theorem, that $F(t)$ and $K(t)$ are bounded operators on $D(A)$.

We shall now write (5.3.1) in the form

(5.3.3)
$$\begin{cases} x'(t) = Ax(t) + \int_0^t F(t-s)Ax(s)\,ds + \int_0^t K(t-s)x(s)\,ds, \\ x(0) = x_0 \in E_1. \end{cases}$$

Then it is easy to verify that $y(t) = e^{-\omega t}x(t)$ satisfies the equation

(5.3.4)
$$\begin{cases} y'(t) = (A - \omega I)y(t) + \int_0^t B(t-s)e^{-\omega(t-s)}y(s)\,ds, \\ y(0) = x_0. \end{cases}$$

We may further assume that A has a bounded inverse, so that $K(t) \equiv 0$. Thus (5.3.1) finally takes the form

(5.3.5)
$$\begin{cases} x'(t) = Ax(t) + \int_0^t F(t-s)Ax(s)\,ds, \\ x(0) = x_0. \end{cases}$$

We shall analyze the form of (5.3.5) in which A generates a C_0-semigroup and $F(t)$ is a bounded operator on E_1 that is strongly continuously differentiable. Since A is closed, it is clear that $D(A)$ endowed with the graph norm $\|x\| = |x| + |Ax|$ yields a Banach space, which will be denoted by $(E_2, \|\cdot\|)$. It is easy to show under the usual conditions (see Section 4.4) that the equation (5.3.5) has a resolvent operator $R(t)$ and that, for $x_0 \in E_2$, the solution $x(t)$ of (5.3.5) is given by

$$x(t) = R(t)x_0.$$

We shall now formulate the concept of wave propagation in the following form.

APPLICATIONS

Definition 5.3.1 Let $\{P_t : t \in R^+\}$ be a family of closed subspaces of E_1. Let $\{S(t) : t \geq 0\}$ be a family of bounded operators on E_1. The family $\{S(t) : t \in R^+\}$ is said to propagate $\{P_t : t \in R^+\}$ if

$$S(t-s)P_s \subset P_t, \quad s < t.$$

We shall now establish the following main result, which indicates that the wave propagation properties of (5.3.2) and (5.3.5) are equivalent.

Theorem 5.3.1 Suppose that $\mathcal{F} = \{P_t : t \in R^+\}$ is a family of closed subspaces of E_1 with $P_s \subset P_t$, $s < t$. Further, assume that

(i) $F(0)(P_t) \subset P_t$ for each $t \in R^+$;

(ii) $F'(t)$ propagates $\mathcal{F}$.

Then the resolvent $R(t)$ of (5.3.5) propagates $\mathcal{F}$ if and only if the semigroup $T(t)$ generated by A propagates $\mathcal{F}$.

Proof Suppose that $T(t)$ propagates $\mathcal{F}$. Then we shall first prove that

$$R(t)P_0 \subset P_t \quad \text{for } t \in R^+.$$

To prove this, we show that if $x(t)$ is a solution of (5.3.5) with $x_0 \in P_0$, then

$$x(t) \in P_t \quad \text{for } t \in R^+.$$

Define

(5.3.6) $$y(t) = \int_0^t Ax(s)\,ds,$$

where $x(t)$ is a solution of (5.3.5).

Using integration by parts, we obtain from (5.3.5) and (5.3.6) that

(5.3.7) $$\begin{cases} x'(t) = Ax(t) + F(0)y(t) + \int_0^t F'(t-s)y(s)\,ds, \\ y'(t) = Ax(t), \\ x(0) = x_0, \quad y(0) = 0. \end{cases}$$

Following the notation of Section 4.3, we note that if $E_1 \times E_1$ is normed with

$$|||(x,y)||| = |x| + |y|$$

then C given by

$$\begin{bmatrix} A & 0 \\ A & 0 \end{bmatrix}$$

generates a semigroup $T_1(t)$ given by

$$\begin{bmatrix} T(t) & 0 \\ T(t) - I & I \end{bmatrix},$$

where I is an identity operator.

A simple calculation now shows that the semigroup $T_1(t)$ propagates the family

$$Q_t = P_t \times P_t$$

of closed subspaces of $E_1 \times E_1$. Thus we can write the equations (5.3.7) as the system

(5.3.8)
$$\begin{cases} z'(t) = Cz(t) + D_1 z(t) + \int_0^t D_2(t-s) z(s) \, ds, \\ z(0) = z_0 \in Q_0, \end{cases}$$

where D_1 and D_2 are bounded operators satisfying

$$D_1(Q_t) \subset Q_t,$$

$$D_2(t-s)(Q_s) \subset Q_t \text{ for } t \geq s.$$

Since C generates a semigroup $T_1(t)$, a solution $z(t)$ of (5.3.8) must be a fixed operator $U \colon C([0, \alpha], E_1 \times E_1)$ defined by

$$(Uy)(t) = T_1(t) z_0 + \int_0^t T_1(t-s) \left[D_1 y(s) + \int_0^s D_2(s-\tau) y(\tau) \, d\tau \right] ds$$

for $0 \leq t \leq \alpha$, where α is to be determined later. Moreover, if $z_0 \in Q_0$, $T_1(t) z_0 \in Q_t$ and $y(t) \in Q_t$, $0 \leq t \leq \alpha$, it is easy to show that

$$D_1 y(s) + \int_0^s D_2(s-\tau) y(\tau) \, d\tau \in Q_s,$$

and hence U maps the set

$$\{y \in C([0,\alpha], E_1 \times E_1) : y(t) \in Q_t\}$$

into itself. Thus if α is chosen small enough, it is indeed true that U is a contraction and therefore by Banach's fixed point theorem, U has a unique fixed point $z(t)$ with $z(t) \in Q_t$, $0 \leq t \leq \alpha$. Define $z_1(t) = z(t+\alpha)$, $0 \leq t \leq \alpha$. Then from (5.3.8), we have

$$z_1'(t) = Cz_1(t) + D_1 z_1(t) + \int_0^t D_2(t-\tau) z_1(\tau) \, d\tau + F_1(t),$$

$$z_1(0) = z(\alpha) \in Q_\alpha,$$

where

$$F_1(t) = \int_0^\alpha D_2(t+\alpha-s) z(s) \, ds \in Q_{t+\alpha}, \quad 0 \leq t \leq \alpha.$$

By the same argument as above, with obvious modifications because of the presence of $F_1(t)$, it follows that

$$z(t) \in Q_t, \quad \alpha \leq t \leq 2\alpha.$$

Hence, by induction, we have

$$z(t) \in Q_t \quad \text{for all } t \in R^+.$$

Since $z(t) = (x(t), y(t))$, it follows that

$$R(t) z_0 = x(t) \in P_t, \quad t \in R^+.$$

To see that $R(t)$ propagates $\{P_t : t \in R^+\}$, let us define for $s > 0$,

$$P_t^s = P_{t+s}.$$

Then it is clear that

$$D_1(P_t^s) \subset P_t^s,$$

$$D_2(t-\tau) P_\tau^s = D_1(t+s-(s+\tau)) P_{s+\tau} \subset P_{t+s} = P_t^s.$$

Similarly, it is easy to see that $T_1(t)$ propagates the family P_t^s. Thus, by the same argument as above, we obtain

$$R(t)(P_0^s) \subset P_t^s,$$

or

$$R(t)P_s \subset P_{t+s}.$$

This implies that $R(t)$ propagates $\{P_t: t \in R^+\}$.

Conversely, suppose that $R(t)$ propagates $\{P_t: t \in R^+\}$. Let $x_0 \in P_0$ and let n be a positive integer. Since $R(t)$ propagates $\{P_t: t \in R^+\}$, it is clear that

$$R\left(\frac{t}{n}\right)^n x_0 \in P_t$$

for each n and each $t \in R^+$. Further, it follows that for each $x \in E_1$ and $t > 0$

$$T(t)x = \lim_{n \to \infty} R\left(\frac{t}{n}\right)^n x,$$

and the convergence is uniform on bounded intervals for fixed x. Since P_t is closed, we have

$$T(t)x_0 \in P_t \quad \text{if } x_0 \in P_0.$$

Now, by defining P_t^s as before, it is easy to verify that $T(t)$ propagates $\{P_t: t \in R^+\}$. This completes the proof. □

It is sometimes convenient to use (5.3.3), because in many applications B will be of the form $B = FA + K$. However, in this case, it is not necessary to assume that K is differentiable, as the following result shows.

Corollary 5.3.1 *Suppose that $\mathcal{F} = \{P_t: t \in R^+\}$ is a family of closed subspaces of E_1 with $P_s \subset P_t$ for $s < t$. Further, assume that*

(i) $F(0)(P_t) \subset P_t$ for each $t \in R^+$;

(ii) $F'(t)$ and $K(t)$ propagate $\mathcal{F}$.

Then the resolvent $R(t)$ of (5.3.3) propagates $\mathcal{F}$ if and only if $T(t)$ generated by A propagates $\mathcal{F}$.

The proof of this corollary is exactly similar to that of Theorem 5.3.1.

Remark 5.3.1 The function $B(t) = a(t)A$, where $a(t)$ is scalar and continuously differentiable, is admissible in Theorem 5.3.1, since the conditions

$$a(0)I(P_t) \subset P_t,$$

APPLICATIONS

$$a'(t-s)I(P_s) \subset P_s \subset P_t \quad \text{for } s < t,$$

are automatically satisfied.

Further, in the case of hyperbolic equations, given P_0, it is reasonable to define P_t by

$$P_t = T(t)P_0.$$

Thus if it is the case that $P_s \subset P_t$ for $s < t$ then Theorem 5.3.1 can be applied in the usual way.

Example 5.3.1 (Maxwell–Hopkinson dielectric model) Let $\Omega \subseteq R^3$ be an open bounded domain. If $D(x,t)$ is the electric displacement field in a Maxwell–Hopkinson dielectric, $x \in \Omega$, $t > 0$, then the components $D_i(x,t)$ ($i = 1, 2, 3$) of $D(x,t)$ satisfy the system of integro-differential equations

$$(5.3.9) \quad \epsilon\mu \frac{\partial^3 D_i}{\partial t^2} = \Delta D_i(x,t) + \int_0^t \Phi(t-\tau)\Delta D_i(x,\tau)\,d\tau, \quad i = 1,2,3,$$

on $\Omega \times [0,T]$, $T > 0$, where $\Phi(t)$ is a scalar function, with initial–boundary conditions (see Bloom [1])

$$D_i(x,t) = 0 \quad \text{for } (x,t) \in \partial\Omega \times [0,T],$$

$$D_i(x,0) = f_i(x), \quad \frac{\partial D_i}{\partial t}(x,0) = g_i(x), \quad x \in \Omega, \quad i = 1,2,3.$$

Further, we assume that the initial functions f_i and g_i have compact support in Ω. We may now rewrite each equation as

$$(5.3.10) \quad \begin{cases} u_{tt} = (\epsilon\mu)^{-1}\Delta u + \int_0^t a(t-\tau)\Delta u(\tau)\,d\tau, \\ u(0) = f, \quad u_t(0) = g, \end{cases}$$

where $a(t) = (a/\epsilon\mu)\Phi(t)$, and consider the solutions u in $L^2(R^3)$. If $B = (I - \Delta)^{1/2}$ and $E_1 = D(B) \times L^2(R^3)$, where $D(B)$ is the domain of B with graph norm, then the operator A given by

$$\begin{bmatrix} 0 & I \\ \frac{1}{\epsilon\mu}\Delta & 0 \end{bmatrix}$$

generates a semigroup $T(t)$ on E_1. In fact, we may take any of

$E_1^k = D(B^{k+1}) \times D(B^k)$, $k = 0, 1, 2, \ldots$, as our space E_1. Let Σ be a compact set in $\Omega \subset R^3$ and let

$$S(\Sigma, t) = \{y + z \in R^3 : y \in \Sigma,\ \|z\| \leq \sigma t\},$$

where $\sigma = (\epsilon \mu)^{-1/2}$.

Let

$$Q_t = \{u \in L^2(R^3) : \text{the support of } u \text{ is contained in } S(\Sigma, t)\}$$

$$P_t = (D(B^{k+1}) \cap Q_t) \times (D(B^k) \cap Q_t).$$

It is easy to verify that $\{P_t : t \in R^+\}$ is a family of closed subspaces of E_1^k with

$$P_s \subset P_t \quad \text{for } s < t.$$

From the fact that the equation

$$u_{tt} = \tfrac{1}{\epsilon\mu}\Delta u$$

has propagation speed $(\epsilon\mu)^{-1/2}$, it follows that $T(t)$ propagates $\{P_t : t \in R^+\}$.

Finally, let $v = u_t$, then we have

$$u_t = v,$$

$$v_t = \tfrac{1}{\epsilon\mu}\Delta u + \int_0^t \Phi(t-\tau) \tfrac{1}{\epsilon\mu}\Delta u(\tau)\, d\tau.$$

Hence (5.3.9) takes the form

$$\begin{bmatrix} u \\ v \end{bmatrix}_t = \begin{bmatrix} 0 & I \\ \tfrac{1}{\epsilon\mu}\Delta & 0 \end{bmatrix} \begin{bmatrix} u \\ v \end{bmatrix}$$

$$+ \int_0^t \begin{bmatrix} 0 & 0 \\ 0 & \Phi(t-\tau) \end{bmatrix} \begin{bmatrix} 0 & I \\ \tfrac{1}{\epsilon\mu}\Delta & 0 \end{bmatrix} \begin{bmatrix} u \\ v \end{bmatrix} d\tau,$$

which can be written as

$$x'(t) = Ax(t) + \int_0^t F(t-\tau) Ax(\tau)\, d\tau.$$

APPLICATIONS

Thus if $\Phi(t)$ is continuously differentiable then $F(t)$ satisfies all the assumptions of Theorem 5.3.1. Therefore the application of Theorem 5.3.1 yields that the equation (5.3.9) has the property of propagating waves with finite speed $(\epsilon\mu)^{-1/2}$.

5.4 Nuclear Reactors

Consider the system of integro-differential equations

(5.4.1)
$$\begin{cases} (a) \ u'(t) = -\int_{-\infty}^{\infty} \alpha(x) T(x,t) \, dx, \\ (b) \ T_t(x,t) = T_{xx}(x,t) + \eta(x) \phi(u(t)), \end{cases}$$

with initial conditions

(5.4.2) $\qquad u(0) = u_0, \qquad T(x,0) = f(x), \qquad x \in R.$

In the special case when $\phi(u) = e^u - 1$, the system (5.4.1) may be regarded as a mathematical model describing the behavior of a continuous-medium nuclear reactor, where the unknown functions $u(t)$ and $T(x,t)$ stand for the derivations of the logarithm of the total reactor power and of the temperature from their equilibrium values. The reactor here is modeled for a doubly infinite rod. An interesting special case is to replace the infinite interval of integration in (5.4.1) by a finite integral and assume suitable boundary conditions on $T(x,t)$ (see Remark 5.4.2).

It is well known that under the usual assumptions, the solution of the heat equation

$$T_t(x,t) = T_{tt}(x,t) + F(x,t)$$

with initial condition $T(x,0) = f(x)$, $x \in R$, is given by

(5.4.3) $\qquad T(x,t) = \int_{-\infty}^{\infty} G(x-\tau,t) f(\tau) \, d\tau$

$$+ \int_0^t \int_{-\infty}^{\infty} G(x-\tau, t-s) F(\tau,s) \, d\tau \, ds,$$

where

$$G(x,t) = \frac{1}{(4\pi t)^{1/2}} \exp\left(-\frac{x^2}{4t}\right), \quad t > 0.$$

If we assume that $u(t)$ is known then, by taking $F(x,t) = \eta(x)\phi(u(t))$, we obtain from (5.4.3) that

$$(5.4.4) \quad T(x,t) = \int_{-\infty}^{\infty} G(x-\tau,t)f(\tau)\,d\tau$$

$$+ \int_0^t \int_{-\infty}^{\infty} G(x-\tau,t-s)\eta(\tau)\phi(u(s))\,d\tau\,ds.$$

Let

$$(5.4.5) \quad K(t) = \int_{-\infty}^{\infty} \alpha(x)\,dx \int_{-\infty}^{\infty} G(x-\tau,t)\eta(\tau)\,d\tau,$$

$$(5.4.6) \quad b(t) = \int_{-\infty}^{\infty} \alpha(x)\,dx \int_{-\infty}^{\infty} G(x-\tau,t)f(\tau)\,d\tau.$$

Then (5.4.1a), (5.4.2) and (5.4.4) – (5.4.6) give

$$(5.4.7) \quad \begin{cases} u'(t) = -b(t) - \int_0^t K(t-s)\phi(u(s))\,ds, \\ u(0) = u_0. \end{cases}$$

In order to obtain a more convenient form for $b(t)$ and $K(t)$, we shall use Parseval's formula for Fourier transforms. The Fourier transform of $G(x,t)$, as a function of x, is given by

$$(5.4.8) \quad \widehat{G}(\tau,t) = \exp(-\tau^2 t)$$

for any $t > 0$.

Set

$$(5.4.9) \quad \rho(x) = \int_{-\infty}^{\infty} G(x-\xi,t)\eta(\xi)\,d\xi.$$

APPLICATIONS

Then from (5.4.5) and (5.4.9), we have

$$K(t) = \int_{-\infty}^{\infty} \alpha(x)\rho(x)\,dx.$$

This implies by Parseval's formula that

$$K(t) = \frac{1}{2\pi} \int_{-\infty}^{\infty} \overline{\hat{\alpha}(x)}\hat{\rho}(x)\,dx.$$

Since $\alpha(x)$ is real-valued (i.e. $\overline{\hat{\alpha}(x)} = \hat{\alpha}(-x)$), we obtain

(5.4.10) $$K(t) = \frac{1}{\pi} \int_{0}^{\infty} \mathrm{Re}[\hat{\alpha}(-x)\hat{\rho}(x)]\,dx.$$

In view of (5.4.8) and (5.4.9), we get

(5.4.11) $$\hat{\rho}(x) = \hat{\eta}(x)\exp(-x^2 t),$$

for any $t > 0$.

Thus (5.4.10) and (5.4.11) lead to

(5.4.12) $$K(t) = \frac{1}{\pi} \int_{0}^{\infty} h_1(x)\exp(-x^2 t)\,dx,$$

where

$$h_1(x) = \mathrm{Re}[\hat{\alpha}(-x)\hat{\eta}(x)].$$

Similarly, from (5.4.6), we obtain

(5.4.13) $$b(t) = \frac{1}{\pi} \int_{0}^{\infty} h_2(x)\exp(-x^2 t)\,dx,$$

where

(5.4.14) $$h_2(x) = \mathrm{Re}[\hat{\alpha}(-x)\hat{f}(x)].$$

In order to obtain the asymptotic behavior of solutions of the nuclear reactor system (5.4.1), (5.4.2) as $t \to \infty$, we need the following hypotheses and a known result.

(H_1) The functions $\alpha(x)$ and $\eta(x)$ are measurable such that

$$\alpha(x), \eta(x) \in L^2(R, R)$$

and $\eta(x)$ is locally Hölder-continuous;

(H_2) $h_1(x) \geq 0$ and $\int_0^\infty h_1(x) dx > 0$;

(H_3) $f(x) \in L^2(R,R) \cap C(R,R)$.

Remark 5.4.1 (i) From the hypothesis (H_1), it is clear that both $\widehat{\alpha}(x)$ and $\widehat{\eta}(x)$ are defined as generalized Fourier transforms (see Corduneanu [1]) and belong to L^2. Moreover

$$h_1(x) \in L^1(R,R).$$

(ii) The hypothesis (H_3) implies that there exists a $\widehat{f}(x) \in L^2$, and $h_2(x)$ defined by (5.4.14) satisfies

$$h_2(x) \in L^1(R,R).$$

(H_4) There exists a measurable function $h_3(x)$ with $h_3(x) \geq 0$ on R^+ and $h_3(x) \in L^1(R^+, R)$ and a number $\lambda > 0$ such that

$$h_2^2(x) \leq h_1(x) h_3(x), \quad x \in R^+,$$

$$h_1(x)\xi^2 + 2h_2(x)\xi + h_3(x) \geq \lambda [\,|\widehat{\eta}(x)|^2 \xi^2 + 2\mathrm{Re}[\widehat{f}(x)\widehat{\eta}(-x)]\xi + |\widehat{f}(x)|^2\,],$$

for $x \in R^+$ and $\xi \in R$.

Lemma 5.4.1 (Corduneanu [1], p. 188) *Assume that*

(i) $K(t) \in C(R^+, R)$, $K(t) \neq K(0)$, and $(-1)^j K^{(j)}(t) \geq 0$ for $0 < t < \infty$, $j = 0,1,2,3$;

(ii) $b(t) \in C(R^+, R)$ and is twice continuously differentiable for $t > 0$;

(iii) there exists $C(t)$ with the same properties as $b(t)$ in (ii) and such that

$$[b^{(j)}(t)]^2 \leq K^{(j)}(t) C^{(j)}(t), \quad j = 0,1,2;$$

(iv) $\phi(u) \in C(R,R)$ such that $u\phi(u) > 0$ for $u \neq 0$ and

$$\Phi(u) = \int_0^u \phi(s)\, ds \to \infty \quad \text{as } |u| \to \infty.$$

Then for each $u_0 \in R$, there exists a solution $u(t)$ for the IVP (5.4.7) for $t \in R^+$ and

$$\lim_{t \to \infty} u^{(j)}(t) = 0 \quad \text{for } j = 0,1,2.$$

Theorem 5.4.1 *Suppose that all the assumptions of Lemma 5.4.1 are satisfied. Further, assume that the hypotheses $(H_1)-(H_4)$ hold. Then there exists a solution $(u(t), T(x,t))$ for the IVP (5.4.1), (5.4.2) on $0 \leq t < \infty$, $x \in R$, satisfying*

$$\lim_{t \to \infty} u^{(j)}(t) = 0, \quad j = 0,1,2,$$

$$\lim_{t \to \infty} T(x,t) = 0 \quad \text{uniformly in } x \in R.$$

Proof From (5.4.12) and (5.4.13), it follows that

(5.4.15) $$K^{(j)}(t) = \frac{(-1)^j}{\pi} \int_0^\infty x^{2j} h_1(x) \exp[-x^2 t] \, du,$$

(5.4.16) $$h^{(j)}(t) = \frac{(-1)^j}{\pi} \int_0^\infty x^{2j} h_2(x) \exp[-x^2 t] \, du$$

for $j = 0,1,2,\ldots$ and $0 < t < \infty$.

In view of the exponential factor, the hypotheses (H_2) and (H_3) yield the convergence of both the integrals (5.4.15) and (5.4.16) for any nonnegative integer j. Further, (5.4.8) and the hypothesis (H_2) imply

$$K(t) \in C(R^+, R), \quad (-1)^j K^{(j)}(t) \geq 0$$

for $j = 0,1,2,\ldots$ and $0 < t < \infty$.

Define

$$C(t) = \frac{1}{\pi} \int_0^\infty h_3(x) \exp(-x^2 t) \, dt$$

where $h_3(x)$ is as defined in (H_4). Then we see that

(5.4.17) $$C^{(j)}(t) = \frac{(-1)^j}{\pi} \int_0^\infty x^{2j} h_3(x) \exp(-x^2 t) \, dt.$$

From (5.4.15) – (5.4.17), using Schwartz's inequality and the hypothesis (H_4), it is easy to verify that the hypothesis (iii) of Lemma 5.4.1 holds for $j = 0,1,2,\ldots$. Thus (5.4.7) satisfies all the assumptions of Lemma 5.4.1, and

hence there exists a solution $u(t)$ of the IVP (5.4.7) on R^+ such that
$$\lim_{t \to \infty} u^{(j)}(t) = 0 \quad \text{for } j = 0, 1, 2.$$
This proves the first part of the theorem. To prove the second part, we consider the function $T(x,t)$ given by (5.4.4).

In view of the classical result on the heat equation, it is clear that the function $T(x,t)$ defined by (5.4.4) satisfies the equation (5.4.1b) with initial condition $T(x,0) = f(x)$. We shall now verify that the pair $(u(t), T(x,t))$ satisfies (5.4.1a).

From the fact that L^2 is an invariant subspace of the convolution operator with integrable kernel (see Corduneanu [1], pp. 42 – 44), it follows for each fixed t that $T(x,t)$ is square integrable on R. From (5.4.4) and (5.4.8), we obtain

(5.4.18) $$\widehat{T}(x,t) = f(x)e^{-x^2 t} + \widehat{\eta}(x) \int_0^t \phi(u(s))e^{-x^2(t-s)} ds.$$

Thus (5.4.18), (5.4.5), (5.4.6) and Parseval's formula lead to
$$\int_{-\infty}^{\infty} \alpha(x) T(x,t) dx = \frac{1}{2\pi} \int_{-\infty}^{\infty} \widehat{\alpha}(-x) \widehat{T}(x,t) dx$$
$$= \int_0^t K(t-s)\phi(u(s)) ds + b(t)$$
$$= -u'(t).$$

This verifies (5.4.1a). Further, it is easy to show from (5.4.3) and (5.4.4) that
$$T_x(x,t), \quad T_{xx}(x,t) \in L^2$$
for any $t > 0$.

From the identity
$$T^2(x,t) - T^2(\xi,t) = 2 \int_\xi^x T(v,t) T_x(v,t) dv,$$
for any $x, \xi \in R$ and $t > 0$, we obtain

(5.4.19) $$|T^2(x,t) - T^2(\xi,t)| \leq 4 \int_{-\infty}^{\infty} T^2(x,t)\,dx \int_{-\infty}^{\infty} T_x^2(x,t)\,dx.$$

Since $T(x,t) \in L^2$ for any $t > 0$, there exists a sequence $\{x_n\}$ such that

$$T(x_n, t) \to 0 \quad \text{as } n \to \infty.$$

Now letting ξ pass through such a sequence, the inequality (5.4.19) leads to

(5.4.20) $$\sup_{x \in R} T^4(x,t) \leq 4 \int_{-\infty}^{\infty} T^2(x,t)\,du \int_{-\infty}^{\infty} T_x^2(x,t)\,dx.$$

Using Fubini's theorem, from (5.4.7), (5.4.12) and (5.4.13), we obtain

(5.4.21) $$u'(t) = -\frac{1}{\pi} \int_0^{\infty} [h_1(x)\gamma(x,t) + h_2(x)] \exp(-x^2 t)\,dx,$$

where

(5.4.22) $$\gamma(x,t) = \int_0^t \phi(u(s)) \exp(x^2 s)\,ds.$$

We now consider the energy function

(5.4.23) $$V(t) = \Phi(u(t))$$
$$+ \frac{1}{2\pi} \int_0^{\infty} [h_1(x)\gamma^2(x,t) + 2h_2(x)\gamma(x,t) + h_3(x)] \exp(-2x^2 t)\,dx.$$

It is clear from the hypothesis (H_4) and Lemma 5.4.1 (iv) that

$$V(t) \geq 0.$$

Further, the time derivative of $V(t)$ along the solutions of (5.4.21) is given by

(5.4.24) $$\frac{dV}{dt} = -\frac{1}{\pi} \int_0^{\infty} [h_1(x)\gamma^2(x,t) + 2h_2(x)\gamma(x,t) + h_3(x)] x^2 \exp(-2x^2 t)\,dx.$$

This together with hypothesis (H_4) implies

(5.4.25) $$\frac{dV}{dt} \leq 0.$$

Using the hypothesis (H_4), it follows from (5.4.22), (5.4.23) and (5.4.25) that

$$\Phi(u(t)) \leq V(t) \leq V(0) = \Phi(u_0) + \frac{1}{2\pi}\int_0^\infty h_3(x)\,dx.$$

Since $h_3 \in L^1(R^+, R)$, we have

(5.4.26) $$|u(t)| \leq K(u_0)$$

with $K(u_0) = \Phi_0^{-1}[\Phi(u_0) + (2\pi)^{-1}\int_0^\infty h_3(x)\,dx]$, where $\Phi_0(u)$ is an increasing function for $u \geq 0$ such that $\Phi_0(u) \leq \Phi(\pm u)$, $u \in R^+$. Thus (5.4.22) and (5.4.26) give

(5.4.27) $$|x^2 \gamma(x,t)| \leq K_0 \exp(x^2 t) \quad \text{for } t, x \in R^+,$$

where K_0 depends on u_0.

Further, from (5.4.24) on differentiation and using (5.4.22), we obtain

(5.4.28)
$$\begin{cases}\dfrac{d^2 V}{dt^2} = -\dfrac{2}{\pi}\phi(u(t))\int_0^\infty [h_1(x)\gamma(x,t) + h_2(x)]x^2 \exp(-x^2 t)\,dx \\ \qquad + \dfrac{2}{\pi}\int_0^\infty [h_1(x)\gamma^2(x,t) + 2h_2(x)\gamma(x,t) + h_3(x)]x^4 \exp(-2x^2 t)\,dx.\end{cases}$$

Using the fact that $h_1, h_2, h_3 \in L^1(R^+, R)$, it follows from (5.4.26)–(5.4.28) that d^2V/dt^2 is bounded. This together with $V(t) \geq 0$ and $dV/dt \leq 0$ implies

(5.4.29) $$\lim_{t\to\infty} \frac{dV}{dt} = 0.$$

Moreover, using Parseval's formula twice and the fact that the Fourier transform of $T_x(x,t)$ is $-ix\widehat{T}(x,t)$, the inequality (5.4.20) can be written as

(5.4.30) $$\sup_{x \in R} T^4(x,t) \leq \frac{1}{\pi^2}\int_{-\infty}^\infty |\widehat{T}(x,t)|^2\,dx \int_{-\infty}^\infty x^2 |\widehat{T}(x,t)|^2\,dx.$$

The relations (5.4.18) and (5.4.22) give

(5.4.31) $$\begin{aligned}|\widehat{T}(x,t)|^2 &= \{|\widehat{\eta}(x)|\gamma^2(x,t) + 2\operatorname{Re}[\widehat{f}(-x)\widehat{\eta}(-x)]\gamma(x,t) \\ &\quad + |\widehat{f}(x)|^2\}\exp(-2x^2 t).\end{aligned}$$

Since $dV/dt \leq 0$ and $\widehat{T}(-x,t) = \overline{\widehat{T}(x,t)}$, the hypothesis ($H_4$), (5.4.23) and (5.4.31) lead to

$$2\pi V(0) \geq 2\pi V(t)$$

$$\geq \int_0^\infty [h_1(x)\gamma^2(x,t) + 2h_2(x)\gamma(x,t) + h_3(x)]e^{-x^2 t}\, dx$$

$$\geq \lambda \int_0^\infty \{|\widehat{\eta}(x)|^2 \gamma^2(x,t) + 2\mathrm{Re}[\widehat{f}(-x)\widehat{\eta}(-x)]\gamma(x,t)$$

$$+ |\widehat{f}(x)|^2\}e^{-2x^2 t}\, dx$$

$$= \lambda \int_0^\infty |\widehat{T}(x,t)|^2\, dx$$

$$= \frac{\lambda}{2}\int_{-\infty}^\infty |\widehat{T}(x,t)|^2\, dx,$$

and hence

(5.4.32) $$\int_{-\infty}^\infty |\widehat{T}(x,t)|^2\, dx \leq \frac{4\pi}{\lambda}V(0) \quad \text{for } t > 0.$$

Similarly, we obtain from (5.4.24) that

$$-\pi\frac{dV}{dt} \geq \lambda \int_0^\infty x^2|\widehat{T}(x,t)|^2\, dx = \frac{\lambda}{2}\int_{-\infty}^\infty x^2|\widehat{T}(x,t)|^2\, dx.$$

This implies that

(5.4.33) $$\int_{-\infty}^\infty x^2|\widehat{T}(x,t)|^2\, dx \leq -\frac{2\pi}{\lambda}\frac{dV}{dt} \quad \text{for } t > 0.$$

Therefore, from (5.4.30), (5.4.32) and (5.4.33), we get

$$\sup_{x \in R} T^4(x,t) \leq -\frac{8}{\lambda^2}V(0)\frac{dV}{dt}.$$

This together with (5.4.29) gives

$$\lim_{t\to\infty} T(x,t) = 0 \quad \text{uniformly in } x \in R.$$

Hence the proof of the theorem is complete. □

Remark 5.4.2 Levin and Nohel [4] have discussed the system

$$u'(t) = -\int_0^\pi \alpha(x) T(x,t)\, dx,$$

$$T_t(x,t) = T_{xx}(x,t) + \eta(x)\phi(u(t))$$

on $0 \leq x \leq \pi$, $0 < t < \infty$, with initial boundary data $u(0) = u_0$, $T(x,0) = f(x)$, $0 \leq x \leq \pi$, and $T_x(0,t) = T_x(\pi,t) = 0$ for all $t > 0$. The natural technique in this case would be some convenient series instead of Fourier transforms.

Remark 5.4.3 In the special case $\phi(u) = 1 - e^u$, letting $u(t) = \log(P(t)/\tau)$, the system (5.4.1) can be replaced by a delay system

$$(5.4.34) \begin{cases} P'(t) = -P(t) \int_{-\infty}^{\infty} \alpha(x) T(x,t)\, dx - \frac{\beta}{\rho} P(t) + \sum_{i=1}^{m} \lambda_i C_i(t), \\ \\ C'_i(t) = \frac{\beta_i}{\rho} P(t) - \lambda_i C_i(t), \quad i = 1, 2, 3, \ldots, m \\ \\ T_t(x,t) = T_{xx}(x,t) + \eta(x)[P(t) - \tau], \quad x \in R, \quad 0 < t < \infty, \end{cases}$$

where $\beta = \sum_{i=1}^{m} \beta_i$, and the asymptotic behavior of solutions of (5.4.34) can be examined by using Theorem 5.4.1 with minor modifications. The system (5.4.34) represents a nuclear reactor in an infinite continuous medium with m groups of delayed neutrons, where the concentration of the emitter in the ith group, is $C_i(t)$.

5.5 Viscoelasticity

Consider the one-dimensional motion of an elastic bar. Let $u(t,x)$ be the position at time t and let $u_x(t,x)$ be the measure of strain. Nonlinear elasticity assumes the stress σ at time t is given by $\sigma = \sigma(u_x(t,x))$. If the bar has infinite length with unit density then the appropriate dynamical model is

(5.5.1) $$u_{tt} = \sigma(u_x(t,x))_x, \quad 0 < t < \infty, \quad x \in R,$$

with initial data

(5.5.2) $$u(0,x) = u_0, \quad u_t(0,x) = u_1(x).$$

It is well known (Lax [1]) that if σ is genuinely nonlinear then the problem (5.5.1) – (5.5.2) cannot have global smooth solutions for any nonzero data. One of the remedies for this paradox is to give up the requirement of smooth solutions and look for weak (or shock) solutions. The latter idea leads to a slight change in the problem by introducing a "viscosity" term. Thus (5.5.1) assumes the form

(5.5.3) $$u_{tt} = \sigma(u_x(t,x))_x + \frac{\partial}{\partial x}[\psi(u_x)u_{xt}], \quad \psi(\xi) > 0.$$

It is shown in MacCamy [1] that the initial value problem (5.5.3)-(5.5.2) always has a global smooth solution that is asymptotically stable, no matter how large the initial data.

The above situation is somewhat similar to the following simple case. Consider the equation

$$u_t + uu_x = 0, \quad x \in R, t > 0,$$

with

$$u(0,x) = \frac{\alpha}{1+x^2}.$$

It is easy to verify that this simple problem has no smooth solutions, but there does exist a weak solution. However, for the problem

(5.5.4) $$u_t + uu_x - \lambda u_{xx} = 0, \quad \lambda > 0,$$

with

$$u(0,x) = \frac{\alpha}{1+x^2},$$

there will be a unique global smooth solution for any α. On the other hand, if (5.5.4) is replaced by

$$u_t + uu_x + \lambda u_{xx} = 0, \quad \lambda > 0,$$

then there will be a smooth solution for small α but not for large α.

In this section, we shall consider a more general problem of one-dimensional motion of an elastic material with memory,

(5.5.5)
$$\begin{cases} u_{tt}(t,x) = \sigma(u_x(t,x))_x + \int_0^t a'(t-s)\sigma(u_x(s,x))_x\,ds + g(t,x), \\ u(0,x) = u_0, \quad u_t(0,x) = u_1(x), \quad 0 < t < \infty, \quad x \in R, \end{cases}$$

and give sufficient conditions on σ, $a'(t)$ and g that guarantee the existence of a unique global solution $u(x,t)$ for the IVP (5.5.5) that tends to zero as $t \to \infty$ uniformly in $x \in R$. The main technique we adopt in our discussion is the method of energy estimates. To this end, we make the following hypotheses:

(H_1) $\sigma \in C^3(R)$, $\sigma(0) = 0$ and $\sigma'(0) > 0$;

(H_2) $g, g_t \in L^1(R^+, L^2(R))$ and $g_x, g_{tt}, g_{tx} \in L^2(R^+, L^2(R))$;

(H_3) the initial functions $u_0(x)$ and $u_1(x)$ satisfy $u_{0x}, u_{0xx}, u_{0xxx} \in L^2(R)$ and $u_{1x}, u_{1xx}, u_{1xx} \in L^2(R)$;

(H_4) the function $a(t)$ satisfies the following conditions:
(i) $a(t) \in C^3[0,\infty)$, $a(t), a'(t), a''(t)$ and $a'''(t)$ are bounded on R^+;
(ii) $a(t) = a_\infty + A(t)$, $a_\infty > 0$ and $a(0) = 1$;
(iii) $(-1)^m A^{(m)}(t) \geq 0$, $0 \leq t < \infty$, $m = 0,1,2$, and $A'(t) \neq 0$;
(iv) $t^n A^{(m)}(t) \in L^1(0,\infty)$, $n = 0,1,2,3$ and $m = 0,1,2,3$.

We shall now define the resolvent kernel $K(t)$ associated with $a'(t)$ by the equation

$$K(t) + (a' * K)(t) = -a'(t), \quad 0 \leq t < \infty,$$

where $*$ denotes the convolution

$$(a' * K)(t) = \int_0^t a'(t-s)K(s)\,ds.$$

From the standard theory of integral equations (Bellman and Cooke [1], Theorem 7.4), it is clear that if a in C^3, smooth then $K(\cdot)$ can be determined uniquely and is C^3 smooth on R^+. Further, for any $\phi \in L^1_{loc}(0,\infty)$, the unique solution of the Volterra equation

(5.5.6) $$y(t) + (a' * y)(t) = \phi(t), \quad 0 \leq t < \infty,$$

is given by
$$y(t) = \phi(t) + (K*\phi)(t), \quad 0 \le t < \infty.$$

Now by considering (5.5.5) as a Volterra equation of the form (5.5.6) with $y(t) = \sigma(u_x)_x$, we obtain from (5.5.5) that

(5.5.7) $\quad u_{tt} + (K*u_{tt})(t,x) = \sigma(u_x(t,x))_x + g(t,x) + (K*g)(t,x).$

Integration by parts with respect to t in the convolution term on the left-hand side of (5.5.7) shows that (5.5.5) is equivalent to

(5.5.8) $\begin{cases} u_{tt} + \dfrac{\partial}{\partial t}\displaystyle\int_0^t K(t-s)u_t(s,x)\,ds = \sigma(u_x(t,x))_x + \Phi(t,x), \quad 0 < t < \infty, \ x \in R, \\[6pt] u(0,x) = u_0(x), \quad u_t(0,x) = u_1(x), \quad x \in R, \end{cases}$

where

(5.5.9) $\quad \Phi(t,x) = g(t,x) + (K*g)(t,x) + K(t)u_1(x).$

The following well-known result (MacCamy [2]) is useful in our subsequent discussion.

Lemma 5.5.1 *Suppose (i) − (iv) of the hypothesis (H_4) are satisfied. Let $K(t)$ be the resolvent kernel corresponding to $a'(t)$. Then the following properties hold for $K(t)$:*

(i) $K(t) \in C^3([0,\infty)), K(t), K'(t)$ and $K''(t)$ are bounded on R^+;
(ii) $K^{(m)}(t) \in L^1(0,\infty), m = 0, 1, 2;$
(iii) *for any $T > 0$ and every $v(t) \in L^2(0,T)$,*

(5.5.10) $\quad \displaystyle\int_0^T v(t)\dfrac{d}{dt}(K*v)(t)\,dt \ge 0.$

Remark 5.5.1 It is clear from the hypothesis (H_2) together with $K''(t) \in L^\infty(0,\infty)$ and $K(t), K'(t), K''(t) \in L^1(0,\infty)$ that $\Phi(t,x)$ defined by (5.5.9) satisfies

(H_5) $\Phi, \Phi_t \in L^1(R^+, L^2(R))$ and $\Phi_x, \Phi_{tt}, \Phi_{tx} \in L^2(R^+, L^2(R)).$

The inequality (5.5.10) indicates a weak dissipative mechanism. In fact, the dissipative mechanism for the viscoelastic equations is quite subtle, and it will reveal itself through a device of MacCamy [1] that involves another form of (5.5.5). Indeed, let us define a function $r: R^+ \to R$ by

$$(5.5.11) \qquad r(t) = \beta + K(t) + \beta \int_0^t K(s)\,ds,$$

where $\beta > 0$ is a constant and $K(t)$ is the resolvent kernel corresponding to $a'(t)$. It is clear that the solution $y(t)$ of (5.5.6) satisfies the equation

$$(5.5.12) \qquad y(t) + \beta \int_0^t y(s)\,ds = \phi(t) + (r*\phi)(t), \quad 0 \leq t < \infty.$$

Take $y = \sigma(u_x)_x$ in (5.5.5). Then it satisfies the equation of the form (5.5.6). Thus (5.5.12) leads to

$$(5.5.13) \quad u_{tt} + (r*u_{tt})(t,x) = \sigma(u_x(t,x))_x + \beta \int_0^t \sigma(u_x(s,x))]_x\,ds$$
$$+ g(t,x) + (r*g)(t,x).$$

Hence the IVP (5.5.5) is equivalent to

$$(5.5.14) \begin{cases} u_{tt}(t,x) + (r*u_{tt})(t,x) = \sigma(u_x(t,x))_x + \beta \int_0^t \sigma(u_x(s,x))_x\,ds + \psi(t,x), \\ u(0,x) = u_0(x), \quad u_t(0,x) = u_1(x), \quad 0 \leq t < \infty, \quad x \in R, \end{cases}$$

where

$$(5.5.15) \qquad \psi(t,x) = g(t,x) + (r*g)(t,x).$$

The justification for considering the variant (5.5.14) of (5.5.5) is provided in the following well-known result (MacCamy [2]).

Lemma 5.5.2 *Assume that $(i) - (iv)$ of the hypothesis (H_4) hold. Let $r(t)$ be defined by (5.5.11). Then we have the following results:*

(i) $r(t) \in C^2[0,\infty)$ and $r(t), r'(t)$ and $r''(t)$ are bounded on R^+;

(ii) $r(t) = r_\infty + R(t), r_\infty = \beta/a_\infty$ and $R^{(m)}(t) \in L^1(0,\infty), m = 0,1,2;$

(iii)   for any $T > 0$, there exist constants $\gamma, q > 0$ with $\beta q < 1$ such that

(5.5.16)
$$q \int_0^t v(t)\frac{d}{dt}(r*v)(t)\,dt - \int_0^T v(t)(R*v)(t)\,dt$$
$$\geq (1+\gamma) \int_0^T v^2(t)\,dt,$$

for every $v(t) \in L^2(0,T)$.

Remark 5.5.2 It is the estimate (5.5.16) that reveals the dissipative mechanism induced by the memory term in the viscoelastic equations and plays a crucial role in the analysis of our main result on asymptotic behavior of solutions of the IVP (5.5.5).

The following local existence result (Dafermos and Nohel [1], Theorem 3.1) whose proof is based on application of Banach's fixed point theorem, will be useful in our subsequent discussion.

Theorem 5.5.1 Assume that
(i)   the hypotheses (H_4) and (H_5) hold;
(ii) $\sigma \in C^3(R), \sigma(0) = 0$ and $\sigma'(\omega) \geq p_0 > 0, -\infty < \omega < \infty$;
(iii) $K', K'' \in C(R^+) \cap L^1(0,\infty)$.

Then there exists a unique solution $u(t,x) \in C^2([0,T_0) \times R)$ of (5.5.8) defined on a maximal interval $[0, T_0)$, $T_0 \leq \infty$, such that $T \in [0, T_0)$ and

$$u_t, u_x, u_{tt}, u_{tx}, u_{xx}, u_{ttt}, u_{ttx}, u_{txx}, u_{xxx} \in L^\infty([0,T], L^2(R)).$$

Further, if $T_0 < \infty$ then

$$\int_{-\infty}^{\infty} [u_t^2(t,x) + u_x^2(t,x) + u_{tt}^2(t,x) + u_{tx}^2(t,x) + u_{xx}^2(t,x)$$
$$+ u_{ttt}^2(t,x) + u_{ttx}^2(t,x) + u_{txx}^2(t,x) + u_{xxx}^2(t,x)]\,dx \to \infty$$

as $t \to T_0$.

Basically we work with the problem (5.5.8), which is equivalent to the problem (5.5.5). In order to ensure that the problem is well-posed, we restrict the range of $u_x(t,x)$ to the set on which $\sigma' > 0$. To this end, we introduce a constant $k_0 > 0$ such that

(5.5.17) $$\sigma'(w) \geq p_0 > 0 \quad \text{for } w \in [-k_0, k_0].$$

Define
$$W(\eta) = \int_0^\eta \sigma(\xi)\,d\xi.$$

Then it is clear from (5.5.17) that

(5.5.18) $$W(\eta) \geq \tfrac{1}{2}p_0\eta^2 \quad \text{for } \eta \in [-k_0, k_0].$$

Finally, we say that a quantity is "sufficiently small" if it can be made arbitrarily small by a proper choice of the initial functions $u_0(x)$ and $u_1(x)$ and is the forcing function $\Phi(t,x)$ is appropriately small in the L^2-norm. For example, in view of the hypotheses $(H_2)-(H_4)$ and (5.5.9), the $L^2(R^+, L^2(R))$-norms of $\Phi, \Phi_t, \Phi_x, \Phi_{tt}, \Phi_{tx}$ and the $L^2(R)$-norms of u_1, u_{1x}, u_{1xx} are "sufficiently small". Our strategy here is to show there exists a positive number $\mu(0 < \mu < k_0)$ depending on β, γ, q (see Lemma 5.5.2), p_0 (see (5.5.18)), the bounds of $|\sigma'(\cdot)|, |\sigma''(\cdot)|$ and $|\sigma'''(\cdot)|$ on $[-k_0, k_0]$, and the $L^1(0, \infty)$-norm of $k'(t)$ such that if the local solution $u(t,x)$ of (5.5.8) in the sense of Theorem 5.5.1 satisfies the estimates

(5.5.19) $$|u_x(t,x)|, |u_{tx}(t,x)|, |u_{xx}(t,x)| \leq \mu$$

for $0 \leq t < T$ and $x \in R$ then certain functionals of the solution are "sufficiently small". Thus our main approach to the problem of investigating the asymptotic behavior of solutions of (5.5.5) is to establish "energy" estimates of the form

(5.5.20)
$$E(t) - E(0) \leq \int_0^t \int_{-\infty}^\infty Q[u,u]\,dx\,ds + \int_0^t \int_{-\infty}^\infty P[u,u]\,dx\,ds$$
$$+ \int_0^t \int_{-\infty}^\infty H[u,\Phi]\,dx\,ds,$$

where

(i) $E(t)$ is an "energy" that controls the growth of the solution;

(ii) $Q[u,u]$, the dissipative term induced by the memory term, is a positive definite quadratic form in a set of derivatives of $u(t,x)$;

(iii) $P[u,u]$, the remainder term due to the nonlinearity of the problem, is a quadratic form in the same derivatives as $Q[u,u]$

APPLICATIONS

and with coefficients that are small whenever the "energy" E is small;

(iv) $H[u,\Phi]$ is a bilinear form in the set of derivatives of $u(t,x)$ involved in $Q[u,u]$ and in $\Phi(t,x)$ and some of its derivatives.

The basic idea here is that, as long as $E(t)$ is small, $P[u,u]$ is dominated by $-Q[u,u]$. Moreover, the Cauchy–Schwartz inequality allows us to dominate the u part in $H[u,\Phi]$ by $-Q[u,u]$. Then if $E(0)$ and Φ are small, (5.5.20) shows that $E(t)$ remains sufficiently small for large t. To this end, we shall state and prove the following main result.

Theorem 5.5.2 *Let the hypotheses* $(H_1)-(H_5)$ *hold. If the* $L^1(R^+,L^2(R))$-*norms of* g *and* g_t, *the* $L^2(R^+,L(R))$-*norms of* g_x, g_{tt} *and* g_{tx} *and the* $L^2(R)$-*norms of* $u_{0x}, u_{0xx}, u_{0xxx}, u_1, u_{1x}$ *and* u_{1xx} *are sufficiently small then there exists a unique global solution* $u(t,x) \in C^2(R^+ \times R)$ *of the IVP* (5.5.5) *with the following properties:*

(a) $u_t, u_x, u_{tt}, u_{tx}, u_{xx}, u_{ttt}, u_{ttx}, u_{txx}, u_{xxx} \in L^\infty(R^+, L^2(R))$;

(b) $u_{tt}, u_{tx}, u_{xx}, u_{ttt}, u_{ttx}, u_{txx}, u_{xxx} \in L^2(R^+, L^2(R))$;

(c) $u_{tt}(t,\cdot), u_{tx}(t,\cdot), u_{xx}(t,\cdot) \to 0$ as $t \to \infty$ in $L^2(R)$;

(d) $u_t(t,x), u_x(t,x), u_{tt}(t,x), u_{tx}(t,x), u_{xx}(t,x) \to 0$ as $t \to \infty$ uniformly for all $x \in R$.

Proof Introduce a constant k_0 such that (5.5.17) and (5.5.18) hold. It follows from Theorem 5.5.1 that there exist a unique local solution $u(t,x)$ of the IVP (5.5.5) satisfying the estimates (5.5.19) for some T, $0 < T \leq \infty$, and a small positive number μ ($0 < \mu < k_0$). Since the initial value problems (5.5.5) and (5.5.8) are equivalent, for convenience, we derive all our results for the solutions of (5.5.8). Multiplying (5.5.8) by $u_t(t,x)$ and integrating over $[0,s] \times R$, $0 < s < T$, integration by parts with respect to x and (5.5.10) leads to

$$\frac{1}{2}\int_{-\infty}^{\infty} u_t^2(s,x)\,dx + \int_{-\infty}^{\infty} W(u_x(s,x))\,dx \leq \frac{1}{2}\int_{-\infty}^{\infty} u_t^2(0,x)\,dx + \int_{-\infty}^{\infty} W(u_x(0,x))\,dx$$

$$+ \int_0^s \int_{-\infty}^{\infty} \Phi u_t\,dx\,dt.$$

Thus, using the fact that

$$\int_0^s \int_{-\infty}^\infty \Phi u_t \, dx \, dt \geq \tfrac{1}{4} \max_{[0,s]} \int_{-\infty}^\infty u_t^2(t,x) \, dx + \left[\int_0^\infty \left(\int_{-\infty}^\infty \Phi^2 \, dx \right)^{1/2} dt \right]^2$$

and the inequality (5.5.18), we obtain

(5.5.21) $\quad \tfrac{1}{2} \int_{-\infty}^\infty u_t^2(s,x) \, dx + p_0 \int_{-\infty}^\infty u_x^2(s,x) \, dx$

$$\leq \int_{-\infty}^\infty u_1^2(x) \, dx + 2 \int_{-\infty}^\infty W(u_{0x}(x)) \, dx + 2 \left[\int_0^\infty \left(\int_{-\infty}^\infty \Phi^2 \, dx \right)^{1/2} \right]^2, \quad 0 \leq s \leq T.$$

From the hypotheses of the theorem and (5.5.9), it is easy to see that the $L^1(R^+, L^2(R))$-norm of Φ is sufficiently small. Hence, (5.5.21) yields that

$$\int_{-\infty}^\infty u_t^2(s,x) \, dx \quad \text{and} \quad \int_{-\infty}^\infty u_x^2(s,x) \, dx$$

are sufficiently small, uniformly in $[0, T]$.

In the next estimate, we make use of the dissipative mechanism introduced by the inequality (5.5.16). To this end, we consider the equivalent form (5.5.14) of the original problem (5.5.5).

Differentiating (5.5.14) with respect to t and multiplying the resulting equation, first by $u_{tt}(t,x)$ and then by $u_t(t,x)$, and integrating over $[0,s] \times R$, $0 < s < T$, we get the following two equations:

(5.5.22) $\quad \tfrac{1}{2} \int_{-\infty}^\infty u_{tt}^2(s,x) \, dx + \tfrac{1}{2} \int_{-\infty}^\infty \sigma'(u_x(s,x)) u_{tx}^2(s,x) \, dx$

$$+ \int_0^s \int_{-\infty}^\infty u_{tt}(r * u_{tt})_t \, dx \, dt - \beta \int_0^s \int_{-\infty}^\infty \sigma'(u_x) u_{tx}^2 \, dx \, dt$$

$$= \tfrac{1}{2} \int_{-\infty}^\infty u_{tt}^2(0,x) \, dx + \tfrac{1}{2} \int_{-\infty}^\infty \sigma'(u_x(0,x)) u_{tx}^2(0,x) \, dx$$

$$+ \int_0^s \int_{-\infty}^\infty \tfrac{1}{2} \sigma''(u_x) u_{tx}^3 \, dx \, dt - \beta \int_{-\infty}^\infty \sigma(u_x(s,x)) u_{tx}(s,x) \, dx$$

$$+ \beta \int_{-\infty}^{\infty} \sigma(u_x(0,x))u_{tx}(0,x)\,dx + \int_0^s \int_{-\infty}^{\infty} \Psi_t u_{tt}\,dx\,dt,$$

(5.5.23) $\quad \dfrac{\beta}{2a_\infty} \displaystyle\int_{-\infty}^{\infty} u_t^2(s,x)\,dx + \beta \int_{-\infty}^{\infty} W(u_x(s,x))\,dx$

$$- \int_0^s \int_{-\infty}^{\infty} u_{tt}(R*u_{tt})\,dx\,dt - \int_0^s \int_{-\infty}^{\infty} u_{tt}^2\,dx\,dt$$

$$+ \int_0^s \int_{-\infty}^{\infty} \sigma'(u_x)u_{tx}^2\,dx\,dt$$

$$= \dfrac{\beta}{2a_\infty} \int_{-\infty}^{\infty} u_t^2(0,x)\,dx + \beta \int_{-\infty}^{\infty} W(u_x(0,x))\,dx$$

$$- \int_{-\infty}^{\infty} u_t(s,x)u_{tt}(s,x)\,dx + \int_{-\infty}^{\infty} u_t(0,x)u_{tt}(0,x)\,dx$$

$$- \int_{-\infty}^{\infty} u_t(s,x)(R*u_{tt})(s,x)\,dx + \int_0^s \Psi_t u_t\,dx\,dt.$$

Multiplying (5.5.22) by q (see Lemma 5.5.2), adding it to (5.5.23) and then using (5.5.16), we obtain

(5.5.24) $\quad \dfrac{\beta}{2a_\infty} \displaystyle\int_{-\infty}^{\infty} u_t^2(s,x)\,dx + \beta \int_{-\infty}^{\infty} W(u_x(s,x))\,dx$

$$+ \tfrac{1}{2}q \int_{-\infty}^{\infty} u_{tt}^2(s,x)\,dx + \tfrac{1}{2}q \int_{-\infty}^{\infty} \sigma'(u_x(s,x))u_{tx}^2(s,x)\,dx$$

$$+ \gamma \int_0^s \int_{-\infty}^{\infty} u_{tt}^2\,dx\,dt + (1-q\beta) \int_0^s \int_{-\infty}^{\infty} \sigma'(u_x)u_{tx}^2\,dx\,dt$$

$$\leq \dfrac{\beta}{2a_\infty} \int_{-\infty}^{\infty} u_t^2(0,x)\,dx + \beta \int_{-\infty}^{\infty} W(u_x(0,x))\,dx$$

$$+ \tfrac{1}{2}q \int_{-\infty}^{\infty} u_{tt}^2(0,x)\,dx + \tfrac{1}{2}q \int_{-\infty}^{\infty} \sigma'(u_x(0,x))u_{tx}^2(0,x)\,dx$$

$$- \int_{-\infty}^{\infty} u_t(s,x)u_{tt}(s,x)\,dx + \int_{-\infty}^{\infty} u_t(0,x)u_{tt}(0,x)\,dx$$

$$- q\beta \int_{-\infty}^{\infty} \sigma(u_x(s,x))u_{tx}(s,x)\,dx$$

$$+ q\beta \int_{-\infty}^{\infty} \sigma(u_x(0,x))u_{tx}(0,x)\,dx - \int_{-\infty}^{\infty} u_t(s,x)(R*u_{tt})(s,x)\,dx$$

$$+ q \int_0^s \int_{-\infty}^{\infty} \tfrac{1}{2}\sigma''(u_x)u_{tx}^3\,dx\,dt + \int_0^s \int_{-\infty}^{\infty} \Psi_t u_t\,dx\,dt$$

$$+ q \int_0^s \int_{-\infty}^{\infty} \Psi_t u_{tt}\,dx\,dt.$$

In view of the fact (see (5.5.8)) that

$$u_{tt}(0,x) = \sigma(u_{0x}(x))_x + \Phi(0,x) - k(0)u_1(x)$$

and the hypotheses, it is clear that the $L^2(R)$-norm of $u_{tt}(0,x)$ is sufficiently small.

Further, as long as (5.5.19) holds for sufficiently small μ, we have the following estimates:

$$(5.5.25) \qquad -\int_{-\infty}^{\infty} u_t(s,x)u_t(s,x)\,dx \le \tfrac{1}{q}\int_{-\infty}^{\infty} u_t^2(s,x)\,dx + \tfrac{1}{4}q \int_{-\infty}^{\infty} u_{tt}^2(s,x)\,dx,$$

$$(5.5.26) \qquad -\int_{-\infty}^{\infty} u_t(s,x)(R*u_{tt})(s,x)\,dx$$

$$\le \int_{-\infty}^{\infty} |u_t(s,x)| \left[\int_0^s R^2(\tau)\,d\tau\right]^{1/2}\left[\int_0^s u_{tt}^2(\tau,x)\,d\tau\right]^{1/2} dx$$

$$\leq \tfrac{1}{\gamma} \sup_{t \in R^+} |R(t)| \left[\int_0^\infty |R(t)|\, dt \right] \left[\int_{-\infty}^\infty u_t^2(s,x)\, dx \right] + \tfrac{1}{4}\gamma \int_0^s \int_{-\infty}^\infty u_{tt}^2\, dx\, dt;$$

(5.5.27) $\quad q \int_0^s \int_{-\infty}^\infty \tfrac{1}{2}\sigma''(u_x) u_{tx}^3\, dx\, dt \leq \tfrac{1}{2}\mu q \max_{[-k_0, k_0]} |\sigma''(\cdot)| \int_0^s \int_{-\infty}^\infty u_{tx}^2\, dx\, dt;$

(5.5.28) $\quad \int_0^s \int_{-\infty}^\infty \Psi_t u_t\, dx\, dt \leq \tfrac{1}{4}\max_{[0,s]} \int_{-\infty}^\infty u_t^2(t,x)\, dx + \left[\int_0^s \left(\int_{-\infty}^\infty \Psi_t^2\, dx \right)^{1/2} dt \right]^2,$

(5.5.29) $\quad q \int_0^s \int_{-\infty}^\infty \Psi_t u_{tt}\, dx\, dt \leq \tfrac{q^2}{\gamma} \int_0^s \int_{-\infty}^\infty \Psi_t^2\, dx\, dt + \tfrac{1}{4}\gamma \int_{-\infty}^\infty u_{tt}^2\, dx\, dt.$

Moreover, for sufficiently small μ, the inequalities (5.5.18) and (5.5.19) yield

(5.5.30) $$\mu q \max_{[-k_0, k_0]} |\sigma''(\cdot)| \leq (1 - q\beta) p_0.$$

From the estimates (5.5.25) − (5.5.30), it follows, in view of (5.5.24), that

$$\int_{-\infty}^\infty u_{tt}^2(s,x)\, dx, \quad \int_{-\infty}^\infty u_{tx}(s,x)\, dx, \quad \int_0^s \int_{-\infty}^\infty u_{tt}^2\, dx\, dt$$

and

$$\int_0^s \int_{-\infty}^\infty u_{tx}^2\, dx\, dt$$

are "sufficiently small", uniformly on $[0, T]$.

Similarly, it is easy to verify (see Dafermos and Nohel [1]) that

$$\int_{-\infty}^\infty u_{xx}^2(s,x)\, dx \quad \text{and} \quad \int_0^s \int_{-\infty}^\infty u_{xx}^2\, dx\, dt$$

are "sufficiently small", uniformly on $[0, T]$. To obtain the estimates of higher-order derivatives such as $u_{ttt}, u_{ttx}, u_{txx}$ and u_{xxx}, we follow exactly the same procedure as in Dafermos and Nohel [1] (pp. 259 − 263) by differentiating (5.5.14) with respect to t and x, and finally show that

$$\int_{-\infty}^{\infty} u_{ttx}^2(s,x)\,dx, \quad \int_{-\infty}^{\infty} u_{txx}^2(s,x)\,dx, \quad \int_{-\infty}^{\infty} u_{xxx}^2(s,x)\,dx,$$

$$\int_{-\infty}^{\infty} u_{ttt}^2(s,x)\,dx, \quad \int_0^s\int_{-\infty}^{\infty} u_{ttx}^2\,dx\,dt, \quad \int_0^s\int_{-\infty}^{\infty} u_{txx}^2\,dx\,dt,$$

$$\int_0^s\int_{-\infty}^{\infty} u_{xxx}^2\,dx\,dt \quad \text{and} \quad \int_0^s\int_{-\infty}^{\infty} u_{ttt}^2\,dx\,dt$$

are "sufficiently small" for small μ.

Thus, in view of all the above estimates of the integrals of the derivatives of $u(t,x)$, a proper choice of initial data and the forcing function, as long as (5.5.19) holds, it follows that

(5.5.31) $\quad \displaystyle\int_{-\infty}^{\infty} \left(u_t^2 + u_x^2 + u_{tt}^2 + u_{tx}^2 + u_{xx}^2 + u_{ttt}^2 + u_{ttx}^2 + u_{txx}^2 + u_{xxx}^2\right) dx$

$$+ \int_0^s \left[\int_{-\infty}^{\infty}\left(u_{tt}^2 + u_{tx}^2 + u_{xx}^2 + u_{ttt}^2 + u_{ttx}^2 + u_{txx}^2 + u_{xxx}^2\right) dx\right] dt \le \mu^2$$

for $0 \le s \le T$.

Since (5.5.31) implies (5.5.19), we conclude, by Theorem 5.5.1, that the maximal interval of existence of $u(t,x)$ is R^+ and that (5.5.31) is satisfied for $s \in R^+$. In particular, assertions (a) and (b) of the theorem hold. The result (c) follows from (a) and (b), and finally (d) holds as a consequence (a) and (c). This completes the proof of the theorem. □

Remark 5.5.3 If $a(t) = \frac{1}{2}(1+e^{-t})$, then (5.5.5) reduces to Cauchy problem

(5.5.32) $\quad \begin{cases} u_{ttt} + u_{tt} = \sigma(u_x)_{tx} + \frac{1}{2}\sigma(u_x)_x + g + g_t, \\ u(0,x) = u_0(x), \; u_t(0,x) = u_1(x), \; u_{tt}(0,x) = \sigma(u_{0x}(x)) + g(0,x). \end{cases}$

It is clear that Theorem 5.5.2 can be applied to the IVP (5.5.32).

Remark 5.5.4 A mathematical model for heat flow in unbounded one dimensional bodies of materials with memory of the form

(5.5.33) $\quad u_t(t,x) = \displaystyle\int_0^t a(t-s)\sigma(u_x(s,x))_x\,ds + f(t,x)$

APPLICATIONS

is a special case of (5.5.5) with $a(0) = 1$, $g(t,x) = f_t(t,x)$ and $f(0,x) = u_1(x)$. Further, if $a(t) \equiv 1$ then both (5.5.5) and (5.5.33) reduce to a nonlinear damped equation

(5.5.34) $\quad u_{tt} = \sigma(u_x)_x + g(t,x), \quad u(0,x) = u_0(x), \quad u_t(0,x) = u_1(x).$

If $g \equiv 0$ then it is known (see Lax [1]) that the IVP (5.5.34) does not generally have global smooth solutions, no matter how smooth the initial data are. However, for the wave equation with frictional damping

$$u_{tt} + u_t = \sigma(u_x)_x,$$

it is shown in Nishida [1] that when the initial values are small, the dissipation precludes the development of shocks, and as a result global smooth solutions exist. In MacCamy [2, 3], by using energy integrals with a Riemann-invariants argument, it was shown under natural conditions on $a(t)$ (see, the hypothesis (H_4)), that the memory terms in (5.5.5) and (5.5.33) induce dissipative mechanisms that guarantee the existence of global smooth solutions whenever the initial values and the forcing function are sufficiently small.

Remark 5.5.5 It is clear (see Chapter 4) that the initial value problem (5.5.5) can be written in the abstract form

$$(5.5.35) \begin{cases} u''(t) + a(0)Au(t) + \int_0^t a'(t-s)Au(s)\,ds = F(t), & 0 < t < \infty, \\ u(0) = u_0, \quad u'(0) = u_1, \end{cases}$$

where A is a nonlinear maximal monotone operator in a Hilbert space H, $F(t)$ takes values in H while $u(t)$ takes values in a reflexive Banach space dense in H. For the IVP (5.5.35), global existence results are discussed in Londen [1, 2] in the case where the kernel $a(t)$ is positive, smooth, decreasing and convex on R^+ with $a'(0^+) = -\infty$. However, the condition $a'(0^+) = -\infty$ is not generally satisfied for memory functionals occurring in (5.5.5) or (5.5.33), and thus the results of Londen cannot be applied to these equations.

Remark 5.5.6 It follows from Theorem 5.5.1 and a proof similar to that of Theorem 5.5.2 (see Dafermos [2], Remark 4.1) that the solution $u(t,x)$

of (5.5.33) has a finite speed of propagation.

5.6 Large-Scale Systems

In this section, we shall discuss the L_2-stability and asymptotic stability in the sense of Lyapunov for a class of large-scale dynamical systems (also known as composite systems, interconnected systems and multiloop systems) described by Volterra integro-differential equations. We also show that when a large-scale system is bounded-input bounded-output stable ($BIBO$ stable) on L_2, it is asymptotically stable in the sense of Lyapunov. The results presented in this section also cover multiple-input and multiple-output systems ($MIMO$ systems) with interconnections.

Let $C = (C_{ij})$ be an $m \times n$ matrix and let C^T be the transpose of C. If C and D are real $m \times n$ matrices then by $C \leq D$ we mean $C_{ij} \leq D_{ij}$ for all i and j. Let $L_p = L_p^n$ be the set of all Lebesgue-measurable functions $f: R^+ \to R^n$ such that

$$\|f\|_p + \left[\int_0^\infty |f(t)|^p\, dt\right]^{1/p} \quad \text{is finite.}$$

When $p = 2$, L_2 is the Hilbert space with inner product

$$(f, g) = \int_0^\infty f^T(t) g(t)\, dt,$$

and

$$(f, f) = \|f\|_2^2.$$

Given a subspace $X \subset L_2$, $X^\perp = \{f: (f, g) = 0 \text{ for all } g \in X\}$. Also, let L_∞ be the space of essentially bounded functions $f: R^+ \to R^n$ and let

$$\|f\|_\infty = \operatorname*{ess\,sup}_{t \geq 0} |f(t)|.$$

Let $C(R^+)$ be the set of all continuous functions $f: R^+ \to R^n$ and

$$C_0 = \{f \in C(R^+): f(t) \to 0 \text{ as } t \to \infty\}.$$

For any given function $f \in L_p$, we shall denote the truncation of f at time T by f_T; that is,

$$f_T(t) = \begin{cases} f(t) & \text{for } 0 \le t \le T \\ 0 & \text{for } t > T. \end{cases}$$

Also, let L_{p_e} be the extended space of L_p (see Zames [1, 2]), in the sense

$$L_{p_e} = \{f : f_T \in L_p \text{ for all } T > 0\}.$$

Thus it is clear that L_{p_e} is the space of all locally L_p-functions $f(t)$ defined on R^+.

Definition 5.6.1

(i) The operator $H: L_{2e}^k \to L_{2e}^l$ is said to be L_2-stable if H maps L_2^k into L_2^l.

(ii) The gain of H, written as $g(H)$, is the smallest number μ such that

$$\|(Hx)_T\|_2 \le \mu \|x_T\|_2$$

for all $x \in L_{2e}^k$ and all $T > 0$.

(iii) If the stable set of H,

$$S(H) = \{x \in L_2^k : Hx \in L_2^l\},$$

is a proper subset of L_2^k then H is said to be L_2-unstable.

(iv) The conditional gain of H, written as $g_c(H)$, is the smallest number μ such that

$$\|(Hx)_T\|_2 \le \mu \|x_T\|_2$$

for all $T > 0$ and all x in the stable set $S(H)$.

In the special case where $S(H) = L_2$, H is stable and $g_c(H) = g(H)$. H is the interior conic (C, r) if

$$\|(Hx)_T - Cx_T\| \le r \|x_T\|$$

for some real constant $r \ge 0$ and some matrix C.

Let $\mathcal{F}_e$ denote the class of all linear time-invariant operators on L_{p_e} having the following properties. If $H \in \mathcal{F}_e$ then there is a function $h \in L_{1e}$ and two sequences $\{h_i\}$ and $\{t_i\}$ such that $t_i < t_{i+1}$ with $t_1 = 0$, $t_i \to \infty$ as $i \to \infty$ and

(5.6.1) $$(Hx)(t) = \sum_{i=1}^{n} h_i x(t-t_i) + \int_0^t h(s) x(t-s) \, dx$$

for all $x \in L_{p_e}$. The class $\mathcal{F}$ will consist of all $H \in \mathcal{F}_e$ such that the corresponding function $h(t)$ and the sequence $\{h_i\}$ satisfy the conditions

(5.6.2) $$\sum_{i=1}^{\infty} |h_i| < \infty, \qquad \int_0^{\infty} |h(t)| \, dt < \infty.$$

Let $\widehat{H}(s)$ and $\widehat{h}(s)$ be the Laplace transforms of the operator H and of $h(t)$ respectively, where $s = \sigma + j\tau$, $\sigma, \tau \in R$, $j = \sqrt{-1}$ (see Akcasu, Lillonehe and Shotkin [1]). If $H \in \mathcal{F}$, then the representation $\widehat{H}(s)$ is guaranteed to converge for all s with $\operatorname{Re} s = \sigma \geq 0$. The function $\widehat{H}(j\omega)$ is essentially the Fourier transform or the frequency response of H. The resolvent of H, denoted by R, is an operator in $\mathcal{F}_e$ and is given by

$$\widehat{R}(s) = [sI - \widehat{H}(s)]^{-1}.$$

For a given $H \in \mathcal{F}$, we assume the following properties of H:

(P_1) $\det[j\omega I - \widehat{H}(j\omega)] \neq 0$ for all $\omega \in R$;

(P_2) $\det[sI - \widehat{H}(s)] \neq 0$ for all s such that $\operatorname{Re} s \geq 0$.

It is known (see Grossman and Miller [2] and Jordon and Wheller [1]) that if $h_i = 0$ for all $i > 1$ then H has the property (P_2) if and only if the resolvent $R \in \mathcal{F}$. Further, H has the property (P_1) if and only if there is a finite sequence of points $\{s_1, s_2, \ldots, s_N\}$ in the half-plane $\operatorname{Re} s > 0$ such that

$$\det[s_i I - \widehat{H}(s_i)] = 0, \qquad i = 1, 2, \ldots, N.$$

For such an H, one can find a set of matrices M_{jk}, nonnegative integers N_j and an operator $S \in \mathcal{F}$ such that

(5.6.3) $$\widehat{R}(s) = \widehat{S}(s) + \sum_{i=1}^{N} \sum_{k=0}^{N_j} M_{jk}/(s-s_j)^k, \qquad \operatorname{Re} s_j > 0.$$

The operator S is called the residual resolvent of H. If $R \in \mathcal{F}$ then there is no such s_j, and thus in this case the resolvent and the residual resolvent are identical. The resolvent R and the residual resolvent S are assumed to be of the special forms

$$(Rx)(t) = \int_0^t r(t-s)x(s)\,ds,$$

$$(Sx)(t) = \int_0^t s(t-\tau)x(\tau)\,d\tau$$

respectively. The matrix functions $r(t) = (r_{ij}(t))$ and $s(t) = (s_{ij}(t))$ are called the kernels of the operators R and S respectively. If $H \in \mathcal{F}$ and has the property (P_2) (or (P_1)) then

$$r_{ij} \in C_0 \cap L_2 \quad (\text{or } s_{ij} \in C_0 \cap L_2)$$

for all i and j.

It is well known (see Desoer and Vidyasagar [1]) that the gain $g(H)$ on the space L_2 is

$$g(H) = \operatorname*{ess\,sup}_{-\infty < \omega < \infty} |\widehat{H}(j\omega)|.$$

If $H \in \mathcal{F}$ and has the property (P_2) then it follows by the same reasoning that the gain of the resolvent operator R is

$$g(R) = \operatorname*{ess\,sup}_{-\infty < \omega < \infty} |(j\omega I - \widehat{H}(j\omega))^{-1}|.$$

Definition 5.6.2 An operator H is said to be regular if $H \in \mathcal{F}_e$ such that the corresponding sequence $\{h_i\}$ and $h_i = 0$ for all $i > 1$.

If H is regular then the resolvent kernel $r(t) = (r_{ij}(t))$ has continuously differentiable elements $r_{ij}(t)$ on R^+ (see Grossman and Miller [2]). Also, $r'_{ij} \in L_1 \cap L_2 \cap C_0$ when H and $R \in \mathcal{F}$.

We shall now consider a linear integro-differential equation

(5.6.4) $\qquad y'(t) = Hy(t) + f(t), \quad y(0) = y_0, \quad t \in R^+,$

and a nonlinear integro-differential equation

(5.6.5) $\qquad x'(t) = Hx(t) + B(t, x_t) + f(t), \quad x(0) = x_0, \quad t \in R^+,$

where $H \in \mathcal{F}_e$, $f \in L_{2e}$, x_t is the truncation of x at time t and B is a continuous functional from $R^+ \times C(R^+)$ into R^n.

By a solution $x(t)$ of (5.6.5) we mean a function that is absolutely continuous on any finite interval $0 \le t \le T$ with derivative $x'(t)$ satisfying (5.6.5) with $x(0) = x_0$ at almost all points $t \in R^+$. Since $H \in \mathcal{F}_e$ and $f \in L_{2e}$, it is clear (see Grossman and Miller [2]) that the linear equation (5.6.4) has a unique solution $y(t)$ and is given by

$$(5.6.6) \qquad y(t) = r(t)y_0 + \int_0^t r(t-s)f(s)\,ds,$$

where $r(t)$ is the kernel of the resolvent R.

The equation (5.6.6) can be written in the abstract form

$$(5.6.7) \qquad y = R[\delta y_0 + f],$$

where δ is the delta function.

Treating $B(t, x_t) + f(t)$ as the forcing function, it follows from the linear variation of parameters formula that the unique solution $x(t)$ of (5.6.5) with $x(0) = x_0$ (under a Lipschitz condition on $B(t, x_t)$) is given by

$$(5.6.8) \qquad x = R[\delta x_0 + f] + R[B(t, x_t)].$$

Remark 5.6.1 Suppose that $r(t)$ is the resolvent kernel of the regular resolvent R (see Definition 5.6.2) and B has the regular form

$$(5.6.9) \qquad B(t, x_t) = B_1(t, x(t)) + \int_0^t h(t-s)B_2(s, x(s))\,ds.$$

Since

$$r'(t) = r(t)h_1 + \int_0^t r(t-s)h(s)\,ds$$

$$= r(t)h_1 + (r*h)(t),$$

where $*$ is the convolution, it follows that

$$(r*h)(t) = r'(t) - r(t)h_1.$$

Hence the last term on the right-hand side of (5.6.8) can be written as

$$R[B_1 + h*B_2] = R[B_1] + r*h*B_2$$
$$= R[B_1 - h_1 B_2] + r'*B_2.$$

Thus (5.6.8) can be put into the alternative form

(5.6.10) $x = R[\delta x_0 + f] + R[B_1(t, x(t)) - h_1 B_2(t, x(t))] + r'*B_2(t, x(t))$.

We shall now list a number of definitions regarding the stability of (5.6.5).

Definition 5.6.3

(i) Equation (5.6.5) is said to be L_2-stable (i.e. *BIBO* stable on L_2) if for each $f \in L_2$ and initial conditions $x_0 \in R^n$, all the solutions of (5.6.5) are in L_2; otherwise, (5.6.5) is called L_2-unstable.

(ii) Equation (5.6.5) is said to be stable in the sense of Lyapunov if for any $\epsilon > 0$ there is a $\delta = \delta(\epsilon) > 0$ such that $|x_0| \leq \delta$ and $\|f\|_2 \leq \delta$ imply $|x(t)| < \epsilon$ for all $t \geq 0$.

(iii) Equation (5.6.5) is said to be asymptotically stable in the sense of Lyapunov if it is stable and there is a $\delta_0 > 0$ such that $x(t) \in C_0$ whenever $|x_0| \leq \delta_0$ and $\|f\|_2 \leq \delta_0$.

Basically we shall consider, in this section, systems that can be described by a set of integro-differential equations

(5.6.11) $$x'_k(t) = \sum_{j=1}^{N} [H_{kj} x_j(t) + B_{kj}(t, x_{jt})] + f_k(t)$$

for $k = 1, 2, \ldots, N$ and $t \in R^+$ with initial conditions $x_k(0) = x_{0k}$, in which $x_k: R^+ \to R^{n_k}$, $f_k \in L_2^{n_k}$, $H_{kj}: L_{2e}^{n_l} \to L_{2e}^{n_k}$ is in $\mathcal{F}_e$, $x_{jt} = (x_j)_t$ is the truncation of x_j and B_{kj} is a continuous nonlinear functional. Letting $x = (x_1, x_2, \ldots, x_N)^T$, $f = (f_1, f_2, \ldots, f_N)^T$, $B = (\Sigma B_{1j}, \Sigma B_{2j}, \ldots, \Sigma B_{Nj})$ and $H = (H_{jk})$, then the system (5.6.11) assumes the form (5.6.5) with initial conditions $x(0) = (x_{10}, x_{20}, \ldots, x_{N0})^T \in R^m$, where $m = \sum_{j=1}^{N} n_j$. The matrices of gains $g(B)$ and $g(C)$ are defined as $g(B) = (g(B_{ij}))$ and $g(C) = (g(C_{ij}))$. When the system (5.6.5) is of the form (5.6.11), the system (5.6.5) is referred to as a large-scale system with decomposition (5.6.11).

We shall now give the main results on stability.

Theorem 5.6.1 *Assume that*

(i) $g(B_{kj}) < \infty$ for all k and j;

(ii) $H = (H_{ij}) \in \mathcal{F}$ has the property (P_2), with resolvent $R = (R_{ij})$;

(iii) all the successive principal minors of the test matrix $(I - g(R)g(B))$ are positive.

Then the system (5.6.11) is L_2-stable.

Proof By the variation of parameters formula (see (5.6.8)), the solution $x(t) = (x_1(t), x_2(t), \ldots, x_N(t))^T$ of (5.6.11) can be expressed as

$$(5.6.12) \qquad x(t) = r(t)x_0 + Rf(t) + RB(t, x_t)$$

for $t \in R^+$. For any $T > 0$, define

$$\|z\|_T = (\|x_{iT}\|_2)^T, \quad \|z(0)\| = (|x_{i0}(0)|)^T$$

and $\|F\|_2 = (\|f_i\|_2)^T$. Then it follows from (5.6.12) that

$$\|z\|_T \leq (\|r_{ij}\|_2) \|z_0\| + g(R) \|F\|_2 + g(R)g(B) \|z\|_T.$$

This implies that

$$(I - g(R)g(B)) \|z\|_T \leq (\|r_{ij}\|_2) \|z_0\| + g(R) \|F\|_2.$$

From the assumption (iii), it is clear that the test matrix $(I - g(R)g(B))$ has an inverse ρ whose entries are all nonnegative. Hence we have

$$(5.6.13) \qquad \|z\|_T \leq \rho[(\|r_{ij}\|_2) \|z_0\| + g(R) \|F\|_2].$$

Since (5.6.13) is true for all $T > 0$, the result follows and hence the proof is complete. $\square$

Theorem 5.6.2 Suppose that the following conditions are satisfied:
(i) B_{ij} is regular in the sense of (5.6.9);
(ii) $H = (H_{ij}) \in \mathcal{F}$ is regular, satisfies the property (P_2) and has resolvent $R = (R_{kj})$ with kernel (r_{kj});
(iii) $T = (T_{kj})$ is the operator defined by

$$T_{kj}x = r'_{kj} * x;$$

(iv) the successive principal minors of the test matrix $(I - g(R)g(B_1 - h_1 B_2) - g(T)g(B_2))$ are all positive.

Then the system (5.6.11) is L_2-stable.

Proof By the variation of parameters formula (5.6.10), we obtain as in Theorem 5.6.1 that

$$x(t) = r(t)x_0 + Rf(t) + R[B_1(t, x(t)) - h_1 B_2(t, x(t))] + r' * B_2(t, x(t))$$

for $t \in R^+$. Then for any $T > 0$, we have

$$\|z\|_T \le (\|r_{ij}\|_2)\|z_0\| + g(R)\|F\|_2$$
$$+ g(R)g(B_1 - h_1 B_2)\|z\|_T$$
$$+ g(r')g(B_2)\|z\|_T.$$

Using the assumptions (iii) and (iv), we obtain as in Theorem 5.6.1 that

$$\|z\|_T \le \rho_1[(\|r_{ij}\|_2)\|z_0\| + g(R)\|F\|_2]$$

for all $T > 0$, where ρ_1 is an inverse of the test matrix defined in the assumption (iv), whose entries are all nonnegative. Since the above inequality holds for all $T > 0$, the assertion of the theorem follows. This completes the proof. □

Theorem 5.6.3 *Suppose that all the assumptions of Theorem 5.6.1 are satisfied and H is regular (or the hypotheses of Theorem 5.6.2 hold). Then the system (5.6.10) is asymptotically stable in the sense of Lyapunov.*

Proof Since the proofs of Theorems 5.6.1 and 5.6.2 are essentially same, we shall follow the proof of Theorem 5.6.1.

From Schwartz's inequality, it follows that if $s, \phi \in L_2$ then the convolution

$$|(s*\phi)(t)| \le \|s\|_2 \|\phi\|_2,$$

and hence $s*\phi \in L_\infty$.

Moreover, it is clear that $r*\phi \in C_0$. Since R is L_2-stable, the kernel $r \in L_2 \cap C_0$. Thus (5.6.12) implies that

$$(5.6.14) \quad |x(t)| \le (\|r_{ij}\|_\infty)\|x_0\| + (\|r_{ij}\|_2)\|F\|_2 + (\|r_{ij}\|_2)g(B)\|z\|_T.$$

Therefore, from (5.6.13) and (5.6.14), we obtain

$$|x(t)| \le (\|r_{ij}\|_\infty)\|x_0\| + (\|r_{ij}\|_2)\|F\|_2$$
$$+ (\|r_{ij}\|_2)g(B)\rho[(\|r_{ij}\|_2)\|z_0\| + g(R)\|F\|_2].$$

This implies that (5.6.11) is stable in the sense of Lyapunov.

To show that $x \in C_0$, we use the fact that $r \in C_0$ and $s*\phi \in C_0$ whenever $s, \phi \in L_2$. Therefore (5.6.12) implies that $x \in C_0$ for all $x_0 \in R^m$ and all $f \in L_2$. This completes the proof. □

Remark 5.6.2 Suppose that $B_1 = h_1 B_2$ in (5.6.9). Then the term $g(B_1 - h_1 B_2)$ in Theorem 5.6.2 is zero. This case is interesting in itself, and is useful in some applications.

As a special case, we shall consider the system of equations described by (5.6.11) with $H_{ij} = 0$ for all $i \neq j$; that is, $MIMO$ systems of the form

$$(5.6.15) \qquad x'_k(t) = H_k x_k(t) + \sum_{j=1}^{N} B_{kj}(t, x_{jt}) + f_k(t),$$

$k = 1, 2, \ldots, N$, $x_k(0) = x_{k0}$. The equation (5.6.15) may be regarded as an interconnected system of N isolated subsystems of the form

$$(5.6.16) \qquad z'_k(t) = H_k z_k(t) + f_k(t), \qquad z_k(0) = z_{k0},$$

$k = 1, 2, \ldots, N$, with interconnecting structure specified by B_{kj}, $k, j = 1, 2, \ldots, N$.

The following result is an immediate consequence of Theorem 5.6.1.

Corollary 5.6.1 *For the system (5.6.14), assume that*
(i) $H_k \in \mathcal{F}$ *with stable resolvent* R_k, $k = 1, 2, \ldots, N$;
(ii) $g(B_{kj}) < \infty$ *for* $k, j = 1, 2, \ldots, N$;
(iii) $f_k \in L_2$, $k = 1, 2, \ldots, N$;
(iv) *the successive principal minors of the test matrix* $M = (m_{ij})$ *are all positive, where*

$$m_{ij} = \begin{cases} 1 - g(R_i)g(B_{ij}), & i = j \\ -g(R_i)g(B_{ij}), & i \neq j. \end{cases}$$

Then the system (5.6.15) is L_2-stable. Further, if H is regular then it is asymptotically stable in the sense of Lyapunov.

The next result yields sufficient conditions for L_2-instability of interconnected systems described by equations of the form

$$(5.6.17) \qquad \begin{cases} \xi_k = f_k - \sum_{j=1}^{N} B_{kj} x_j, \\ x_k = R_k(\xi_k + v_k), \quad k = 1, 2, \ldots, N \end{cases}$$

where $f_k, v_k \in L_2^{n_k}$, $\xi_k, x_k \in L_{2e}^{n_k}$, $R_k \in \mathcal{F}_e^{n_k}$ and $B_{kj}: L_{2e}^{n_l} \to L_{2e}^{n_k}$.

Define
$$X_k = \{x \in L_2^{n_k}: R_k x \in L_2^{n_k}\},$$

the stable manifold of R_k. Let $M = (m_{ij}) = (g(B_{ij}), g_c(R_j))$ and let $\rho(M)$ be the spectral radius of M.

Theorem 5.6.4 *Assume that*

(i) $(X_k)^\perp \neq 0$ *for at least one* $k = k_0$ *and* $g(B_{kj}) < \infty$ *for all k and j;*

(ii) $\rho(M) \leq 1$ *and* $m_{ij} > 0$ *for all i and j;*

(iii) $f_k + v_k \in (X_k)^\perp$ *for all k and* $f_k + v_k \neq 0$ *when* $k = k_0$.

Then the system (5.6.17) is L_2-unstable (that is, $x_j \notin L_2$ for at least one value of j).

Proof For the sake of contradiction, let us assume that $x_k \in L_2$ for all k. If $\rho(M) < 1$ then we replace each value $g_c(H_i)$ by $g_c(H_i) + \alpha$ and each value $g(B_{ij})$ by $g(B_{ij}) + \alpha$ where α is small and positive. If α is sufficiently small, the new matrix M will still have $\rho(M) < 1$ but not all the entries of (m_{ij}) will be positive. Let us assume that this replacement has been accomplished. Then $m_{ij} > 0$ for all i and j, and by Perron's theorem (see Fielder and Plak [1]), there exist numbers $\alpha_i > 0$ such that $\alpha = (\alpha_1, \alpha_2, \ldots, \alpha_m)$ is the row eigenvector corresponding to the dominant eigenvalue $\rho(M)$ of M.

Let
$$y_i = \sum_{j=1}^m B_{ij} x_j = \sum_{j=1}^m B_{ij} R_j (\xi_j + v_j).$$

Then it is easy to verify using (5.6.17) that
$$y_i = f_i + v_i - (\xi_i + v_i).$$

Since $f_i + v_i \in X_i^\perp$ and $\xi_i + v_i \in X_i$, it follows that $y_i \in L_2$ and
$$\|y_i\|_2^2 = \|f_i + v_i\|_2^2 + \|\xi_i + v_i\|_2^2$$
$$\geq \|\xi_i + v_i\|_2^2.$$

Further, since $\|f_i + v_i\| > 0$ for at least one $i = k_0$ and since α_i are all positive, it follows that

(5.6.18) $$\sum_{i=1}^m \alpha_i \|y_i\|_2 > \sum_{i=1}^m \alpha_i \|\xi_i + v_i\|_2.$$

From the definition of y_i and the assumptions of the theorem, we obtain an estimate

$$(5.6.19) \quad \| y_i \|_2 \leq \sum_{j=1}^{m} \| B_{ij} x_j \|_2$$

$$\leq \sum_{j=1}^{m} g(B_{ij}) g_c(R_j) \| \xi_j + v_j \|_2$$

$$= \sum_{j=1}^{m} m_{ij} \| \xi_j + v_j \|.$$

Since α is a row eigenvector of M, it follows that

$$\sum_{i=1}^{m} \alpha_i m_{ij} = \rho(M) \alpha_j$$

for all j. Thus the inequalities (5.6.19) and $\rho(M) \leq 1$ imply that

$$\sum_{j=1}^{m} \alpha_j \| y_j \|_2 \leq \sum_{i=1}^{m} \sum_{j=1}^{m} \alpha_i m_{ij} \| \xi_j + v_j \|_2$$

$$= \sum_{j=1}^{m} \rho(M) \alpha_j \| \xi_j + v_j \|_2$$

$$\leq \sum_{j=1}^{m} \alpha_j \| \xi_j + v_j \|_2,$$

which is a contradiction to (5.6.18). Hence $x_k \notin L_2$ for at least one value of k. This completes the proof. □

Corollary 5.6.2 *Suppose that the following conditions hold for the system (5.6.15):*

(i) $H_k \in \mathcal{F}$ for $k = 1, 2, \ldots, N$;

(ii) $g(B_{ij}) < \infty$, $i, j = 1, 2, \ldots, N$;

(iii) $f_j \in L_2$ for $j = 1, 2, \ldots, N$;

(iv) *for each k, H_k satisfies property (P_1) and/or property (P_2), and for at least one k_0 it satisfies property (P_1) only; and*

(v) $\rho(M) \leq 1$ *and* $m_{ij} > 0$ *for all* $i, j = 1, 2, \ldots, N$.

Then the system (5.6.15) is L_2-unstable.

Proof Equation (5.6.15) can be written in the equivalent form

$$(5.6.20) \quad \xi_k(t) = f_k(t) + \sum_{j=1}^{m} B_{kj}(t, x_{jt}),$$

(5.6.21) $$x'_k(t) = H_k x_k(t) + \xi_k(t),$$

$x_k(0) = x_{k0}, k = 1, 2, \ldots, N$. Choosing $x_{k0} = 0$ for all k and integrating (5.6.21), we obtain

(5.6.22) $$x_k = R_k \xi_k.$$

Equations (5.6.22) and (5.6.20) are of the form (5.6.17). Thus the application of Theorem 5.7.4 yields the stated result. This completes the proof. □

Example 5.6.1 Consider the system of integro-differential equations

(5.6.23) $$x'_k(t) = f_k(t) + \sum_{j=1}^{N}\left[G_{kj}(t, x_j(t)) + \int_0^t M_{kj}(t-s) N_{kj}(s, x_j(s)) \, ds \right],$$

where G_{kj} is the interior conic (m_{kj}, r_{kj}) and N_{kj} is the interior conic (n_{kj}, ω_{k_j}). Then (5.6.23) can be written as

$$x'_k(t) = f_k(t) + \sum_{j=1}^{N}\left[m_{kj} x_j(t) + \int_0^t M_{kj}(t-s) n_{kj} x_j(s) \, ds \right]$$

$$+ \sum_{j=1}^{N}\left[(G_{kj}(t, x_j(t)) - m_{kj} x_j(t)) \right.$$

$$\left. + \int_0^t M_{kj}(t-s)\left[N_{kj}(s, x_j(s)) - n_{kj} x_j(s) \right] ds \right].$$

Take $B_1 = (G_{kj}(t, x_j(t)) - m_{kj} x_j(t))$, $g(B_1) \leq (r_{ij})$, $B_2 = (M_{kj}*(N_{kj}(t, n_j(t)) - n_{kj} x_j(t))$, $g(B_2) \leq g(M_{kj})(T_{kj})$, $h_1 = (m_{kj})$ and $h(t) = (M_{kj}(t) n_{kj})$. Then, under the assumptions of Theorem 5.6.2, the system (5.6.23) is L_2-stable.

Example 5.6.2 Consider the point kinetics model for a reactor with N cores, as described in Plaza and Kohler [1], and given by

(5.6.24) $$p'_j(t) = \frac{\rho_j - \epsilon_j - \beta_j}{\Lambda_j} p_j(t) + \frac{\rho_j}{\Lambda_j}$$

$$+ \sum_{i=1}^{6} \frac{\beta_{ij}}{\Lambda_j} c_{ij}(t) + \frac{1}{\Lambda_j} \sum_{k=1}^{N} \epsilon_{jk} \frac{P_{k0}}{P_{j0}} \int_0^t h_{kj}(t-s) p_k(s) \, ds,$$

(5.6.25) $$c'_{ij}(t) = \lambda_{ij}[p_j(t) - c_{ij}(t)]$$

for $j = 1,2,\ldots, N$, $i = 1,2,\ldots,6$, in which
$$p_j(t) = \frac{P_j(t) - P_{j0}}{P_{j0}}, \qquad c_{ij}(t) = \frac{\lambda_{ij}\Lambda_j}{P_{j0}\beta_{ij}}(C_{ij} - C_{ij0}),$$

where P_j is the power in the jth core, C_{ij} is the effective concentration of the ith precursor in the jth core, $\beta_{ij}, \lambda_i, \epsilon_j, \epsilon_{jk}$ and Λ_j are positive constants, $\beta_j = \sum_{i=1}^{6} \beta_{ij}$ and $h_{kj}(t)$ is the coupling function relating to neutron migration from kth to the jth core; P_{j0} and C_{ij0} are the equilibrium power and precursor concentrations in the jth core, while ρ_j is the reactivity in the jth core. Assume that

$$\rho_j(t) = \int_0^t W_j(t-s) p_j(s)\, ds$$

is correct at least to linear terms, where the feedback function $W_j \in L_1$.

Solving (5.6.25) for $c_{ij}(t)$, we get

$$c_{ij}(t) = c_{ij}(0) e^{-\lambda_{ij} t} + \lambda_{ij}\left(e^{\lambda_{ij} t} * p_i\right).$$

Substituting $c_{ij}(t)$ into (5.6.24) and linearizing, we obtain

$$(5.6.26) \quad p'_j(t) = f_j(t) - \frac{\epsilon_j + \beta_j}{\Lambda_j} p_j(t) + \frac{W_j * p_j}{\Lambda_j}$$
$$+ \left(\sum_{i=1}^{6} \frac{\beta_j \lambda_{ij}}{\Lambda_j} e^{-\lambda_{ij} t}\right) * p_j$$
$$+ \frac{1}{\Lambda_j} \sum_{k=1}^{N} \epsilon_{kj} \frac{P_{k0}}{P_{j0}} (h_{kj} * p_k),$$

with $p_j(0)$, $j = 1,2,\ldots, N$, given.

Equation (5.6.26) can be written in the form of (5.6.15) with $x_j = p_j$ and $n_j = 1$ for all j (see, (5.6.5)), where

$$(H_j x)(t) = -\frac{\epsilon_j + \beta_j}{\Lambda_j} x(t) + \frac{1}{\Lambda_j} \int_0^t [W_j(t-s) + \sum_{i=1}^{6} \beta_{ij}\lambda_{ij} e^{-\lambda_{ij}(t-s)}] x(s)\, ds,$$

$$f_j(t) = \sum_{i=1}^{6} c_{ij}(0) e^{-\lambda_{ij} t},$$

$$B_{jk}(t, x_t) = \frac{\epsilon_{jk}}{\Lambda_j} \frac{P_{k0}}{P_{j0}} (h_{kj} * x)(t).$$

Thus Corollary 5.6.1 can be applied to (5.6.26). Indeed, the resolvent R_j has Laplace transform $\widehat{R}_j(s) = 1/D_j(s)$ and is stable if and only if

$$D_j(s) = s + \frac{\epsilon_j + \beta_j}{\Lambda_j} - \sum_{i=1}^{6} \frac{\beta_{ij}\lambda_{ij}}{\Lambda_j(s+\lambda_{ij})}$$

$$-\frac{\widehat{W}_j(s)}{\Lambda_j} \neq 0,$$

in the half-plane $\operatorname{Re} s \geq 0$. This condition can be verified through the graph of $D_j(s)$. Further, $1/g(R_j)$ is equal to the minimum distance from the graph of $D_j(j\omega)$, $-\infty < \omega < \infty$, to the origin in the complex plane. Therefore if the successive principal minors of the test matrix $M = (I - g(R_j)g(B_{ij}))$ are all positive then, for all initial values $p_j(0)$ and $c_{ij}(0)$, the solutions of (5.6.24) and (5.6.25) are in $L_2 \cap C_0$. Moreover, they are stable in the sense that, given $\epsilon > 0$, there exists a $\delta = \delta(\epsilon) > 0$ such that $|p_j(0)| \leq \delta$ and $|c_{ij}(0)| \leq \delta$ for all i and j imply

$$|p_j(t)| < \epsilon \quad \text{and} \quad |c_{ij}(t)| < \epsilon,$$

for all i and j and for all $t \geq 0$.

On the other hand, if $D_j(j\omega) \neq 0$ for $-\infty < \omega < \infty$ and $j = 1, 2, \ldots, N$ but $D_j(s_0) = 0$ for some s_0, $\operatorname{Re} s_0 > 0$ and some j, and if the matrix $M = (g(B_{ij})g_c(R_j))$ has spectral radius $\rho(M) < 1$ (i.e. weak interconnections between cores), then by Corollary 5.6.2, it follows that the system (5.6.26) is L_2-unstable. In view of the fact for this type of linear system, L_2-stability and asymptotic stability in the sense of Lyapunov are equivalent, it is clear that the system (5.6.26) is Lyapunov-unstable. Since the instability of the linearized system will carry over to the corresponding nonlinear system, it is obvious that the system (5.6.24) is unstable in this case.

5.7 Notes and Comments

In his study of growth of biological populations, Volterra [1, 2] postulated various mathematical models that are in the form of Volterra integro-differential equations with unbounded delay. The contents of Section 5.1 are taken from Miller [7]. An introduction to such work with references is found in Goel et al. [1], Goh [1], Clark [1], Cushing [2], Lotka [1], MacDonald [1],

Smith [1], Pearl [1] and Pielou [1]. For further study on management and analysis of prey − predator and competing problems, see Ahmad [1], Ahmad and Lazer [1], Ahmad and Rama Mohana Rao [1, 2], Brauer [3], Freedman and Sree Hari Rao [1], Kapur [1], Seifert [4], Stepan [1, 2] and the references therein.

Section 5.2 is based on the general model suggested by Rama Mohana Rao and Pal [1] for grazing grassland on the pattern of prey − predator systems. A nonlinear model is discussed using Krasovskii's method (cf. Hahn [1], p. 270 and Krasovskii [1]). Examples 1 − 3 illustrate the generality of the model. Special cases of this model are found in Noy and Meir [1, 2]. The results covered in Section 5.3 are taken from Desch, Grimmer and Schappacher [1], while Example 5.3.1 is from Davis [1]. Similar results are found in Desch and Grimmer [2], Coleman and Gurtin [1, 2], Gurtin and Pipkin [1], Grimmer and Zeman [1], and Staffans [1].

Section 5.4 deals with the dynamics of nuclear reactors. The first few papers to be devoted to this topic in the spirit of this section are those of Levin and Nohel [1 − 3, 5]; see also Corduneanu [1], Akcasu et al. [1], Nohel [1], and Plaza and Kohler [1]. More complicated systems describing various phenomena in reactor dynamics are discussed in Londen [1, 2]. The theory of nonlinear viscoelastic equations and the energy method presented in Section 5.5 are contained in Dafermos and Nohel [1, 2]. For a general description of viscoelastic models and their basic study, see Bloom [1], Christenen [1], Hrusa and Renardy [1], Dafermos [1, 2], MacCamy [1, 2, 3], Miller [8] and Staffans [1]. The material on large-scale systems covered in Section 5.6 is due to Miller and Michel [1]. The notion of *BIBO* stability is motivated by that of Zames [1, 2] for single loop systems while the Lyapunov stability is taken from Desoer and Vidyasagar [1], Lasley [1, 2] and Lasley and Michel [1]. Various contributions on the stability analysis of feedback and time-lag control systems are found in Willems [1], Corduneanu [1], Chen [1], Fielder and Ptak [1], Paley and Weiner [1], Nohel and Shea [1], Oğuztoreli [1], Bellman and Cooke [1], Halanay [1], Ostrowskii [1], Levin and Nohel [4] and the references therein.

REFERENCES

Aftabizadeh, A.R. and S. Leela
[1] Existence results for boundary value problems of integro-differential equations, *Proc. Colloquia on Qualitative Theory of Differential Equations* at Bolyai Institute, Szeged, Hungary 1984.

Ahmad, S.
[1] On almost periodic solutions of the competing species problems, *Proc. Amer. Math. Soc.* **102** (1988) 855 – 861.

Ahmad, S. and A.C. Lazer
[1] On the nonautonomous N-competing species problem, *J. Math. Anal. Appl.* (to appear).

Ahmad, S. and M. Rama Mohana Rao
[1] Asymptotically periodic solutions of N-competing species problem with time delays, (to appear).

[2] Stability criteria for N-competing species problem with time delays, (to appear).

Akcasu, Z., G.S. Lillonche and M. Shotkin
[1] *Mathematical Methods in Nuclear Reactor Dynamics*, Academic Press, New York 1971.

Alekseev, V.M.
[1] An estimate for perturbations of solutions of ordinary differential equations, *Vestnik Moskov, Univ. Ser. Mat. Mech.* **2** (1961) 28 – 36.

Ambrosetti, A. and G. Prodi
[1] On the inversion of some differentiable mappings with singularities between Banach spaces, *Ann. Math. Pura Appl.* **93** (1972) 231 – 247.

Angelova, D.Ts. and D.D. Bainov
[1] On the behavior of solutions of Volterra second order integro-differential equations, *Q.J. Math.* **39** (1988) 255 – 268.

Bainov, D.D. and A.B. Dishliev
[1] Sufficient conditions for absence of beating in systems of differential equations with impulses, *Applicable Anal.* **18** (1984) 67 – 73.

Barbu, V.
[1] Nonlinear Volterra equations in Hilbert space, *SIAM J. Math. Anal.* **6** (1975) 728 – 741.

Becker, L.C.
[1] *Stability Conditions for Volterra Integro-differential Equations*, Ph.D. Dissertation, Southern Illinois University, 1979.

Becker, L.C., Burton, T.A., and T. Krisztin
[1] Floquet theory for a Volterra equation, *J. Lond. Math. Soc.* **37** (1988) 141 – 147.

Beesack, P.R.
[1] On the variation of parameters methods for integro-differential, integral and quasilinear partial integro-differential equations, *J. Appl. Math. Comput.* **22** (1987) 189 – 215.

Bellman, R. and K.L. Cooke
[1] *Difference – Differential Equations*, Academic Press, New York, 1963.

Bernfeld, S.R. and V. Lakshmikantham
[1] *An Introduction to Nonlinear Boundary Value Problems*, Academic Press, New York, 1974.

Bernfeld, S.R. and M.E. Lord
[1] A nonlinear variation of constants method for integro-differential and integral equations, *J. Appl. Math. Comput.* **4** (1978) 1 – 14.

Bloom, F.
[1] *Ill-posed Problems for Integro-differential Equations in Mechanics and Electromagnetic Theory*, SIAM, Philadelphia, 1981.

Brauer, F.
[1] Asymptotic stability of a class of integro-differential equations, *J. Diff. Eqns* **28** (1975) 180 – 188.
[2] A nonlinear variation of constants formula for Volterra equations, *Math. Systems Theory* **6** (1972) 226 – 234.
[3] Stability of some population models with delay, *Math. Biosci.* **33** (1978) 345 – 358.

Brezis, H.
[1] Opérateurs maximaux monotones et semigroups de contractions dans les espaces de Hilbert, *Math. Stud.* **5** North-Holland, Amsterdam, 1973.

Burton, T.A.
[1] Periodic solutions of linear Volterra equations, *Funkcial. Ekvac.* **27** (1984), 229 – 253.
[2] *Volterra Integral and Differential Equations*, Academic Press, New York, 1983.
[3] An integro-differential equation, *Proc. Amer. Math. Soc.* **79** (1980) 303 – 399.
[4] Construction of Lyapunov functionals for Volterra equations, *J. Math. Anal. Appl.* **85** (1982) 90 – 105.
[5] Perturbed Volterra equations, *J. Diff. Eqns* **43** (1982) 168 – 183.
[6] Boundedness in functional differential equations, *Funkcial Ekavac.* **25** (1982) 51 – 57.

Burton, T.A. and L. Hatvani
[1] Stability theorems for nonautonomous functional differential equations, *Tokohu Math. J.* **41** (1989) 65 – 104.

Burton, T.A. and R.H. Hering
[1] Boundedness in infinite delay systems, *J. Math. Anal. Appl.* **144** (1989) 486 – 502.

Burton, T.A. and W.E. Mahfoud
[1] Stability criterion for Volterra equations, *Trans. Amer. Math. Soc.* **279** (1983) 143 – 174.

Cassago, Jr., H. and C. Corduneanu
[1] The ultimate behavior of certain nonlinear integro-differential equations, **9** (1985) 113 – 124.

Chen, G.
[1] Control and stabilization for the wave equation in a bounded domain, *SIAM J. Control* **17** (1979) 66 – 81.

Chen, G. and R. Grimmer
[1] Integral equations as evolution equations, *J. Diff. Eqns* **45** (1982) 53 – 74.
[2] Semi-groups and integral equations, *J. Integral Eqns* **2** (1980) 133 – 154.

Chen, Yu-Bo and W. Zhuang
[1] The existence of maximal and minimal solution of the nonlinear integro-differential equations in Banach space, *Applicable Anal.* **22** (1980) 139 – 147.

Christenen, R.M.
[1] *Theory of Viscoelasticity: An Introduction*, 2nd edn, Academic Press, New York, 1982.

Clark, C.W.
[1] *Mathematical Bioeconomics*, Wiley, New York, 1976.

Coleman, B.D. and M.E. Gurtin
[1] Waves in materials with memory II: On the growth and decay of one dimensional acceleration waves, *Arch. Rat. Mech. Anal.* **19** (1965) 239 – 265.
[2] Euipresence and constitutive equations for rigid heat conductors, *Z. Angew. Math. Phys.* **18** (1967) 199 – 208.

Coppel, W.A.
[1] *Stability and Asymptotic Behavior of Differential Equations*, Heath, Boston, 1965.

Corduneanu, C.
[1] *Integral Equations and Stability of Feedback Systems*, Academic Press, New York, 1973.
[2] Almost periodic solutions for infinite delay systems, *Spectral Theory of Differential Operators*, North-Holland, Amsterdam, 1981, pp. 99 – 106.
[3] *Integral Equations and Applications*, Cambridge University Press, 1991.
[4] Ultimate behavior of solutions of some nonlinear integro-differential equations, *Libertas Mathematica* **4** (1984) 61 – 72.

Corduneanu, C. and V. Lakshmikantham
[1] Equations with infinite delay: A survey, *Nonlinear Analysis – TMA* **4** (1980) 831 – 877.

Crandall, M.G., Londen, S.O. and J.A. Nohel
[1] An abstract nonlinear Volterra integro-differential equation, *J. Math. Anal. Appl.* **64** (1978) 701 – 735.

Crandall, M.G. and J.A. Nohel
[1] An abstract functional differential equation and a related Volterra equation, *Israel J. Math.* **29** (1978) 313 – 328.

Cushing, J.M.
[1] An operator equation and boundedness of solutions of integro-differential equations, *SIAM J. Math. Anal.* **6** (1975) 433 – 445.
[2] *Integro-differential Equation Models in Population Dynamics*, Lecture Notes in Biomathematics **20**, Springer-Verlag, Berlin, 1977.

Dafermos, C.M.
[1] Asymptotic stability in viscoelasticity, *Arch. Rational Mech. Anal.* **37** (1970) 297 – 308.
[2] An abstract Volterra equation with applications to linear viscoelasticity, *J. Diff. Eqns* **7** (1970) 554 – 569.

Dafermos, C.M. and J.A. Nohel
[1] Energy methods for nonlinear hyperbolic Volterra integro-differential equations, *Commun. Partial Diff. Eqns* **4** (1979) 219 – 278.
[2] Nonlinear hyperbolic Volterra equations in viscoelasticity, *Amer. J. Math. Supp.* (1981) 87 – 116.

Dannan, F.M. and S. Elaydi
[1] Lipschitz stability of nonlinear system of differential equations, *J. Math. Anal. Appl.* **113** (1986) 562 – 577.

DaPrato, G. and M. Iannelli
[1] Linear integro-differential equations of hyperbolic type in Hilbert spaces, *Rend. Sem. Math. Univ. Padova* **62** (1980) 191 – 206.
[2] Linear integro-differential equations in Banach spaces, *Rend. Sem. Mat. Univ. Padova* **62** (1980) 207 – 219.

Davis, P.L.
[1] Hyperbolic integro-differential equations arising in the electromagnetic theory of dielectronics, *J. Diff. Eqns* **18** (1975) 170 – 178.

Deo, S.G. and S.G. Pandit
[1] *Differential Systems Involving Impulses*, Lecture Notes **954**, Springer-Verlag, New York 1982.

Desch, W. and R. Grimmer
[1] Some considerations for linear integro-differential equations, *J. Math. Anal. Appl.* **104** (1984) 219 – 234.
[2] Propagation of singularities for integro-differential equations, *J. Diff. Eqns* **65** (1986) 411 – 426.
[3] Initial boundary value problems for integro-differential equations, *J. Integral Eqns* **10** (1985) 73 – 97.

Desch, W., R. Grimmer and W. Schappacher
[1] Wave propagation for a class of integro-differential equations in Banach space, *J. Diff. Eqns* **74** (1988) 391 – 411.

Desch, W. and W. Schappacher
[1] The semi-group approach to integro-differential equations, (to appear).

Desoer, C.A. and M. Vidyasagar
[1] *Feedback Systems: Input – Output Properties*, Academic Press, New York, 1975.

Driver, R.D.
[1] Existence and stability of solutions of a delay-differential system, *Arch. Rat. Mech. Anal.* **10** (1962) 401 – 426.
[2] *Ordinary and Delay Differential Equations*, Springer-Verlag, New York, 1977.

Elaydi, S. and M. Rama Mohana Rao
[1] Lipschitz stability for nonlinear integro-differential systems, *J. Appl.Math. Comput.* **27** (1988) 191 – 199.

Elaydi, S. and S. Sivasundaram
[1] A unified approach to stability in integro-differential equations via Lyapunov functions, *J. Math. Anal. Appl.*

Erbe, L.H. and X. Liu
[1] Boundary value problems for nonlinear impulsive integro-differential equations, *Appl. Math. Comp.* **36** (1990) 31 − 50.

Faheem, M. and M. Rama Mohana Rao
[1] A boundary value problem for functional differential equations of delay type, *J. Math. Anal. Appl.* **109** (1985) 258 − 278.
[2] A boundary problem for nonlinear nonautonomous functional differential equations of delay type with L_p-initial functions, *Funkcial. Ekvac.* **30** (1987) 237 − 249.
[3] Functional differential equations of delay type and nonlinear evolution operators in L_p-spaces, *J. Math. Anal. Appl.* **123** (1987) 73 − 103.

Feller, W.
[1] An integral equation of renewal theory, *Ann. Math. Stab.* **12** (1941) 243-267.

Fielder, M. and V. Ptak
[1] On matrices with nonpositive off-diagonal elements and positive principal minors, *Czechoslovak Math. J.* **12** (1962) 382 − 400.

Freedman, H.I. and V. Sree Hari Rao
[1] The trade-off between mutual interference and time-lag in a predator-prey system, *Bull. Math. Biol.* **45** (1983) 991 − 1004.

Friedman, A.
[1] On integral equations of Volterra type, *J. Analyse Math.* **11** (1963) 381 − 413.

Friedman, A. and M. Shinbrot
[1] Volterra integral equations in Banach space, *Trans. Amer. Math. Soc.* **126** (1967) 131 − 179.

Goel, N.S., S.C. Maitra and E.W. Montroll
[1] *On Volterra and Other Nonlinear Models of Interacting Populations*, Academic Press, New York, 1977.

Goh, B.S.
[1] *Management and Analysis of Biological Populations*, Elsevier, New York, 1980.

Grimmer, R.
[1] Resolvent operators for integral equations in Banach spaces, *Trans. Amer. Math. Soc.* **273** (1982) 333 − 349.
[2] Existence of periodic solutoins of functional differential equations, *J. Math. Anal. Appl.* **72** (1979) 666 − 673.

Grimmer, R. and F. Kappel
[1] Series expansion for resolvents of Volterra integro-differential equations in Banach space, *SIAM J. Math. Anal.* **15** (1984) 595 − 604.

Grimmer, R. and R.K. Miller
[1] Well-posedness of Volterra integral equations in Hilbert space, *J. Integral Eqs* **1** (1979) 201 – 216.
[2] Existence, uniqueness and continuity for integral equations in a Banach space, *J. Math. Anal. Appl.* **57** (1977) 429 – 447.

Grimmer, R. and A.J. Prichard
[1] Analytic resolvent operator for integral equations in Banach spaces, *J. Diff. Eqs* **50** (1983) 234 – 259.

Grimmer, R. and J. Press
[1] On linear Volterra equations in Banach spaces, *Comput. Math. Appl.* **11** (1985) 189 – 205.

Grimmer, R. and G. Seifert
[1] Stability properties of Volterra integro-differential equations, *J. Diff. Eqns* **19** (1975) 142 – 166.

Grimmer, R. and M. Zeman
[1] Wave propagation for linear integro-differential equations in Banach space, *J. Diff. Eqns* **54** (1984) 274 – 282.

Grossman, S.I. and R.K. Miller
[1] Nonlinear Volterra integro-differential equations with L^1 – kernels, *J. Diff. Eqns* **13** (1973) 551 – 566.
[2] Perturbation theory for Volterra integro-differential equations, *J. Diff. Eqns* **8** (1971) 457 – 474.

Gurtin, M.E. and A.C. Pipkin
[1] A general theory of heat conduction with finite wave speeds, *Arch. Rat. Mech. Anal.* **31** (1968) 113 – 126.

Gustafson, G.B. and K. Schmitt
[1] Periodic solutions of hereditary differential systems, *J. Diff. Eqns* **13** (1973) 567 – 587.

Haddock, J.R. and J. Terjeki
[1] Lyapunov – Razumikhin functions and an invariance principle for functional differential equations, *J. Diff. Eqns* **48** (1983) 95-122.

Hahn, W.
[1] *Stability of Motion*, Springer-Verlag, New York, 1967.

Halanay, A.
[1] *Differential Equations: Stability, Oscillations, Time-lag*, Academic Press, New York, 1966.

Hale, J.K.
[1] Sufficient conditions for stability and instability of autonomous functional differential equations, *J. Diff. Eqns* **1** (1965) 452 – 482.
[2] *Theory of Functional Differential Equations*, Springer-Verlag, New York, 1977.

[3]   Periodic and almost periodic solutions of functional differential equations, *Arch. Rat. Mech. Anal.* **15** (1964) 289 – 304.

Hale, J. and J. Kato
[1]   Phase space for retarded differential equations with infinite delay, *Funkcial. Ekvac.* **21** (1978) 11 – 41.

Hannsgen, K.B.
[1]   The resolvent kernel of an integro-differential equation in Hilbert space, *SIAM J. Math. Anal.* **7** (1976) 431 – 490.

Hara, T., T. Yoneyama and T. Itoh
[1]   Characterization of stability concepts of Volterra integro-differential equations, *J. Math. Anal. Appl.* **142** (1989) 558 – 572.
[2]   Asymptotic stability criteria for nonlinear Volterra integro-differential equations, *Funkcial. Ekvac.* **33** (1990) 39 – 57.

Hatvani, L.
[1]   On the stability of zero solution of certain nonlinear second order differential equations, *Acta. Sci. Math.* **32** (1971) 1 – 9.

Hille, G. and R.S. Phillips
[1]   *Functional Analysis and Semi-groups*, Amer. Math. Soc. Publ., Providence, RI, 1957.

Hrusa, W.J. and M. Renardy
[1]   On wave propagation in linear viscoelasticity, *Q. Appl. Math.* **43** (1985) 237 – 254.

Hu, S. and V. Lakshmikantham
[1]   Periodic boundary value problems for integro-differential equations of Volterra type, *Nonlinear Analysis – TMA* **10** (1986) 1203 – 1208.

Hu, S., V. Lakshmikantham and M. Rama Mohana Rao
[1]   Nonlinear variation of parameters formula for integro-differential equations of Volterra type, *J. Math. anal. Appl.* **129** (1988) 223 – 230.

Hu, S., W. Zhuang and M. Khavanin
[1]   On the existence and uniqueness for nonlinear integro-differential equations, *J. Math. Phys. Sci.* **21** (1987) 93 – 103.

Islam, M.N.
[1]   *Periodic Solutions of Volterra Integral Equations*, Ph.D. Dissertation, Southern Illinois University 1980.

Jordan, G.S.
[1]   Asymptotic stability of a class of integro-differential systems, *J. Diff. Eqns.* **31** (1979) 359 – 365.

Kaplan, J.L. and J.A. Yorke
[1]   On the nonlinear differential delay equation, *J. Diff. Eqns.* **23** (1977) 293-314.

Kapur, J.N.
[1] Some problems in biomathematics, *Int. J. Math. Educ. Sci. Tech.* **9** (1978) 287 – 306.

Kato, J.
[1] On Lyapunov – Razumikhin type theorems, *Japan – United States Seminar on Ordinary and Functional Differential Equations, Kyoto, Japan, 1971*, Lecture Notes in Mathematics **243**, Springer-Verlag, Berlin, 1971, pp. 54 – 65.
[2] Razumkhin type conditions for functional differential equations, *Funkcial. Ekvac.* **16** (1973) 225 – 239.

Kato, T.
[1] *Perturbation Theory for Linear Operators*, 2nd edn, Springer – Verlag, New York, 1976.

Khavanin, M. and V. Lakshmikantham
[1] The method of mixed monotone and second order boundary value problems, *J. Math. Anal. Appl.* **120** (1986) 737 – 744.

Krasovskii, N.N.
[1] *Stability of Motion*, Stanford University Press, 1963.

Krisztin, T.
[1] On the convergence of solutions of functional differential equations with infinite delay, *J. Math. Anal. Appl.* **109** (1985) 509 – 521.
[2] Uniform asymptotic stability of a class of integro-differential equations, *J. Integral Eqs*, to appear.

Kuen, S.M. and K.P. Rybakowski
[1] Boundedness of solutions of a system of integro-differential equations, *J. Math. Anal. Appl.* **112** (1985) 378-390.

Ladde, G.S., V. Lakshmikantham and A.S. Vatsala
[1] *Monotone Iterative Techniques for Nonlinear Differential Equations*, Pitman, London, 1985.

Lakshmikantham, V.
[1] Some problems in integro-differential equations of Volterra type, *J. Integral Eqs* **10** (1985) 137-146.
[2] Recent advances in Lyapunov method for delay differential equations, *Proc. Int. Conf.*

Lakshmikantham, V., D.D. Bainov and P.S. Simeonov
[1] *Theory of Impulsive Differential Equations*, World Scientific, Singapore, 1989.

Lakshmikantham, V. and S. Leela
[1] *Differential and Integral Inequalities*, Vol. I, Academic Press, New York, 1969.
[2] *Nonlinear Differential Equations in Abstract Spaces*, Pergamon Press, Oxford, 1981.

[3] *Differential and Integral Inequalities*, Vol. II, Academic Press, New York, 1969.

Lakshmikantham, V., S. Leela and A.A. Martynyuk
[1] *Stability Analysis of Nonlinear Systems*, Marcel Dekker, New York, 1989.

Lakshmikantham, V., S. Leela and M. Rama Mohana Rao
[1] Integral and integro-differential inequalities, *Applicable Analysis* **34** (1987), 157 – 164.
[2] New directions in the method of vector Lyapunov functions, *Nonlinear Analysis – TMA* **16** (1991) 255 – 262.

Lakshmikantham, V., S. Leela and S. Sivasundaram
[1] Lyapunov function on product spaces and stability theory of delay differential equations, *J. Math. Anal. Appl.* **154** (1990) 391-402.

Lakshmikantham, V., X. Liu and S. Sathananthan
[1] Impulsive integro-differential equations and extension of Lyapunov's method, *Applicable Analysis* **32** (1989) 203-214.

Lakshmikantham, V. and M. Rama Mohana Rao
[1] Integro-differential equations and extension of Lyapunov method, *J. Math. Anal. Appl.* **30** (1970) 435 – 447.
[2] Stability in variation for nonlinear integro-differential equations, *Applicable Analysis* **24** (1987) 165 – 173.

Lakshmikantham, V. and D. Trigiante
[1] *Theory of Difference Equations with Applications to Numerical Analysis*, Academic Press, New York, 1988.

Lakshmikantham, V. and X. Liu
[1] Stability criteria for impulsive differential equations in terms of two measures, *J. Math. Anal. Appl.* **137** (1989) 591 – 604.

Langenhop, C.E.
[1] Periodic and almost periodic solutions of Volterra integro-differential equations with infinite memory, *J. Diff. Eqns* **58** (1985) 391 – 403.

Lasley, E.L.
[1] Input – output stability of interconnected systems, *Proc. IEEE Int. Symp. on Circuits and Systems*, Boston 1975, pp. 131-134.
[2] L_2- and L_∞-stability of interconnected systems, *IEEE Trans. Circuits Systems* **23** (1976) 261 – 270.

Lasley, E.L. and A.N. Michel
[1] Input – output stability of large-scale systems, *Proc. 8th Asilomer Conf. on Circuits, Systems and Computers*, Western Periodicals, North Hollywood, CA, 1975, pp. 472 – 482.

Lax, P.D.
[1] Development of singularity solutions of nonlinear hyperbolic differential equations, *J. Math. Phys.* **5** (1964) 611 – 613.

Leela, S. and M.N. Oguztoreli
[1] Periodic boundary value problems for differential equations with delay and monotone iterative method, Preprint.

Leela, S. and M. Rama Mohana Rao
[1] (h_0, h, M_0)-stability for integro-differential equations, *J. Math. Anal. Appl.* **130** (1988) 460 – 468.

Leitman, M.J. and V.J. Mizel
[1] Asymptotic stability and periodic solutions of $x(t) + \int_{-\infty}^{t} a(t-s) g(s, x(s)) ds = f(t)$, *J. Math. Anal. Appl.* **66** (1978) 606 – 625.

Levin, J.J.
[1] The asymptotic behavior of Volterra equations, *Proc. Amer. Math. Soc.* **14** (1963) 534 – 541.
[2] Resolvents and bounds for linear and nonlinear Volterra equations, *Trans. Amer. Math. Soc.* **228** (1977) 207 – 222.

Levin, J.J. and J.A. Nohel
[1] On a system of integro-differential equations occurring in reactor dynamics, *J. Math. Mech.* **9** (1960) 347 – 368.
[2] The integro-differential equations of a class of nuclear reactors with delayed neutrons, *Arch. Rat. Mech. Anal.* **31** (1968) 151 – 172.
[3] On the system of integro-differential equations occurring in reactor dynamics II, *Arch. Rat. Mech. Anal.* **11** (1962) 210 – 243.
[4] A nonlinear system of integro-differential equations, *Mathematical Theory of Control, (Proc. Conf. University of Southern California, Jan 30 – Feb. 1, 1967)*, Academic Press, New York, (1967), pp. 398 – 405.
[5] The integro-differential equations of a class of nuclear reactors with delay neutrons, *Tech. Report, University of Wisconsin, Madison*, 1968.

Levin, J.J. and D.F. Shea
[1] On the asymptotic behavior of some integral equations, I, II, III, *J. Math. Anal. Appl.* **37** (1972) 42 – 82, 288 – 326, 537 – 575.

Londen, S.O.
[1] On some nonlinear Volterra equations, *Ann. Acad. Sci. Fenn. Ser. A* VI No. 317 (1969).
[2] On the asymptotic behavior of a solution of a nonlinear integro-differential equation, *SIAM J. Math Anal.* **2** (1971) 356 – 367.

Lotka, A.J.
[1] *Mathematical Biology*, Dover, New York, 1956.

Luca, N.
[1] The stability of solutions of a class of integro-differential systems with infinite delays, *J. Math. Anal. Appl.* **67** (1979) 323 – 339.

Lunardi, A. and G. DaPrato
[1] Hopf bifurcation for nonlinear integro-differential equations in Banach spaces with infinite delay, *Indiana Univ. Math. J.* **36** (1987).

MacCamy, R.C.
[1] Existence, uniqueness and stability of $u_{tt} = \frac{\partial}{\partial x}[\sigma(u_x) + \lambda(u_x)u_{xt}]$, *Indiana Math. J.* **20** (1970) 231 – 238.
[2] A model for one dimensional nonlinear viscoelasticity, *Q. Appl. Math.* **35** (1977) 21 – 33.
[3] An integro-differential equation in applications to heat flow, *Q. Appl. Math.* **35** (1977) 1 – 19.

MacDonald, N.
[1] *Time-lags in Biological Models*, Lecture Notes in Biomathematics **27**, Springer-Verlag, Berlin, 1978.

Marcus, M. and V.J. Mizel
[1] Limiting equations for problems involving range memory, *Mem. Amer. Math. Soc.* **43** No. 278 (1983).

Miller, R.K.
[1] Volterra equations in Banach space, *Funkcial. Ekavac.* **18** (1975) 163 – 193.
[2] Asymptotic behavior for a linear Volterra integral equation in Hilbert space, *J. Diff. Eqns* **23** (1977) 270-284.
[3] *Nonlinear Volterra Integral Equations*, Benjamin, Menlo Park, California, 1971.
[4] On the linearization of Volterra integral equations, *J. Math. Anal. Appl.* **23** (1968) 198 – 208.
[5] Asymptotic stability properties of linear Volterra integro-differential equations, *J. Diff. Eqns* **10** (1971) 485 – 506.
[6] On Volterra integral equations with nonnegative integrable resolvents, *J. Math. Appl.* **22** (1968) 319-340.
[7] On Volterra population equation, *SIAM J. Appl. Math.* **14** (1966) 446 – 452.
[8] An integro-differential equation for rigid heat conductors with memory, *J. Math. Anal. Appl.* **66** (1978) 313 – 333.

Miller, R.K and A.N. Michel,
[1] L_2-stability and instability of large-scale systems described by integro-differential equations, *SIAM J. Math. Anal.* **8** (1977) 547 – 557.

Miller, R.K. and R.L. Wheeler
[1] Well-posedness and stability of linear Volterra integro-differential equations in abstract spaces, *Funkcial. Ekvac.* **21** (1978) 279 – 305.

Mönch, H. and G.F. Von Harten
[1] On the Cauchy problem for ordinary differential equations in Banach spaces, *Arch. Math.* **39** (1982) 153 – 160.

Moore, J.C.
[1] A new concept of stability-M_0-stability, *J. Math. Anal. Appl.* **112** (1985) 1 – 13.

Morchazo, J.
[1] Integral equivalence of two systems of integro-differential equations, *Proc. Int. Conf. on Differential Equations, Bulgaria*, 1985, pp. 841 – 844.
[2] Integral equivalence of systems of differential equations, *Rend. Acad. Nazion. Dei Lincei*, (to appear).

Moser, J.
[1] On nonoscillatory networks, *Q. Appl. Math.* **25** (1967) 1 – 9.

Murakami, S.
[1] Periodic solutions of some integro-differential equations, *Res. Rep., Hachinohe Nat. College* **21** (1986) 131 – 133.

Nohel, J.A.
[1] A nonlinear conservation law with memory, *Lecture Notes in Pure and Appl.* **81**, Marcel Dekker, New York, 1982, pp. 91 – 123.
[2] Remarks on nonlinear Volterra equations, *Proc. US – Japan Seminar on Differential and Functional Differential Equations*, Benjamin, New York, 1967, pp. 249 – 266.
[3] Asymptotic equivalence of Volterra equations, *Ann. Mat. Pura Appli.* **96**, (1973), 339 – 347.

Nohel, J.A. and D.F. Shea
[1] Frequency domain methods for Volterra equations, *Adv. Math.* **22** (1976) 278 – 304.

Noy-Meir, I.
[1] Stability of grazing systems: An application prey – predator graph, *J. Ecol.* **63** (1975) 459 – 481.
[2] Stability in simple grazing models, effect of explicit functions, *J. Theor. Biol.* **71** (1978) 347 – 380.

Oğuztoreli, M.N.
[1] *Time-lag Control Systems*, Academic Press, New York, 1966.

Ostrowskii, A.
[1] Determinanten mit überwiegender Hauptdiagonale und die absolute konvergenz von linearen iteration prozessen, *Comm. Math. Helv.* **30** (1956) 175 – 210.

Paley, R.E. and N. Weiner
[1] Fourier transforms in the complex domain, *Amer. Math. Soc. Colloq. Publ.* **19** (1934).

Pazy, A.
[1] Semi-group of linear operators and applications to partial differential equations, *Univ. of Maryland Math. Dept Lecture Notes* **10** (1974).

Pearl, R.
[1] *Introduction to Medical Biometry and Statistics*, Saunders, Philadelphia 1940, pp. 459 – 470.

Pielou, E.C.
[1] *An Introduction to Mathematical Ecology*, Wiley, New York, 1969.

Plaza,D.W. and W.H. Kohler
[1] Coupled-reactor kinetics equations, *Nucl. Sci. Engng* **22** (1966) 419 – 422.

Prakasa Rao, B.L.S. and M. Rama Mohana Rao
[1] Stochastic integral equations of mixed type, *An. Sti. Univ. "AL. I. CUZA" Iasi* **28** (1972) 351 – 359.
[2] Perturbations of stochastic integro-differntial equations, *Bull. Math. Soc. (Romania)* **18** (1974) 187 – 201.

Pritchard, A.J. and J. Zabczyk
[1] Stability and stabilizability of infinite dimensional systems, *SIAM Rev.* **23** (1981) 25 – 52.

Rama Mohana Rao, M.
[1] *Ordinary Differential Equations: Theory and Applications*, Edward Arnold, London, 1981.

Rama Mohana Rao, M. and V.M. Pal
[1] Asymptotic stability of grazing systems with unbounded delay, *J.Math. Anal. Appl.* **163** (1992) 60 – 72.

Rama Mohana Rao, M. and V. Raghavendra
[1] Asymptotic stability properties of Volterra integro-differential equations, *Nonlinear Analysis – TMA* **11** (1987) 475 – 480.
[2] On the stability of differential systems with respect to impulsive perturbations, *J. Math. Anal. Appl.* **48** (1974) 515 – 526.
[3] Volterra integral equations with discontinuous perturbations, *Mathematica, Cluj (Romania)* **17** (1975) 89 – 101.

Rama Mohana Rao, M., S. Sathananthan and S. Sivasundaram
[1] Asymptotic behavior of solutions of Volterra systems with impulsive effect, *Appl. Math. Comput.* **34** (1990) 195 – 211.

Rama Mohana Rao, M. and K.S. Sanjay
[1] Lyapunov functions on product spaces and stability of integro-differential equations, *Dynamical Systems Appl.* **1** (1992) 93 – 102.

Rama Mohana Rao, M., K.S. Sanjay. and S. Sivasundaram
[1] Stability of Volterra integro-differential equations with impulsive effect, *J. Math. Anal. Appl.* **163** (1992) 47 – 59.

Rama Mohana Rao, M. and S. Sivasundaram
[1] Asymptotic stability for equations with unbounded delay, *J. Math. Anal. Appl.* **131** (1988) 97 – 105.

Rama Mohana Rao, M. and V. Sree Hari Rao
[1] Stability of impulsively perturbed systems, *Bull. Austral. Math. Soc.* **16** (1977) 99 – 110.
[2] Uniform asymptotic stability of impulsively perturbed nonlinear differential equations, *Nonlinear Analysis – TMA* **4** (1980) 599 – 606.

Rama Mohana Rao, M. and P. Srinivas
[1] Stability of functional differential equations of Volterra type, *Bull. Austral. Math. Soc.* **29** (1984) 93 – 100.
[2] Positivity and boundedness of solutions of Volterra integro-differential equations, *Libertas Mathematica* **3** (1983) 71 – 81.
[3] Asymptotic behavior of solutions of Volterra integro-differential equations, *Proc. Amer. Math. Soc.* **94** (1985) 55 – 60.

Rama Mohana Rao, M. and C.P. Tsokos
[1] Integro-differential equations of Volterra type, *Bull. Austral. Math. Soc.* **3** (1970) 9 – 22.

Razumikhin, B.S.
[1] The application of Lyapunov's method to problems in the stability of systems with delay, *Autom. Rem. Control* **21** (1960) 515 – 520.

Seifert, G.
[1] Almost periodic solutoins of delay differential equations with infinite delays, *J. Diff. Eqns* **41** (1981) 416 – 425.
[2] Lyapunov – Razumikhin conditions for stability and boundedness of functional differential equations of Volterra type, *J. Diff. Eqns* **14** (1973) 424 – 430.
[3] Lypaunov – Razumikhin conditions for asymptotic stability in functional differential equations of Volterra type, *J. Diff. Eqns* **16** (1974) 289 – 297.
[4] On a delay differential equation for single specie population variations, *Nonlinear Analysis – TMA* **10** (1987) 1051 – 1059.

Shendge, G.R.
[1] A new approach to the stability theory of functional differential equations, *J. Math. Anal. Appl.* **95** (1983) 319 – 334.

Simeonov, P.S. and D.D. Bainov
[1] Stability of solutions of integro-differential equations with impulsive effect, *Math. Rep. Toyama Univ.* **9** (1986) 1 – 24.
[2] Stability with respect to part of the variables in systems with impulsive effect, *J. Math. Anal. Appl.* **117** (1986) 247 – 263.

Sivasundaram, S.
[1] The method of upper and lower solutions and interval analytic method for integro-differential equations, *J. Appl. Math. Comput.* (to appear).
[2] The method of upper and lower solutions and interval analytic method for integral equations, *Proc. Nonlinear Anal., Arlington* 1986.

Smith, J.M.
[1] *Models in Ecology*, Cambridge University Press, 1974.

Staffans, O.
[1] On a nonlinear hyperbolic Volterra equation, *SIAM J. Math. Anal.* **11** (1980) 793 – 812.
[2] Nonlinear Volterra integral equations with positive definite kernels, *Proc. Amer. Math. Soc.* **51** (1975) 103 – 108.

Stepan, G.
[1] Great delay in a prey-predator model, *Nonlinear Analysis – TMA* **10** (1986) 913 – 929.
[2] A stability criterion for retarded dynamical systems, *Z. Angew. Math. Mech.* **64** (1984) 345 – 346.

Strauss, A.
[1] On a perturbed volterra integral equation, *J. Math. Anal. Appl.* **30** (1970) 564 – 575.

Tanabe, H.
[1] *Equations of Evolution*, Pitman, London, 1979.

Volterra, V.
[1] *Theory of Functionals and of Integral and Integro-differential Equations*, Dover, New York, 1959.
[2] *Lecons sur la théorie mathématique de la luttle pour la vie* Gauthier – Villars, Paris, 1931.

Wang, Z.C. and J.H. Wu
[1] Neutral functional differential equations with infinite delay, *Funkcial. Ekvac.* **28** (1985) 157 – 170.

Willems, J.C.
[1] *The Analysis of Feedback Systems*, MIT Press, Cambridge, MA, 1971.

Wu, J.H.
[1] The local theory for neutral functional differential equations with infinite delay, *Acta Math. Appl. Sinica* **8** (1985) 427 – 481.
[2] Globally stable periodic solutoins of linear neutral Volterra integro-differential equations, *J. Math. Anal. Appl.* **130** (1988) 474 – 483.

Yoshizawa, T.
[1] *Stability Theory by Lyapunov's Second Method*, Math. Soc. Japan, Tokyo, 1966.

Zames, G.
[1] On the input-output stability of time-varying nonlinear feedback systems – Part I, *IEEE Trans. Autom. Control* **11** (1966) 228 – 238.
[2] On the input – output stability of time-varying nonlinear feedback systems – Part II, *IEEE Trans. Autom. Control* **11** (1966) 465 – 476.

Zhuang, W.
[1] Existence and uniqueness of solutions of nonlienar integro-differential equations of Volterra type in Banach space, *Applicable Anal.* **22** (1986) 157 – 166.

Zouyousefain, M. and S. Leela
[1]   Stability results for difference equations of Volterra type, *Appl. Math. Comput.* **36** (1990) 51 – 61.

INDEX

Adjoint equation 29, 138
Approximations, successive 19, 202, 207
Ascoli theorem 218, 275
Ascoli-Arzela theorem 10, 12, 208, 265
Asymptotic,
    behavior 126, 275, 305, 317
    equivalence 81, 87, 89
    periodic solutions 116, 122, 123
Attraction,
    uniform 146
    (h_0, h, M)-uniform 175

Barbălat lemma 135, 141, 144
Biological population 274
Boundary value problem 36, 41, 48
Boundedness,
    of solutions 59
    uniform 57, 150, 151, 152
    uniform ultimate 57, 150, 151, 152
Bounded variation 236

Cauchy problem 231, 320
Comparison
    result 37, 166, 180
    theorem 12, 157, 178
Cone 211
    normal 212, 217
Converse theorem(s),
    for exponential asymptotic stability 146
    for uniform asymptotic stability 147

Decrescent ... 145
Differential resolvent ... 26, 29
Differentiability with respect to initial values ... 31, 33
Dissipative mechanism ... 315, 316, 319, 324

Elasticity ... 309
Electric networks ... 98
Energy,
    estimates ... 310, 314
    function ... 306
Equation(s)
    adjoint ... 138
    difference ... 102, 103
    differential ... 97, 100, 137, 141, 225
    evolution ... 225, 243
    functional differential ... 141
    hyperbolic ... 292, 297
    impulsive differential ... 182, 185, 197
    impulsive effect ... 289
    impulsive integro-differential ... 289
    infinite delay ... 137, 255, 272, 275
    integral ... 71, 74, 204, 215, 310
    limiting ... 124
    partial integro-differential ... 242, 299, 311
    unbounded delay ... 137
    Verhulst-Pearl ... 272
    Volterra integral ... 28, 311
Equilibrium ... 56, 275, 289
Equivalence,
    asymptotic ... 86, 89
    M-asymptotic ... 91, 92
    (M, p)-integral ... 91
Existence theorem, ... 9, 209
    local ... 2, 3
    global ... 2, 4
    of maximal solution ... 211
    of minimal solution ... 211
Extremal solutions ... 9

Finer ... 174
Fixed points,
    Banach ... 213, 295
    Darbo ... 204
    Schauder ... 2, 3, 41
    Tychonoff ... 4, 5
Frequency response ... 324
Function(s),
    decrescent ... 167
    equicontinuous ... 10
    forcing ... 238, 314, 320

Function(s) (continued)
- h_0-decrescent 167
- initial 66, 127, 137, 314
- measurable 87
- minimal class 157, 169, 183, 184
- positive definite 145, 162
- h-positive definite 172
- weight 282

Functional,
- continuous linear 212
- decrescent 145
- Lipschitz continuous 144
- nonanticipative 144, 147
- response 289
- positive definite 145, 148

Fundamental,
- solution 227, 228, 245
- system 227

Heat,
- conduction 219, 252
- equation 300
- flow 254, 321

Hereditary 272

Impulsive effect 180, 193
- integro-differential equation 180, 186, 194

Inequality,
- difference 98, 103
- differential 181, 183, 192
- Gronwall's 8, 61, 67
- Hölder 91, 92, 96
- impulsive integro-differential 186
- integral 18
- integro-differential 6, 15, 23, 209
- Jenson 176
- Schwartz 98, 304

Initial function 127, 169
Initial value(s) 169, 170
- continuous dependence on 21, 22, 23
- differentiability with respect to 31, 34

Instability (unstable) 131, 132
L_2- 259, 330, 331, 335
Iterative bounds for solution 44

Linear,
- convolution system 54
- independence 54

Logistic growth 302

Lyapunov's
    functions 157, 158, 254, 259
    functions on product spaces 174
    functional 127, 126, 137, 285
    piecewise continuous functions 183, 185

Mapping,
    bicontinuous 83, 87
    contraction 83
Matrix,
    equation 137
    fundamental 141
    positive definite 137
    stable 137
Method,
    energy estimates 310
    integral analytic 43
    Krasovskii's 298, 299, 300
    of Lyapunov function 157
    of Lyapunov functionals 125
    of Lyapunov – Razumikhin 284
    monotone iterative 211
    of reduction 64
Measure of noncompactness,
    Hausdorff's 212, 213
    Kuratowski 204
Mönch and Von Harten lemma 212
Monotone sequences 213
Monotone solutions 274

Nuclear reactors 299, 302
    continuous-medium 300, 308
    system 302

Operator(s)
    bounded 85, 223, 238, 293
    bounded linear 223, 226, 231
    closed 238
    closed linear 85, 220, 224
    evolution 227, 237, 247, 249
    integral 5, 48, 206
    linear 219
    kernel of 325
    monotone 43, 95, 216, 321
    nonanticipative 150
    positive definite 101
    resolvent 230, 239, 242, 256

INDEX

Parseval's formula 301, 304, 307
Partial ordering 311
Periodic solutions 122, 128
    asymptotically 123
Perturbation(s) 149, 155, 249
    result 235

Resolvent 220, 233, 237
    differentiable 138, 238, 246
    kernel 310, 312, 323
    residual 324

Semigroup 222, 228, 232, 236
    C_0 .... 223, 242
    Favard class 235, 237
    stable 228, 292, 293
    translation 220, 222, 244, 250
Sets,
    bounded 2
    closed 2
    convex 2
    equicontinuous 3
    minimal 184, 288
Solution(s),
    asymptotically periodic 119, 124, 125
    boundedness of 56, 79, 83, 263
    lower 35, 212
    M-bounded 95
    maximal 9, 10, 13, 42, 48, 69, 213
    minimal 9, 48, 210, 213
    periodic 117
    stability of 255
    ultimate behavior of 94
    uniform boundedness of 55, 117
    uniform ultimate boundedness of 55, 117
    upper 35, 212
    weak 59, 160, 257, 327
Stability
    asymptotic 126, 143, 329
    bounded − input bounded − output 327
    eventual uniform 143
    exponential 55
    exponential asymptotic 152
    (h_0, h)-uniform 177
    (h_0, h)-uniform asymptotic 178
    (h_0, h, M)-uniform 180
    (h_0, h, M)-uniform asymptotic 181
    L_2- 255, 257, 328, 330
    Lipschitz 75
    uniform 58, 72, 166
    uniform asymptotic 126, 135, 176

System(s)
- convolution 55
- composite 323
- grazing 281
- impulsive integro-differential 109
- interconnected 322, 330
- large-scale 322, 327
- multi − input and multi − output (MIMO) 329
- multi − loop 322
- periodic 117
- perturbed 145
- prey − predator 281
- variational 69, 75

Techniques,
- Laplace transform 29, 220, 244
- Fourier transform 303, 324
- Lyapunov − Razumikhin 284
- monotone iterative 35

Total variation 229

Variation of parameters,
- linear 23, 26, 138, 237, 328
- nonlinear 30, 34

Viscosity 310
Viscoelasticity 310

Wave,
- equation 321
- motion 299
- propagation 292, 294, 297, 299

Wellposedness 219, 220, 223, 227
- uniform 227